VOLUME SIX HUNDRED AND SIXTY NINE

METHODS IN ENZYMOLOGY

Coenzyme B_{12} Enzymes Part B

METHODS IN ENZYMOLOGY

VOLUME SIX HUNDRED AND SIXTY NINE

METHODS IN ENZYMOLOGY

Coenzyme B_{12} Enzymes Part B

Edited by

E. NEIL G. MARSH
Professor of Chemistry and Biological Chemistry, Department of Chemistry, University of Michigan, United States

ELSEVIER

ACADEMIC PRESS
An imprint of Elsevier

Academic Press is an imprint of Elsevier
50 Hampshire Street, 5th Floor, Cambridge, MA 02139, United States
525 B Street, Suite 1650, San Diego, CA 92101, United States
The Boulevard, Langford Lane, Kidlington, Oxford OX5 1GB, United Kingdom
125 London Wall, London, EC2Y 5AS, United Kingdom

First edition 2022

ISBN: 978-0-323-95557-7
ISSN: 0076-6879

For information on all Academic Press publications
visit our website at https://www.elsevier.com/books-and-journals

Publisher: Zoe Kruze
Developmental Editor: Federico Paulo S. Mendoza
Production Project Manager: James Selvam
Cover Designer: Christian Bilbow

Typeset by STRAIVE, India

Contents

Contributors

Squire J. Booker
The Department of Chemistry; The Department of Biochemistry and Molecular Biology; The Howard Hughes Medical Institute, The Pennsylvania State University, University Park, PA, United States

Jennifer Bridwell-Rabb
Department of Chemistry; Program in Chemical Biology, University of Michigan, Ann Arbor, MI, United States

Thomas C. Brunold
Department of Chemistry, University of Wisconsin—Madison, Madison, WI, United States

Jun-Ru Chen
Department of Physics, National Dong Hwa University, Hualien, Taiwan

Taewon Chung
Department of Chemistry, University of Michigan, Ann Arbor, MI, United States

Amanda L. Darbyshire
Department of Chemistry, University of British Columbia - Okanagan Campus, Kelowna, BC, Canada

Paul Dewan Jr.
Biophysics Program, University of Michigan, Ann Arbor, MI, United States

Zerick Dill
Department of Chemistry; Program in Chemical Biology, University of Michigan, Ann Arbor, MI, United States

Tadashi Eguchi
Department of Chemistry, Tokyo Institute of Technology, Tokyo, Japan

Laura D. Elmendorf
Department of Chemistry, University of Wisconsin—Madison, Madison, WI, United States

Arghya P. Ghosh
Department of Chemistry, University of Louisville, Louisville, KY, United States

Samantha J.O. Hardman
Manchester Institute of Biotechnology, The University of Manchester, Manchester, United Kingdom

Joanna A. Hughes
Laboratory of Ultrafast Spectroscopy, ISIC, and Lausanne Centre for Ultrafast Science (LACUS), École Polytechnique Fédérale de Lausanne, Lausanne, Switzerland

Alina Ionescu
Department of Physics, Emory University, Atlanta, GA, United States

Alex R. Jones
Biometrology, Department of Chemical and Biological Sciences, National Physical Laboratory, Middlesex, United Kingdom

Shyue-Chu Ke
Department of Physics, National Dong Hwa University, Hualien, Taiwan

Hayley L. Knox
The Department of Chemistry, The Pennsylvania State University, University Park, PA, United States

Meghan Kohne
Department of Physics, Emory University, Atlanta, GA, United States

Pawel M. Kozlowski
Department of Chemistry, University of Louisville, Louisville, KY, United States

Fumitaka Kudo
Department of Chemistry, Tokyo Institute of Technology, Tokyo, Japan

Ryan M. Lamb
Department of Chemistry, University of Michigan, Ann Arbor, MI, United States

Yu-Hsuan Lee
Department of Chemistry, University of Texas at Austin, Austin, TX, United States

Bin Li
Department of Chemistry, University of Michigan, Ann Arbor, MI, United States

Wei Li
Department of Physics, Emory University, Atlanta, GA, United States

Hung-wen Liu
Department of Chemistry; Division of Chemical Biology and Medicinal Chemistry, College of Pharmacy, University of Texas at Austin, Austin, TX, United States

Valentina Lukinović
Institute of Biochemistry, Carleton University, Ottawa, ON, Canada

Amarendra Nath Maity
Department of Physics, National Dong Hwa University, Hualien, Taiwan

Caitlyn Makins
Department of Chemistry, University of British Columbia - Okanagan Campus, Kelowna, BC, Canada

E. Neil G. Marsh
Department of Chemistry; Department of Biological Chemistry, University of Michigan, Ann Arbor, MI, United States

Taylor P. McClain
Biophysics Program, University of Michigan, Ann Arbor, MI, United States

Saurav Parmar
Department of Chemistry, University of Louisville, Louisville, KY, United States

James E. Penner-Hahn
Department of Chemistry; Biophysics Program, University of Michigan, Ann Arbor, MI, United States

Shusuke Sato
Department of Chemistry, Tokyo Institute of Technology, Tokyo, Japan

Roseanne J. Sension
Department of Chemistry, University of Michigan, Ann Arbor, MI, United States

Erica K. Sinner
Department of Chemistry, Johns Hopkins University, Baltimore, MD, United States

Megan J. Toda
Department of Chemistry, University of Louisville, Louisville, KY, United States

Craig A. Townsend
Department of Chemistry, Johns Hopkins University, Baltimore, MD, United States

Kurt Warncke
Department of Physics, Emory University, Atlanta, GA, United States

Kirsten R. Wolthers
Department of Chemistry, University of British Columbia - Okanagan Campus, Kelowna, BC, Canada

Jonathan R. Woodward
Graduate School of Arts and Sciences, The University of Tokyo, Tokyo, Japan

Chen Zhu
Department of Physics, Emory University, Atlanta, GA, United States

Preface

Vitamin B_{12} (cyanocobalamin) is the precursor to two biologically active forms of coenzyme B_{12}: adenosylcobalamin (AdoCbl) and methylcobalamin (MeCbl). Its structure comprises a highly modified tetrapyrrole macrocycle that chelates a cobalt ion; a further axial ligand to cobalt is provided by a nucleotide-derived ligand (most commonly dimethylbenzimidazole) that is tethered to the macrocycle. This leaves the sixth coordination site open to catalyze biological reactions, which take advantage of cobalt's ability to access the +1, +2, and +3 oxidation states. Coenzyme B_{12} has been called nature's most beautiful cofactor. Whether one shares this view or not, it is certainly one of the most structurally complex cofactors; its biosynthesis requires over 20 steps starting from uroporphyrinogen III, the common intermediate from which other tetrapyrroles such as heme and chlorophyll are derived.

Vitamin B_{12} is the precursor to the catalytically active cofactors for two enzymes in animals. Methylmalonyl-CoA mutase uses AdoCbl to catalyze the rearrangement of methylmalonyl-CoA to succinyl-CoA in the metabolism of odd-chain fatty acids, while methionine synthase uses MeCbl as the intermediate methyl donor in the transfer of a methyl group from methyl-tetrahydrofolate to homocysteine to produce methionine. These are the only known uses of this complex molecule in animals, although both AdoCbl- and MeCbl-dependent enzymes are found far more extensively in bacteria and archaea. By contrast, B_{12} appears to be absent from plants.

Vitamin B_{12} was discovered through efforts to cure pernicious anemia—a disease that we now know is generally caused by defects in the cobalamin uptake system in afflicted individuals. In 1926, Minot and Murphy showed that a diet high in liver and kidney, and low in fat, cured patients of their anemia noting "*prompt, rapid and distinct remission of their anemia and a marked symptomatic improvement, except for pronounced disorders due to spinal cord degeneration.*" It was this observation that would lead, some 96 years later, to this volume! Although liver extracts enriched in "antipernicious anemia factor" became an effective treatment for the disease during the 1920s and 1930s, it was not until 1947 that the compound was isolated in crystalline form. This herculean task, given the very small quantities present in liver, was achieved independently by Folkers' group at Merck in the United States and Smith's group at Glaxo in the United Kingdom.

In 1948, with its structure and mode of action still unknown, the crystalline "antipernicious anemia factor" was renamed vitamin B_{12}—"B" connotes only nutritional significance and the number "12" had not been used previously in vitamin nomenclature.

The structure of vitamin B_{12} was solved by X-ray crystallography in Hodgkin's laboratory in 1955—at the time, it was the largest molecular structure to be solved and a landmark in structural biology. However, the structure gave no immediate clue to the biochemical function of B_{12}; that would have to wait until 1958 for the discovery by Barker's laboratory of the first B_{12} enzyme, glutamate mutase. When the crystal structure of the active coenzyme was determined in 1960, again by Hodgkin, an adenosyl moiety was found to replace the cyanide ligand (cyanide is likely an artifact of the purification procedure) as the sixth ligand. Most unexpectedly, this ligand was bonded to cobalt through the 5′-carbon of adenosine. Shortly afterward, Wood and Lester demonstrated that methylcobalamin was involved in the biosynthesis of methionine. With the discovery of these two active forms of vitamin B_{12}, the stage was set to explore this new class of metalloenzymes that catalyze unusual reactions enabled by the unique chemistry of the B_{12} cobalt-carbon bond.

Homolysis of the AdoCbl cobalt-carbon bond serves to generate the highly reactive 5′-deoxyadenosyl radical. In this respect, AdoCbl-dependent enzymes parallel radical SAM enzymes, which form the same radical by reductive cleavage of *S*-adenosylmethionine. AdoCbl-dependent enzymes likely evolved after radical SAM enzymes, probably as a response to a more aerobic environment—all radical SAM enzymes are very oxygen-sensitive whereas AdoCbl-dependent enzymes are much less so. Historically though, radical SAM enzymes were discovered after B_{12} enzymes, prompting Barker to dub them "a poor man's coenzyme B_{12}." In subsequent decades this sobriquet has reverberated throughout the literature—and has often been challenged. In one respect, however, Barker was the most percipient; for just as there are many more poor people than rich people, there are now many more radical SAM enzymes known than B_{12} enzymes! In the six decades since glutamate mutase was discovered, only about a dozen AdoCbl enzymes have been discovered, all of which catalyze mechanistically similar rearrangements or dehydration/deamination reactions. By contrast, there are now around 100 different radical SAM enzymes known. These have been found in all kingdoms of life and catalyze an astonishingly wide array of reactions.

Despite their relative rarity, AdoCbl-dependent enzymes have fascinated enzymologists since their discovery because they catalyze unusual rearrangement reactions with no obvious counterpart in conventional organic chemistry. Elucidating the mechanisms by which AdoCbl-dependent enzymes catalyze these reactions has provided fundamental insights into how enzymes harness highly reactive free radicals for catalysis, thereby catalyzing reactions that would be hard to achieve by other means. What we have learned from studying AdoCbl-dependent enzymes has provided the foundational knowledge with which to tackle the much larger (and much more technically challenging) family of radical SAM enzymes.

With this brief historical introduction as context, I turn to this volume of *Methods in Enzymology* that propels B_{12} enzymology firmly into the 21st century. First, I express my profound thanks to all the authors who have so generously contributed their expertise and time to produce the chapters collected here. The response from the B_{12} community has been remarkable! Although I originally conceived this collection of methods-focused papers as a single volume, the enthusiasm of the contributors has been such that it has expanded into two volumes. Within these volumes, I have loosely organized the chapters into four sections.

Section I deals with enzymes involved in the biosynthesis of cobalamin and methods for the biosynthesis of cobalamin analogs. The structural complexity of coenzyme B_{12} ensures that the enzymes involved in its biosynthesis are both interesting and diverse! This section includes chapters on the enzymes involved in the construction of the nucleotide "tail" of the coenzyme, which in many microbes varies from the canonical dimethylbenzimidazole ligand, and the enzymes involved in transforming the vitamin form of B_{12} to the enzymatically active form, AdoCbl. In animals, the uptake, transport, and installation of B_{12} into methylmalonyl-CoA mutase and methionine synthase require a dedicated set of chaperone enzymes that are also discussed. Lastly, the synthesis and biological properties of "antivitamin B_{12}" analogs are discussed—these synthetic compounds have potential as mechanistic probes and therapeutic agents.

Section II groups together chapters that focus on the purification and characterization of enzymes that use either AdoCbl or MeCbl as cofactors. The chapters include discussions of enzymes that have been known from the early days, including diol dehydratase and ethanolamine ammonia-lyase, glutamate mutase and 2-methyleneglutarate mutase, and the human enzymes methylmalonyl-CoA mutase and methionine synthase. Also included are more recently recognized functions of B_{12}. Thus, B_{12} is used as a redox cofactor by a class of reductive dehalogenases that initiate the degradation of halogenated hydrocarbons by various anaerobic microorganisms.

And the most recently recognized role for AdoCbl is as a light sensor. AdoCbl has long been known to be light-sensitive, but this property has now been recognized to have a biological role with the discovery of the bacterial light-sensing transcription factor CarH.

Section III is devoted to the emerging group of enzymes that, excitingly for both B_{12} and radical SAM enzymologists, use both radical SAM and cobalamin chemistry. Many of these enzymes catalyze reactions in which methyl groups are added to nonacidic positions on the substrate, e.g., carbon and phosphorus. These methylations proceed by radical mechanisms in which radicals are generated from SAM and the methyl group is supplied by MeCbl. However, these enzymes have proved especially challenging to study due to the difficulty of recombinantly expressing enzymes that have both the iron-sulfur cluster and the cobalamin correctly installed. The chapters here focus on practical methods for expressing, purifying, and characterizing a number of these fascinating enzymes.

Section IV focuses on methods for studying the mechanisms of B_{12} enzymes. These chapters cover a wide range of techniques and describe how they have been used to elucidate the mechanisms of various B_{12} enzymes. In many cases, these techniques and approaches should be readily adaptable to study the mechanisms of other B_{12} enzymes, and related enzymes such as radical SAM enzymes. The rich electronic spectrum of the B_{12} corrin ring is sensitive to the oxidation and coordination state of cobalt. These properties make UV-visible spectroscopy and magnetically coupled circular dichroism particularly useful techniques to study the reactivity of the coenzyme, especially when combined with stopped flow. The mechanisms of many B_{12} enzymes involve radical intermediates, which naturally makes EPR spectroscopy a valuable tool to investigate them. The transfer of hydrogen between the substrate and coenzyme is a hallmark of AdoCbl-dependent reactions; as such, isotope effect measurements are an informative tool. Many of these techniques can be adapted to apply them at very short timescales (in some cases on the picosecond timescale), making it possible to precisely dissect individual steps in the mechanism. Last but not least, computational methods continue to increase in the power and sophistication with which they can model enzyme reactions, and B_{12} enzymes are no exception.

In closing, I hope this collection of chapters will serve as a valuable resource for the current community of scientists working on B_{12} and, most importantly, those entering the field in the future.

E. Neil G. Marsh

SECTION III

B_{12}-dependent radical SAM enzymes

CHAPTER ONE

Structural characterization of cobalamin-dependent radical *S*-adenosylmethionine methylases

Hayley L. Knox[a] and Squire J. Booker[a,b,c,*]
[a]The Department of Chemistry, The Pennsylvania State University, University Park, PA, United States
[b]The Department of Biochemistry and Molecular Biology, The Pennsylvania State University, University Park, PA, United States
[c]The Howard Hughes Medical Institute, The Pennsylvania State University, University Park, PA, United States
*Corresponding author: e-mail address: squire@psu.edu

Contents

Abstract

Cobalamin-dependent radical *S*-adenosylmethionine (SAM) methylases catalyze key steps in the biosynthesis of numerous biomolecules, including protein cofactors, antibiotics, herbicides, and other natural products, but have remained a relatively understudied subclass of radical SAM enzymes due to their inherent insolubility upon overproduction in *Escherichia coli*. These enzymes contain two cofactors: a [4Fe-4S] cluster that is ligated by three cysteine residues, and a cobalamin cofactor typically bound by residues in the N-terminal portion of the enzyme. Recent advances in the expression and purification of these enzymes in their active states and with both cofactors present has allowed for more detailed biochemical studies as well as structure determination by X-ray crystallography. Herein, we use *Ks*TsrM and TokK to highlight methods for the structural characterization of cobalamin-dependent radical SAM (RS) enzymes and describe recent advances in in the overproduction and purification of these enzymes.

Methods in Enzymology, Volume 669
ISSN 0076-6879
https://doi.org/10.1016/bs.mie.2021.12.013

1. Introduction

Methylation is a critical reaction in all forms of life. It is involved in the modification of DNA, RNA, lipids, proteins, and a wide array of medically important secondary metabolites. SAM is the source of the appended methyl groups in the overwhelming majority of biological methylation reactions. Methyl groups are appended most often to N, O, and S atoms, and even carbon atoms that form carbanions. More recently, it has come to light that SAM can also be used to methylate inert carbon centers, meaning those that do not form carbanions under physiological conditions. These methylation reactions take place by mechanisms involving carbon-centered radicals and are catalyzed exclusively by enzymes within the radical SAM (RS) superfamily (Akiva et al., 2014; Frey, Hegeman, & Ruzicka, 2008; Sofia, Chen, Hetzler, Reyes-Spindola, & Miller, 2001). RS enzymes bind a [4Fe-4S] cluster that is used to catalyze a reductive cleavage of SAM to generate a 5′-deoxyadenosyl 5-radical (5′-dA•), which functions most often to abstract hydrogen atoms from enzyme-bound substrates (Broderick, Duffus, Duschene, & Shepard, 2014; Frey & Booker, 2001; Frey et al., 2008; Landgraf, McCarthy, & Booker, 2016; Sofia et al., 2001). RS enzymes, which number in the hundreds of thousands, catalyze over 100 unique reactions, including oxidative decarboxylation, hydroxylation, metallocofactor assembly, oxidation, mutase reactions, sulfur insertion and thioether bond formation, methylation and methylthiolation, among many others (Broderick et al., 2014; Challand, Driesener, & Roach, 2011; Holliday et al., 2018).

RS methylases (RSMTs) can append methyl groups to unactivated carbon and phosphinate phosphorus centers and are organized into distinct classes (Bauerle, Schwalm, & Booker, 2015; Miller et al., 2020). Class A RSMTs, which comprise the two enzymes RlmN and Cfr, use a cysteine dyad to methylate C2 and/or C8 of specific adenosine nucleotides in ribosomal or transfer RNAs (Toh, Xiong, Bae, & Mankin, 2008). Class B enzymes use cobalamin (Cbl) as a secondary cofactor to methylate sp^2- or sp^3-hybridized carbon centers as well as phosphinate phosphorus centers. Finally, class C RS methylases use two simultaneously bound molecules of SAM to methylate sp^2-hybridized carbon centers often found in intermediates during the biosynthesis of a number of complex natural products (Bauerle et al., 2015; Jin, Wu, Xu, Yuan, & Tang, 2019; Zhang, Mahanta, Hudson, Mitchell, & van der Donk, 2017). Class B or

Cbl-binding RS methylases are found in biosynthetic pathways of numerous antibiotics, anticancer agents, enzyme cofactors, and other natural products, including thienamycin, asparenomycin, thiostrepton, gentamicins, pactamycin, mitomycin C, moenomycin A, Fosfomycin, clorobiocin, fortimicin, quinomycin and chondrochloren, among others (Bridwell-Rabb & Drennan, 2017). However, despite their importance, their study has been complicated by solubility issues upon overproduction in *E. coli* (Sinner, Marous, & Townsend, 2021). Recently, strategies for generating soluble enzymes fully reconstituted with their required metallocofactors has led to advances in the understanding of this class of RS methylases, including the determination of several X-ray crystal structures of two of these enzymes. In this chapter, we share strategies that for determining X-ray crystal structures of two Cbl-binding RSMTs: *Ks*TsrM and TokK. *Ks*TsrM methylates C2 of tryptophan—an sp^2-hybridized carbon center—during the biosynthesis of thiostrepton, an antitumor antibiotic. By contrast, TokK methylates C6 of a carbapenam substrate—an sp^3-hybridized carbon center—during the biosynthesis of asparenomycin, a carbapenem antibiotic (Knox, Chen, et al., 2021; Knox, Sinner, Townsend, Boal, & Booker, 2021). A previous chapter in Methods in Enzymology described our strategies for overproducing and characterizing cobalamin-dependent RSMTs (Blaszczyk, Wang, & Booker, 2017). In this chapter, we discuss updates to those methods, but focus primarily on the structural characterization of these enzymes.

2. Overproduction and purification of cobalamin-dependent radical SAM methylases

Initial attempts to purify Cbl-binding RSMTs often resulted in low quantities of protein, which was found almost exclusively in inclusion bodies (Blaszczyk et al., 2017; Lanz et al., 2018). Several strategies designed to ameliorate this issue are detailed in a previous Methods in Enzymology contribution (Blaszczyk et al., 2017). First, we found that overproducing our proteins as a fusion on the C-terminus of SUMO—which in turn contained an N-terminal hexahistidine tag—slightly enhanced the solubility of the protein and allowed the protein to be purified to near homogeneity. The SUMO tag is a small protein-based tag (~12 kDa) that can be cleaved from the fusion construct—using the highly specific Ulp1 protease—in a way that that does not leave any residual amino acids on the target protein (Marblestone et al., 2006). Second, we found that the RosettaBlue (DE3) pLysS strain of *E. coli* combined with expression in an M9-ethanolamine medium supplemented with Cbl yielded low, but reproducible amounts of near-homogeneous protein (Blaszczyk et al., 2017). However, a strategy using a recently developed plasmid, called pBAD42-BtuCEDFB (Btu), has increased the solubility and yield of these Cbl-dependent RSMTs significantly (Lanz et al., 2018). The Btu plasmid encodes five proteins that aid in the uptake of Cbl into the cytoplasm of the cell (Lanz et al., 2018). The expression of the genes on the plasmid along with the gene of interest has led to a dramatic increase in the yield of the encoded protein, which in turn has facilitated structural studies. Below is an updated description of the methods used to overproduce *Ks*TsrM and TokK and purify to homogeneity.

2.1 Gene cloning and expression strategy

As described previously, the *tsrM* gene from *Kitasatospora setae* was cloned into the pSUMO vector (LifeSensors) to yield a construct that encodes a SUMO-TsrM fusion protein with an N-terminal hexahistidine tag (Blaszczyk et al., 2016; Knox, Chen, et al., 2021; Lanz et al., 2018). The pSUMO vector is often used for robust overexpression of proteins due to its ability to enhance the solubility of fused proteins as well as the unique ability to remove the SUMO tag without leaving any nonnative amino acids on the protein of interest (Marblestone et al., 2006; Panavas, Sanders, & Butt, 2009). The pSUMO vector confers kanamycin (Kan) resistance, which allows for a cotransformation of *E. coli* with pDB1282, a plasmid that confers ampicillin (Amp) resistance. Use of the pDB1282 plasmid was described in a previous Methods in Enzymology contribution (Lanz et al., 2012). It contains the *isc* operon from *Azotobacter vinelandii*, which comprises genes that encode proteins involved in the assembly of Fe-S clusters (Cicchillo et al., 2004; Zheng, Cash, Flint, & Dean, 1998). A third plasmid, pBAD42-BtuCEDFB, confers spectinomycin (Spec) resistance. The BL21 (DE3) strain of *E. coli* was transformed with the pDB1282, pSUMO-*Ks*TsrM, and Btu plasmids, yielding a strain that confers Kan, Amp, and Spec resistance.

A construct for the expression of the *tokK* gene has already been described and is the exact construct employed for overproducing the protein for X-ray crystallography, which was done in collaboration with the Townsend laboratory. TokK from *Streptomyces tokunonensis* was cloned into pET26b (Kan resistance), which produces a fusion protein having a

C-terminal hexahistidine tag with a small linker between the end of the *tokK* gene and the tag. This linker contains a TEV protease recognition sequence, which allows the hexahistidine tag to be excised from the protein (Sinner, Lichstrahl, Li, Marous, & Townsend, 2019). The TokK-pET26b, pDB1282, and Btu plasmids were used to transform the BL21 (DE3) strain of *E. coli* as above for gene expression and protein overproduction.

We typically overproduce Cbl-binding RSMTs in an M9-ethanolamine medium supplemented with Cbl (Blaszczyk et al., 2016; Lanz et al., 2018; Sinner et al., 2019). This medium is similar to standard M9-minimal medium but lacks the traditional carbon (glucose) and nitrogen (ammonium chloride) sources (Lanz et al., 2018). The use of ethanolamine as the sole nitrogen and carbon source leads to upregulation of AdoCbl-dependent ethanolamine ammonia lyase, which enhances Cbl uptake in Cbl-supplemented media (Bandarian & Matthews, 2001; Blackwell & Turner, 1978). *Ks*TsrM was overproduced in *E. coli* cultured in M9-ethanolamine media supplemented with hydroxoCbl acetate (Sigma) (Knox, Chen, et al., 2021), which was found to increase the yield of *Ks*TsrM threefold. TokK, however, was overproduced in *E. coli* cultured in Luria-Bertani (LB) medium supplemented with hydroxoCbl acetate (Knox, Sinner, et al., 2021). When overproduced as described for *Ks*TsrM, the isolated TokK did not yield good quality crystals. Because the methods for TsrM have been previously reported in a Methods in Enzymology contribution, we focus herein on overproduction of TokK, which is an adaptation of a previously described protocol by the Townsend laboratory (Blaszczyk et al., 2017). Overall, the methods are highly similar, with the major exception of the culturing medium.

2.2 Protocol for expression of the *tokK* gene

1. Simultaneous transformations are performed to introduce the pDB1282, Btu, and TokK-pET26b plasmids into *E. coli* BL21 (DE3) using standard procedures. Transformants are selected on LB plates containing Kan, Spec, and Amp at final concentrations of 50, 50, and 100 μg/mL, respectively
2. Generate a starter culture by using a single colony to inoculate 100 mL LB medium containing Kan, Spec, and Amp at 50, 50, and 100 μg/mL, respectively. Incubate at 37 °C with shaking at 250 rpm for ~20 h. Place four 6 L flasks, each containing 4 L of LB medium, in an incubator-shaker and equilibrate at 37 °C overnight

Table 1 Composition of 10,000 × micronutrient mixture and Cbl supplement.

10,000 × Micronutrient Mixture	**mg/L**
$(NH_4)_2Mo_7O_2$	37.1
$CoCl_2·6H_2O$	71.4
H_3BO_3	247
$CuSO_4·5H_2O$	25
$MnCl_2·4H_2O$	198
$ZnSO_4·7H_2O$	28.8
500 mL of 10 × Cbl Supplement	
10,000 × Micronutrient Mixture	4 mL
$MgSO_4·7H_2O$	9.2 g
$CaCl_2·2H_2O$	0.6 g
Thiamine	37 mg
OHCbl acetate	120 mg

3. Prepare 10,000 × micronutrient mix by combining the ingredients in Table 1 in 500 mL water. Sterilize the solution by filtration before use. (Blaszczyk et al., 2017)
4. Prepare a 10 × Cbl supplement by combining 4 mL 10,000 × micronutrient mix, 9.2 g $MgSO_4·7H_2O$, 0.6 g $CaCl_2·2H_2O$, 37 mg thiamine, and 120 mg hydroxoCbl acetate (see Table 1) in 500 mL of water and sterilizing by filtration
5. Prior to inoculation, add the following to each flask: 50 μg/mL Kan, 50 μg/mL Spec, 100 μg/mL Amp, and 54 mL 10 × Cbl supplement
6. Inoculate each 6 L flask with ~12 mL of the 100 mL starter culture grown to an OD_{600} of ~2.0
7. At an OD_{600} of 0.3 (~4 h), add L-arabinose to a final concentration of 0.2% (*w/v*) and $FeCl_3$ and L-cysteine to final concentrations of 25 and 150 μM, respectively. L-arabinose initiates expression of genes both on the pDB1282 and on the Btu plasmids
8. At an OD_{600} of 0.6, place flasks in an ice bath for ~1 h. Adjust the temperature of the incubator to 18 °C. Place the flasks back into the

incubator and induce *tokK* gene expression by adding isopropyl-β-D-thiogalactopyranoside (IPTG) to a final concentration of 1 mM. Note: for *Ks*TsrM, IPTG is added to a final concentration of 250 μM

9. Incubate cultures for ~20 h at 18 °C with shaking at 180 rpm
10. Harvest the bacterial cells by centrifugation and freeze the resulting cell paste (~2.5 g/L) in liquid N_2. Store the cell paste in a liquid N_2 dewar

2.3 Purification strategy

The purification strategies for Cbl-binding RS have been detailed previously (Blaszczyk et al., 2017; Lanz et al., 2012). Both *Ks*TsrM and TokK contain hexahistidine tags that are used to purify them by immobilized metal affinity chromatography (IMAC). The his_6-tag on the SUMO-TsrM construct is excised using the Ulp1 protease, while the his_6-tag on the TokK construct is excised using the TEV protease. After protease cleavage, the proteins are reapplied to the IMAC resin (usually nickel nitrilotriacetic acid), which captures the His_6-tags and other impurities that bind nonspecifically to the matrix, resulting in enhanced purity (Marblestone et al., 2006). Note that TEV cleavage of the TokK construct results in an additional six amino acids appended to the C-terminus of TokK. For X-ray crystallography, removal of tags is generally preferred because they can inhibit crystal formation through unfavorable protein interactions. The tagged protein is typically reconstituted immediately after the protease cleavage step (Blaszczyk et al., 2017). The protease (Ulp1 or TEV) is incubated with the protein for 1 h at room temperature, which is immediately followed by reconstitution with iron, sulfide, and hydroxocobalamin overnight on ice, as described previously Blaszczyk et al., 2017). The resulting native (i.e., untagged) protein is applied to an S-200 column connected to an ÅKTA fast-protein liquid chromatography (FPLC) system for further purification by size-exclusion chromatography.

RS enzymes are inherently unstable in the presence of oxygen. Consequently, the greatest challenge in purifying them is maintaining sufficiently anoxic conditions. Therefore, the purification of TsrM and TokK as well as all subsequent manipulations of the proteins are performed in a Coy anaerobic chamber to exclude oxygen, as previously described (Blaszczyk et al., 2017; Lanz et al., 2012). The purification is conducted in 2 days (due to the tag cleavage step) without interruption and while maintaining the proteins on ice. The crystallography is performed entirely within a Coy anaerobic chamber as well, which includes initial screening by hand—or with a crystallography robot—and looping. The following sections will detail all crystallography protocols.

3. Protocols for crystallization and structure determination

Below, we describe two protocols involving different, but related approaches for 3D structure determination of *Ks*TsrM and TokK involving Fe anomalous scattering or molecular replacement for phase determination (Knox, Chen, et al., 2021; Knox, Sinner, et al., 2021). Anomalous scattering requires a heavy atom to be bound to the protein. If a heavy atom is not bound natively either as a cofactor or prosthetic group, one can often be soaked into proteins or crystals. Cbl-binding RSMTs contain bound [4Fe-4S] and Cbl cofactors, both of which can be used for anomalous scattering. We elected to use the Fe K-edge instead of the Co K-edge, because there are more Fe atoms present, resulting in a stronger Fe signal.

3.1 Crystallization and structure determination of *Ks*TsrM

3.1.1 Background

TsrM methylates C2 of the indole ring of Trp, forming 2-methylTrp (MeTrp) as the first step in the formation of the quinaldic acid moiety of thiostrepton A, a thiopeptide antibiotic that exhibits strong potency against Gram-positive pathogens (Fig. 1A) (Frenzel, Zhou, & Floss, 1990; Pierre et al., 2012; Zhou et al., 1989). Although it is annotated as an RS enzyme, unlike all other RS enzymes studied to date, it does not catalyze formation of any SAM-derived radical. The only SAM-related product of its reaction is *S*-adenosylhomocysteine (SAH), which is formed upon generating methylcobalamin (MeCbl) from SAM (Benjdia et al., 2015; Blaszczyk et al., 2016). The proposed mechanism for the TsrM reaction is shown in Fig. 1B, which is proposed to take place by electrophilic aromatic substitution reaction (Blaszczyk, Knox, & Booker, 2019; Knox, Chen, et al., 2021). In this mechanism, a methyl group is transferred from MeCbl to C2 of Trp, which is facilitated by removal of the N1 proton to make C2 sufficiently nucleophilic. The base that removes the N1 proton appears to be the carboxylate group of SAM. In traditional RS reactions, the carboxylate of SAM is coordinated to the unique iron of the [4Fe-4S] cluster throughout the reaction (Knox, Chen, et al., 2021). Given the fundamental difference between TsrM and all other RS enzymes studied to date, it was difficult to

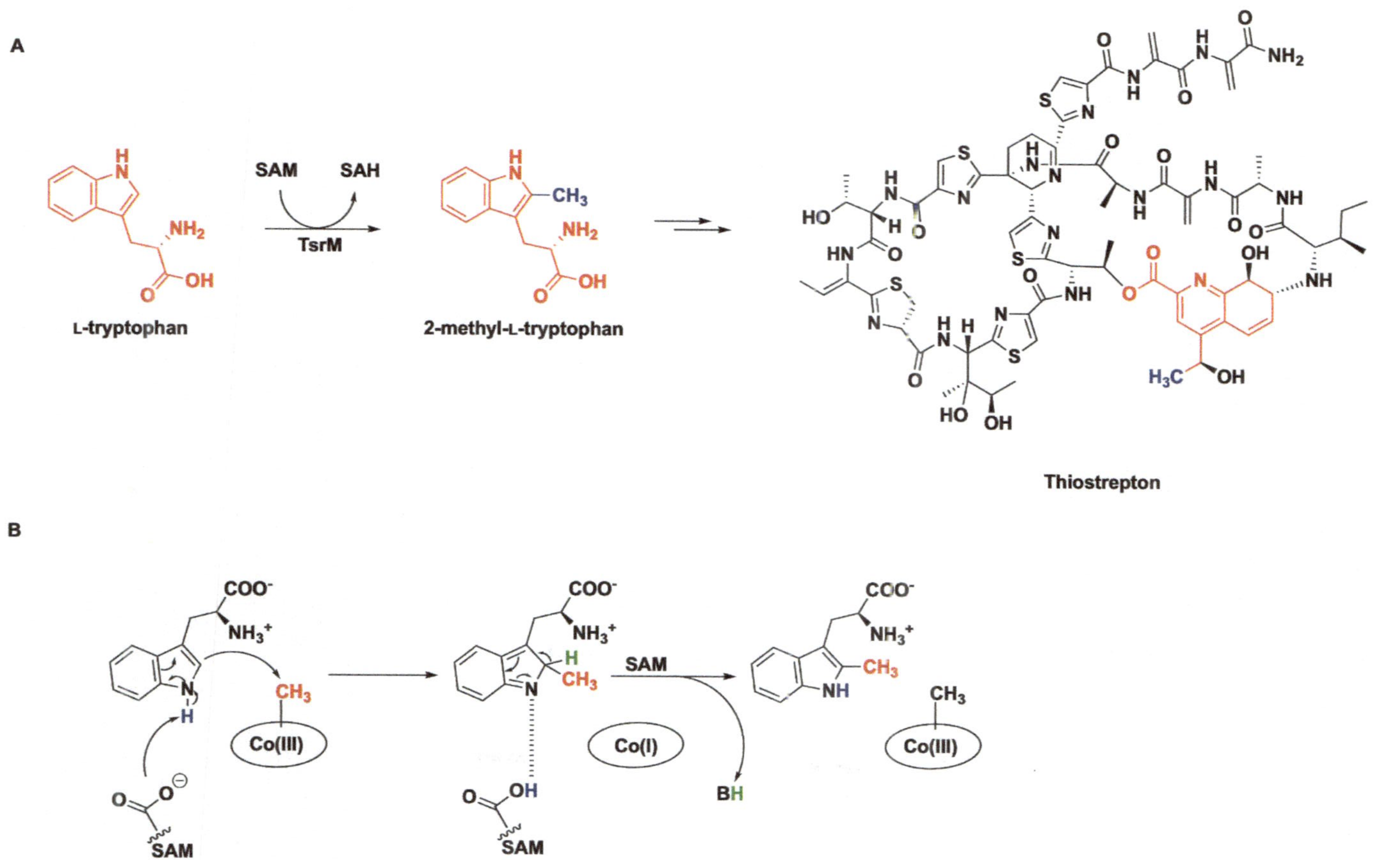

Fig. 1 Reaction catalyzed by TsrM. (A) TsrM methylates C2 of Trp to form MeTrp, a precursor to the quinaldic acid moiety of thiostrepton (red and blue). (B) Proposed non-radical mechanism performed by TsrM, where the carboxylate of SAM is used to as a base to remove the N1 proton during an electrophilic aromatic substitution by the methyl group of MeCbl.

understand, in the absence of a structure, what drives the TsrM reaction and why it is a member of the RS superfamily (Knox, Chen, et al., 2021). However, the TsrM used for most mechanistic studies, from *Streptomyces laurentii* (*Sl*TsrM), was difficult to crystallize. Therefore eight orthologs of *Sl*TsrM were identified, purified, and analyzed for their ability to form crystals (Knox, Chen, et al., 2021). One ortholog, from *K. setae* (*Ks*TsrM), generated crystals of suitable diffraction quality (Knox, Chen, et al., 2021).

3.1.2 Crystallization

Initial screens were set up using purified *Ks*TsrM (aliquoted at ~45 mg/mL) that was diluted in water to a concentration of ~8 mg/mL, resulting in the following components at their indicated concentrations from the storage buffer: 13.3 mM HEPES, pH 7.5, 53.2 mM KCl, 2.6% (*w*/*v*) glycerol, and 0.2 mM DTT. The screens (sitting drop) were set up by hand in a drop ratio of 1 μL protein to 1 μL precipitating solution. Initial crystal hits were obtained after approximately 1 day using 0.1 M KCl, 12% (*w*/*v*) PEG 8000, and 5% (*w*/*v*) glycerol as the precipitating solution. In hanging-drop vapor-diffusion trials with the above precipitating solution, large brown plate-shaped crystals appeared within 15 h (Fig. 2A). Trials were initiated by adding 1 μL protein to 1 μL precipitating solution, which was followed by equilibration against 0.5 mL of 0.5 M LiCl well solution at room temperature. The conditions screened were the Classics and ComPAS Suite from Qiagen (Germantown, MD). The humidity in the glovebox was maintained below 50% (ideal crystal formation at 40%, which we determined empirically). We found that if the humidity of the glovebox were maintained above 50%, crystal formation took more than a week and crystal quality declined. If the humidity were maintained between 40% and 45%, crystals

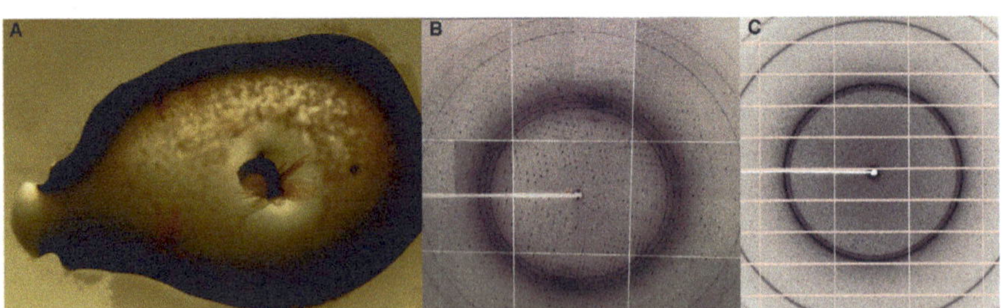

Fig. 2 *Ks*TsrM crystals. (A) *Ks*TsrM crystals that formed overnight as large, clusters of brown plates. (B) Typical diffraction of native *Ks*TsrM crystals. (C) Diffraction of *Ks*TsrM with non-merohedral twinning and small spots.

would form overnight. Crystals were briefly dipped in cryoprotectant (50% (*v*/*v*) precipitating reagent and 50% (*v*/*v*) ethylene glycol), mounted on nylon loops (Hampton Research Corp, Alieso Viejo, CA), and flash-cooled in liquid nitrogen. The crystals that yielded the highest-resolution (HR) *Ks*TsrM dataset formed in sitting-drop vapor diffusion trays set up by hand. Crystals formed overnight using 0.2 M sodium acetate, 0.1 M MES pH 6.5, and 15% (*w*/*v*) PEG 8000 as the precipitating solution. Trials were initiated by adding 1 μL protein to 1 μL precipitating solution, followed by equilibration against 0.5 mL of a 0.5 M LiCl well solution at room temperature. Crystals were briefly dipped in cryoprotectant (80% (*v*/*v*) precipitating reagent and 20% (*v*/*v*) ethylene glycol), mounted on nylon loops, and flash-cooled in liquid nitrogen.

Initial screens for the substrate bound *Ks*TsrM structure were performed as above. *Ks*TsrM was diluted as above and *S*-5′-azamethionine-5′-deoxyadenosine (aza-SAM) and Trp were added to final concentrations of 2 and 4 mM, respectively. The protein solutions were incubated for 15 min prior to setting up crystal screens. Initial screens were set up by hand in a drop ratio of 1 μL protein to 1 μL precipitating solution. Initial crystal hits were obtained after approximately 1 day using 0.1 M trisodium citrate, pH 5.6, 20% (*v*/*v*) isopropanol, and 20% (*w*/*v*) PEG 4000 as the precipitating solution. In hanging-drop vapor-diffusion trials with the above precipitating solution supplemented with 0.05 M nickel (II) dichloride hexahydrate, small brown plate-shaped crystals appeared within 15 h. Trials were initiated by adding 1 μL protein to 1 μL precipitating solution, followed by equilibration against either a 0.5 mL well solution of 0.5 M LiCl or a 0.5 M well solution of calcium acetate at room temperature. As before, humidity in the glovebox was maintained below 50%. Crystals were prepared for data collection by a brief soak in perfluropolyether cryo oil (Hampton Research) for cryoprotection, followed by mounting on nylon loops and flash-freezing in liquid nitrogen.

3.1.3 Structure determination

The single anomalous dispersion (SAD) technique was used to phase *Ks*TsrM diffraction data. Diffraction data for the iron K-edge absorption peak ($\lambda = 1.7433$ Å) with 360° of data measured using a 0.2° oscillation range were collected to 2.50 Å resolution. On the same crystal, a 2.00 Å resolution native dataset was collected at $\lambda = 1.0332$ Å. The anomalous and native *Ks*TsrM datasets were processed using the HKL2000 or HKL3000 package (Minor, Cymborowski, Otwinowski, & Chruszcz, 2006). Phenix Autosol, which uses HySS to identify heavy atom sites, was used to analyze the anomalous dataset (Adams et al., 2010). Eleven heavy atom sites were identified in total. Because TsrM has two molecules in the asymmetric unit, each containing 1 Co ion in the cobalamin and 4 Fe irons in the [4Fe-4S] cluster, we expected that there would be 10 heavy atom sites. Upon further examination, we determined that the eleventh site was an artifact. After solvent flattening using RESOLVE, the initial figure-of-merit (FOM) for refinement of the 10 heavy atom sites (8 Fe and 2 Co sites) was 0.271 to 2.50-Å resolution (Adams et al., 2010). The FOM is a metric used to evaluate how well the program identifies the heavy atom sites. A value of 0.271 is modest, but the density-modified maps were of sufficient quality to build an initial model. Phenix Autobuild was used to construct the initial model (residues 1–170, 172–526, and 532–567 out of 575 in chain A and 1–165, 179–482, 490–519, and 533–564 in chain B), which was subjected to one cycle of rigid body refinement to generate phase information for the native dataset (2.00-Å-resolution) by molecular replacement using Phenix Phaser-MR (Adams et al., 2010). The R_{free} flags were maintained across all datasets, so the same 5% of the reflections were used for the test set. Coot was used to add in cobalamin and the [4Fe-4S] cluster in the early stages of refining the model against the native data (Emsley, Lohkamp, Scott, & Cowtan, 2010). The final structure was refined using iterative rounds of positional refinement and individual *B*-factor refinement in Phenix. Water and small molecules were added in the later stages of refinement. Model building and adjustment were performed in Coot (Emsley et al., 2010). Non-crystallographic symmetry restraints were used throughout.

The crystals for the HR native *Ks*TsrM (PDB ID: 6WTE) and the substrate bound *Ks*TsrM (PDB ID: 6WTF) datasets were both non-merohedrally twinned crystals and thus could not be processed as above (Fig. 2B). Non-merohedrally twinning usually results from two identical crystals that have overlapped, resulting in two overlapping diffraction patterns (Sevvana, Ruf, Uson, Sheldrick, & Herbst-Irmer, 2019). Because of the nature of this twinning, it is important to separate the overlapping lattices. Both datasets were processed using XDS (Kabsch, 2010a, 2010b). The reflections from each lattice were split from each other during the COLSPOT stage of XDS. The lattices were individually indexed, integrated, and scaled, and only the reflections from the cleanest lattice were used for subsequent analysis. Because there was a prevalent ice ring for both datasets, all reflections between the resolution range 3.65–3.68 Å were excluded. The HR native structure (1.67-Å) was collected at $\lambda = 0.97872$ Å. Rigid body refinement from phenix.refine was used to refine the 2.00-Å model into the HR map data with the same R_{free} test sets (Adams et al., 2010). As before, the structure was refined through iterative rounds of positional and individual *B*-factor refinements performed in Phenix with model building and adjustment performed in Coot. In the later stages of refinement, Translation/Libration/Screw (TLS) parameters were used. The substrate-bound structure was processed as above, using the same R_{free} test set and refining the HR model into the map.

The refinement constraints for the cobalamin were generously provided by O. Smart at Global Phasing. The [4Fe-4S] cluster restraints were based on the values for the structure of *Moorella thermoacetica* carbon monoxide dehydrogenase/acetyl-CoA synthase (PDB ID: 3I01). The aza-SAM restraints were generated from the Grade Web Server (Global Phasing). The coordination distance between the carboxylate side chain of Glu273 and the unique Fe of the cluster was set to 2.0 Å based on the F_o-F_c map and previously observed values (Harding, Nowicki, & Walkinshaw, 2010).

3.2 Crystallization and structure determination of TokK

3.2.1 Background

Because of their potency and broad-spectrum activity, carbapenems are critical in the fight against bacterial infection and are considered antibiotics of last resort (Bodner et al., 2011; Coulthurst, Barnard, & Salmond, 2005). The clinically-used carbapenem antibiotics are distinct from other known classes of β-lactam antibiotics due to a C6 alkyl side chain that provides resistance to β-lactamases (Bodner et al., 2011). The assembly of the C6 alkyl side chain is accomplished by a Cbl-dependent RS enzyme through serial methyl transfers with stereochemical control for each step. ThnK from *Streptomyces cattleya* is a Cbl-dependent RS enzyme responsible for performing two sequential methylations in the biosynthesis of thienamycin, with the first appended methyl group being stereoselective for the *R*-configuration (Fig. 3A) (Marous et al., 2015). A highly similar enzyme, TokK, performs three sequential methylations to form the C6 isopropyl chain of asparenomycin A, with the first methylation also being stereoselective for the *R*-configuration (Fig. 3B) (Sinner et al., 2019). The ability of these enzymes to perform iterative sequential methylations while controlling the stereochemical outcome of the first methylation was of much interest, as well as how Cbl-dependent RS methylases function in general

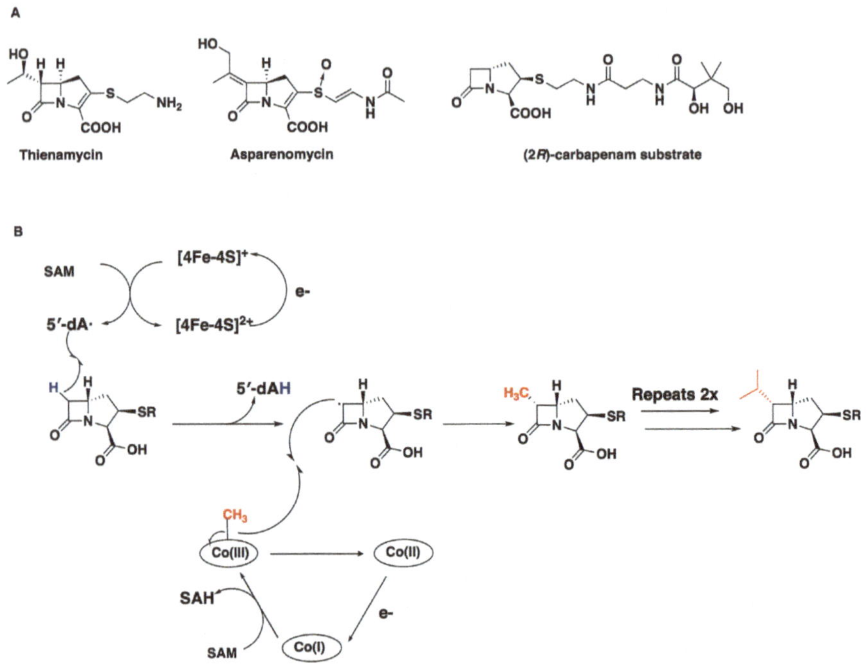

Fig. 3 TokK substrate and reaction scheme. (A) Chemical structures of thienamycin, asparenomycin, and (2*R*)-carbapenam substrate for TokK. (B) Proposed reaction catalyzed by TokK.

(Knox, Sinner, et al., 2021; Sinner et al., 2019). While ThnK was resistant to crystallization, TokK crystallized quite readily.

3.2.2 Crystallization

Preliminary crystallization screens were set up using an Oryx8 robot from Douglas Instruments (Berkshire, UK) (Fig. 4). Initial screens were performed with purified TokK (aliquoted at ~38 mg/mL) that was diluted to 8 mg/mL in 34 mM HEPES, pH 7.5. 5′-dAH and Met (Millipore Sigma, St. Louis, MO) were added to the solution to final concentrations of 2 mM each. The mixture was incubated for approximately 30 min at room temperature before screening with the Oryx8. Initial screens were set up in a drop ratio of 0.3 μL protein to 0.3 μL precipitating solution. Ten different screens were utilized: Classics, ComPAS, PEGs Suite I, PEGs Suite II, Natrix, and ProComplex screens from NeXtal Biotech (Holland, OH); the Berkeley and JCSGTop96 screens from Rigaku Reagents (Bainbridge Island, WA); and the XP and PACT++ screens from JenaBioscience

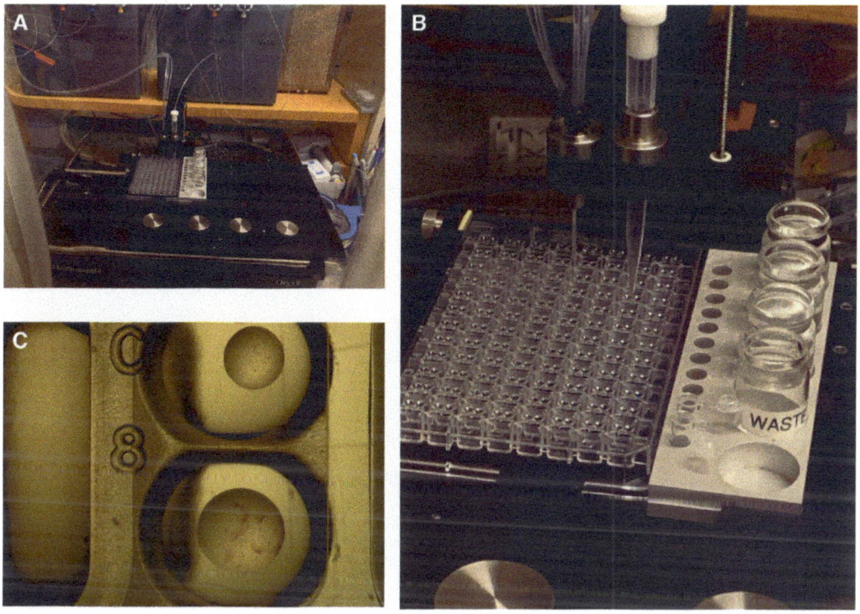

Fig. 4 Use of the Oryx8 in an anaerobic chamber. (A) Overview of the Oryx8 crystallography robot in a Coy anaerobic chamber. (B) Close-up of the Oryx8 pipetting system with two-drop tray. (C) Example of drops pipetted by the Oryx8. The top drop has no seed solution added while the bottom drop had seed solution added, resulting in brown plate crystals.

(Jena, Germany). Initial crystal hits were obtained after about a week using 100 mM magnesium chloride, 100 mM calcium chloride, 20% PEG 8000, and 10% 1,6-hexanediol as the precipitating solution (also referred to as the mother liquor). Those crystals (small brown plates) were used to make a seed stock by looping 10–20 crystals into a 1 μL drop of the mother liquor. The drop was then transferred to 50 μL of mother liquor in an Eppendorf tube with glass seed beads (Hampton Research Corp, Alieso Viejo, CA). The Eppendorf tube was vortexed for approximately 2 min to form the seed stock. Prior to use, the seed stock was was diluted 1:100 in mother liquor solution. In hanging-drop vapor-diffusion trials set up by hand with the above precipitating solution, brown plate-shaped crystals appeared within 3 days. Trials were initiated by adding 1 μL protein to 1 μL precipitating solution and 0.5 μL of seed solution, followed by equilibration against 0.5 mL of a 0.5 M LiCl well solution at room temperature. The humidity in the glovebox was kept between 45% and 55%. The humidity did not appear to have the same effect on the rate of crystal formation as in the case of *Ks*TsrM, but it was kept consistent for reproducibility. Crystals were prepared for data collection by a brief soak in cryoprotectant solution (50% (*v*/*v*) precipitating reagent and 50% (*v*/*v*) ethylene glycol), which was followed by mounting on nylon loops and flash-cooling in liquid nitrogen (Fig. 5).

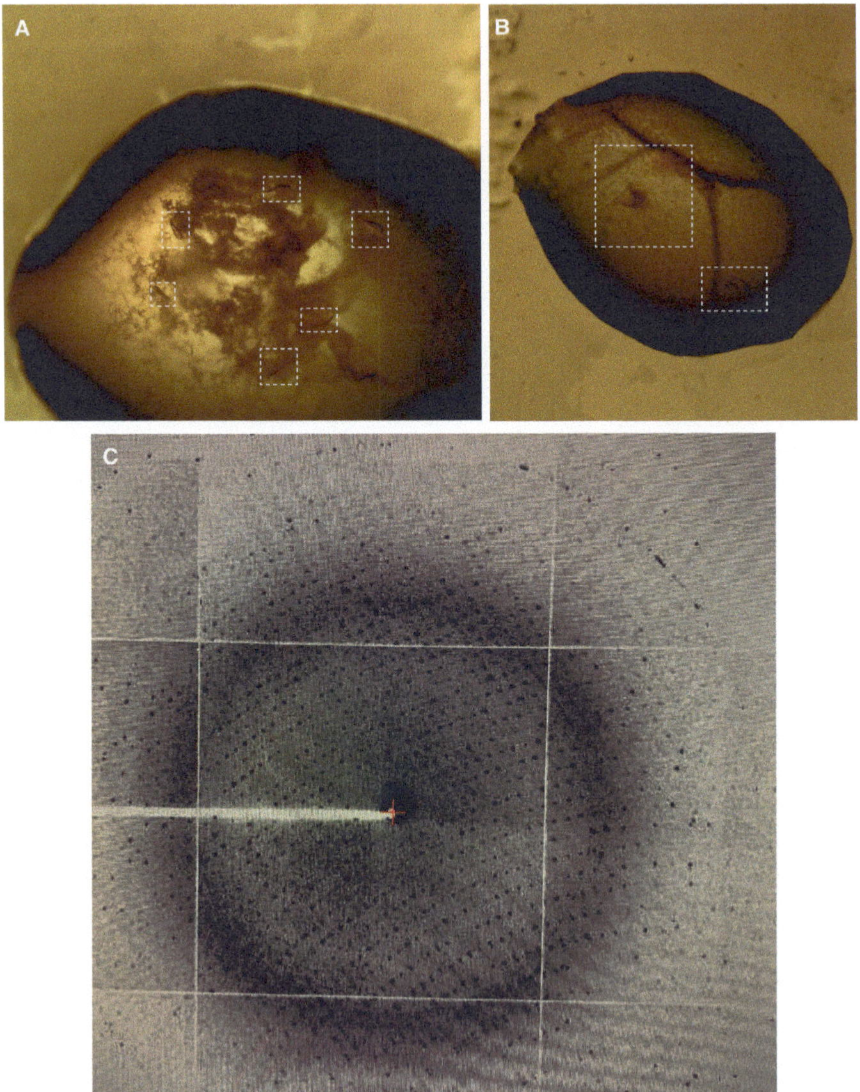

Fig. 5 TokK crystals. (A) Example of TokK crystals with 5′-dAH and Met bound. (B) Example of TokK crystals with 5′-dAH, Met, and (2*R*)-carbapenam substrate bound. (C) Typical TokK diffraction pattern.

Initial screens in the presence of substrate were set up as described above. Purified TokK (aliquoted at ~38 mg/mL) was diluted to 8 mg/mL in 34 mM HEPES, pH 7.5. 5′-dAH, Met, and (2*R*)-carbapenam substrate were added to the solution to a final concentration of 2 mM each. The mixture was incubated for ~15 min at room temperature before initiating screening with the Oryx8. Preliminary screens were set up in a drop ratio of 0.3 μL protein to 0.2 μL precipitating solution to 0.1 μL seed solution. Initial hits were obtained after approximately a week. In hanging-drop vapor-diffusion trays with 0.2 M lithium sulfate, 0.1 M Tris-HCl, pH 8.5, and 18% (*w*/*v*) PEG 8000 as the precipitant, brown plate-shaped crystals appeared within 2 weeks. Trials were initiated by mixing 1 μL protein with 1 μL precipitant and 0.5 μL seed solution followed by equilibration at room temperature against a 0.5 mL reservoir of the precipitating solution. Humidity of the glovebox was kept between 45% and 55%. Prior to looping, substrate concentration was increased in the drop to 2.7 mM and allowed to sit for ~1 h. Crystals were prepared for data collection by a brief soak in perfluropolyether cryo oil (Hampton Research) for cryoprotection, followed by mounting on nylon loops and flash-cooling in liquid nitrogen.

3.2.3 Structure determination

The SAD technique was also used to phase TokK diffraction datasets. A diffraction dataset for the iron K-edge absorption peak ($\lambda = 1.72194$ Å) with 360° of data measured using a 0.5° oscillation range was collected to 2.52 Å resolution. Additionally, a 1.79 Å-resolution native dataset was collected at $\lambda = 1.03313$ Å. All datasets were processed using either HKL2000 or HKL3000 (Minor et al., 2006). As before, heavy atom sites were identified by HySS implemented within Phenix Autosol (Adams et al., 2010). In this case, HySS identified 13 sites total, which is close to the expected 10 sites for the 2 molecules of TokK within the asymmetric unit (each TokK containing 4 Fe ions in a [4Fe-4S] cluster and 1 Co ion in the Cbl molecule). Two sites were likely artifacts based on the output data (Adams et al., 2010). Nine of the sites were identified as Fe sites. HySS had assigned three Fe sites to an entire [4Fe-4S] cluster for two of the clusters in the asymmetric unit while the remaining three Fe sites were from an adjacent asymmetric unit. The remaining two sites belonged to Co. The initial overall FOM was 0.265–2.50 Å resolution after solvent flattening in RESOLVE. Again, despite the modest FOM, the density-modified experimental maps were of sufficient quality for Phenix Autobuild to construct an initial model (residues 11–138, 148–412, 437–509, 535–566, 573–592, 600–616, 642–650 out of 682 total in chain A and residues 10–155, 165–208, 216–412, 414–427, 433–509, 512–530, 536–567, 573–595, 640–665 in chain B). Unlike in the case of *Ks*TsrM, a less complete model was built by Autobuild, and more of this structure was built manually. This model was subjected to several iterative rounds of positional refinement and group B-factor refinement, which was performed in Phenix, while model building and model adjustment were performed in Coot (Emsley et al., 2010). This model was used to obtain phase information for the 1.79 Å resolution native dataset by molecular replacement using Phenix Phaser-MR (Adams et al., 2010). R_{free} flags were maintained so that the same 5% of the reflections were used as the test set. Coot was used to add the [4Fe-4S] cluster and the Cbl in the early stages of refinement of the model against the native dataset, whereas water and small molecules were added in the later stages. The structure was refined using iterative rounds of positional refinement and individual B-factor refinement, as described for the Fe anomalous dataset (Adams et al., 2010; Emsley et al., 2010). Non-crystallographic symmetry restraints and TLS parameters were used. The substrate-bound structure was processed the same as above and Phaser-MR was used to place the model into the map. The structure was then manually adjusted with Coot, fitting in substrates if appropriate and building in loops. Restraints for the (2*R*)-carbapenam substrate were created by the PRODRG 2 server (Schuttelkopf & van Aalten, 2004). Restraints for 5′-dAH were obtained from the Grade Web Server (Global Phasing).

3.2.4 Identification of cobalamin form from the crystal

Use of the pBAD42-BtuCEDFB plasmid results in an increase in the cellular Cbl content, which is predominantly AdoCbl and OHCbl, with small amounts of MeCbl (Lanz et al., 2018). While MeCbl is a key intermediate in Cbl-dependent RSMT reactions, AdoCbl and OHCbl have been observed to bind opportunistically to these enzymes, especially when overproduced in the presence of the pBAD42-BtuCEDFB plasmid. While as-isolated *Ks*TsrM contains only AdoCbl and OHCbl, as isolated TokK contains a mixture of all three Cbl forms. Because it can be difficult to distinguish between OHCbl and MeCbl *in crystallo*, we used high-resolution mass spectrometry (HR-MS) to identify the form of Cbl bound (Knox, Sinner, et al., 2021).

The protocol was adapted from previously published studies on Cbl RSMTs (Fig. 6) (Lanz et al., 2018). Approximately 15 crystals of TokK with 5′-dAH and Met were looped from the crystallization drop and placed into a

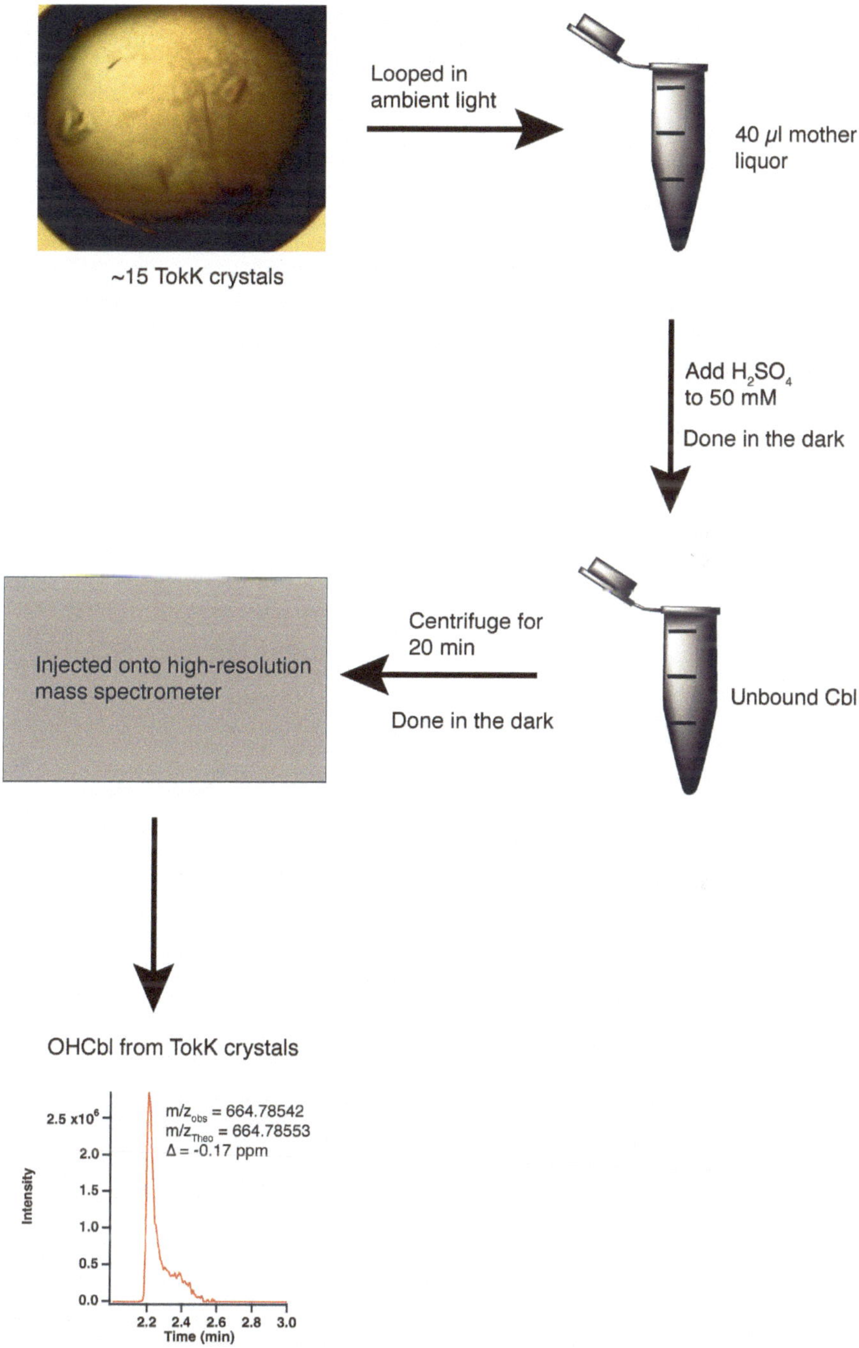

Fig. 6 Workflow for extracting Cbl from TokK crystals. The crystals are looped from the drop into the mother liquor in a dark or light-sensitive Eppendorf tube. The protein is precipitated by acid in the dark prior to centrifugation and mass spectrometry analysis. Mass spectrometry analysis shows that OHCbl (observed $m/z = 664.78542$; theoretical $m/z = 664.78553$) is the overwhelming cobalamin species in TokK crystals.

darkly colored Eppendorf tube containing 40 µL mother liquor. In the dark, H_2SO_4 was added to the tube at a final concentration of 50 mM. The tube was vortexed vigorously for 1 min to solubilize any remaining crystals. The sample was centrifuged for approximately 20 min at 4 °C. 5 µL of the solution was injected into a Thermo Fisher Scientific UHPLC/QExactive HF-X mass spectrometer equipped with a (2.1 × 100 mm) C18 column equilibrated in 5% solvent A (0.1% formic acid) and 95% solvent B (0.1% formic acid in acetonitrile). From 1 to 7 min, solvent B composition was increased linearly to 98%. Cbl forms were detected by ESI^+ by scanning from m/z 150–1700 with a resolution of 120,000. A calibration curve of Cbl standards (0.1–5 µM) was run concurrently to determine the relative Cbl forms in the sample (Knox, Sinner, et al., 2021).

4. Structure analysis

Despite the differences in substrate and mode of mechanism, the structures of *Ks*TsrM and TokK are highly similar (Fig. 7). They comprise the same three modular domains: an N-terminal Rossmann fold that binds Cbl; a shortened TIM barrel that contains the three Cys residues that ligate

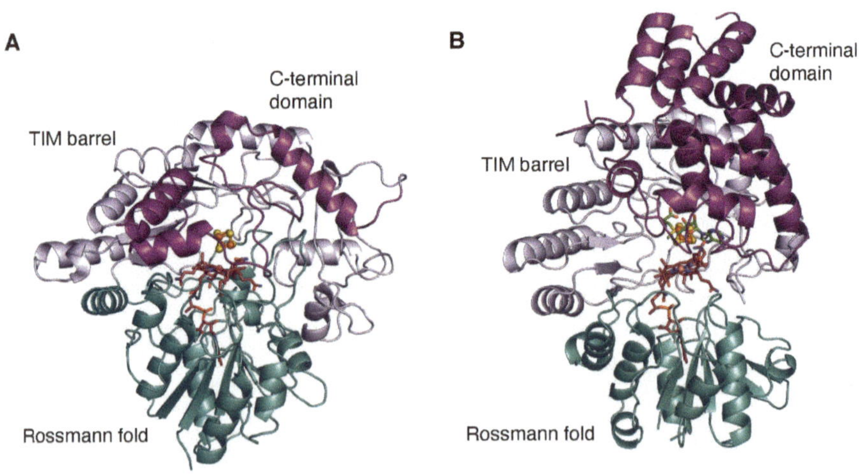

Fig. 7 Structural comparison of *Ks*TsrM (A) and TokK (B) in ribbon diagram. Both are comprised of the same three domains: the Rossmann fold that binds Cbl (teal); the RS domain, which is a shortened TIM barrel (light blue); and a C-terminal domain (purple). The Cbl is shown in red and the [4Fe-4S] cluster is shown in yellow and orange spheres. TokK has 5′-dAH and Met bound, shown as green sticks.

to the cluster, also referred to as the RS domain; and a C-terminal domain, which is a bit different for each structure (Knox, Chen, et al., 2021; Knox, Sinner, et al., 2021). The overall folds are highly similar except for the C-terminal domain. *Ks*TsrM has a mobile loop in the C-terminal domain that moves in the presence of substrate to close off the active site, while TokK has a more expansive domain comprised of several helices (Knox, Chen, et al., 2021; Knox, Sinner, et al., 2021). For TokK, the C-terminal domain helps to create a channel through which substrate can accesses the active site (Knox, Sinner, et al., 2021). While the overall folds are similar, the interactions with the cofactors are different and may play a role in controlling the reactivity. In the case of *Ks*TsrM, there is a conserved Arg residue on the proximal face of the Cbl (Knox, Chen, et al., 2021). Substitution of this residue results in near complete loss of activity, indicating the importance of this residue in catalysis. This residue does not coordinate to the cobalt ion of Cbl but is believed to prevent the Cbl cofactor from attaining a hexacoordinate geometry by blocking access to the cobalt ion by small molecules. Pentacoordinate alkylcobalamins are much less stable (susceptible to heterolysis) than hexacoordinate alkylcobalamins. Moreover, the positive charge on the Arg residue is believed to destabilize the pentacoordinate methylcobalamin further—which contains a cobalt ion with a + 3 charge—by charge-charge repulsion. Indeed, these characteristics likely explain how the weakly nucleophilic C2 carbanion of Trp can displace cob(*I*)alamin, the strongest nucleophile in biology (Knox, Chen, et al., 2021). For TokK, there is a conserved Trp residue on the proximal face, and substitution of this residue is more tolerated, with the exception of a Lys substitution, which results in a 50-fold loss in activity (Knox, Sinner, et al., 2021). Although Trp likely plays a similar role to Arg in *Ks*TsrM by blocking access of small molecules to the cobalt ion of Cbl, pentacoordinate alkylcobalamins are not significantly more destabilized toward homolytic cleavage of the cobalt-carbon bond than hexacoordinate alkylcobalamins. The [4Fe-4S] cluster is the other key cofactor in Cbl-dependent RSMTs. While both TokK and *Ks*TsrM bind the cluster with the expected three Cys residues in the canonical RS motif, *Ks*TsrM has a fourth ligand to the cluster, a conserved Glu (Knox, Chen, et al., 2021; Knox, Sinner, et al., 2021). For *Ks*TsrM, this fully ligated cluster occludes SAM binding to the cluster, preventing the reductive cleavage of SAM to a 5′-dA•. From these two structures, which offer the first glimpses into the Cbl-binding RSMT family, we begin to see how a nearly identical scaffold can be used to promote very different reactivities.

5. Conclusions

Cbl-binding RSMTs methylate both sp^2- and sp^3-hybridized carbon centers as well as phosphinate phosphorus centers in the biosynthesis of many biomolecules that play crucial roles in human health. The difficulties in overproduction and purification of these enzymes have complicated their study for some time, but recent advances have led to breakthroughs. Structural studies KsTsrM and TokK have led to a deeper understanding of this diverse class of enzymes and how the same domain architecture can be used to catalyze very different reactions. Continued structural characterization of the Cbl-binding RSMTs will likely yield more insight into the structural diversity that dictates mechanism.

Acknowledgments

This work was supported by NIH (GM122595 to S.J.B.) and the Eberly Family Distinguished Chair in Science (S.J.B.). S.J.B. is an investigator of the Howard Hughes Medical Institute. This research used resources of the Advanced Photon Source, a US Department of Energy (DOE) Office of Science User Facility operated for the DOE Office of Science by Argonne National Laboratory under contract no. DE-AC02-06CH11357. Use of GM/CA@APS has been funded in whole or in part with federal funds from the National Cancer Institute (ACB-12002) and the National Institute of General Medical Sciences (AGM-12006). The Eiger 16M detector at GM/CA-XSD was funded by NIH grant S10 OD012289. Use of the LS-CAT Sector 21 was supported by the Michigan Economic Development Corporation and the Michigan Technology Tri-Corridor (grant 085P1000817). This research also used the resources of the Berkeley Center for Structural Biology supported in part by the Howard Hughes Medical Institute. The Advanced Light Source is a DOE Office of Science User Facility under contract no. DE-AC02-05CH11231. The ALS-ENABLE beamlines are supported in part by the National Institutes of Health, National Institute of General Medical Sciences grant P30 GM124169.

References

Adams, P. D., Afonine, P. V., Bunkoczi, G., Chen, V. B., Davis, I. W., Echols, N., et al. (2010). PHENIX: A comprehensive python-based system for macromolecular structure solution. *Acta Crystallographica, Section D: Biological Crystallography, 66*, 213–221. https://doi.org/10.1107/S0907444909052925.

Akiva, E., Brown, S., Almonacid, D. E., Barber, A. E., Custer, A. F., Hicks, M. A., et al. (2014). The structure-function linkage database. *Nucleic Acids Research, 42*(D1), D521–D530. https://doi.org/10.1093/nar/gkt1130.

Bandarian, V., & Matthews, R. G. (2004). Measurement of energetics of conformational change in cobalamin-dependent methionine synthase. *Methods in Enzymology, 380*, 152–169. https://doi.org/10.1016/S0076-6879(04)80007-7.

Bauerle, M. R., Schwalm, E. L., & Booker, S. J. (2015). Mechanistic diversity of radical S-adenosylmethionine (SAM)-dependent methylation. *The Journal of Biological Chemistry, 290*(7), 3995–4002. https://doi.org/10.1074/jbc.R114.607044.

Benjdia, A., Pierre, S., Gherasim, C., Guillot, A., Carmona, M., Amara, P., et al. (2015). The thiostrepton a tryptophan methyltransferase TsrM catalyses a cob(II)alamin-dependent methyl transfer reaction. *Nature Communications*, *6*. https://doi.org/10.1038/ncomms9377.

Blackwell, C. M., & Turner, J. M. (1978). Microbial metabolism of amino alcohols. Purification and properties of coenzyme B12-dependent ethanolamine ammonia-lyase of Escherichia coli. *The Biochemical Journal*, *175*(2), 555–563. https://doi.org/10.1042/bj1750555.

Blaszczyk, A. J., Knox, H. L., & Booker, S. J. (2019). Understanding the role of electron donors in the reaction catalyzed by Tsrm, a cobalamin-dependent radical S-adenosylmethionine methylase. *Journal of Biological Inorganic Chemistry*, 1–9. https://doi.org/10.1007/s00775-019-01689-8.

Blaszczyk, A. J., Silakov, A., Zhang, B., Maiocco, S. J., Lanz, N. D., Kelly, W. L., et al. (2016). Spectroscopic and electrochemical characterization of the iron sulfur and cobalamin cofactors of TsrM, an unusual radical S-adenosylmethionine methylase. *Journal of the American Chemical Society*, *138*(10), 3416–3426. https://doi.org/10.1021/jacs.5b12592.

Blaszczyk, A. J., Wang, R. X., & Booker, S. J. (2017). TsrM as a model for purifying and characterizing cobalamin-dependent radical S-adenosylmethionine methylases. *Methods in Enzymology*, *595*, 303–329. https://doi.org/10.1016/bs.mie.2017.07.007.

Bodner, M. J., Li, R., Phelan, R. M., Freeman, M. F., Moshos, K. A., Lloyd, E. P., et al. (2011). Definition of the common and divergent steps in carbapenem beta-lactam antibiotic biosynthesis. *Chembiochem*, *12*(14), 2159–2165. https://doi.org/10.1002/cbic.201100366.

Bridwell-Rabb, J., & Drennan, C. L. (2017). Vitamin B-12 in the spotlight again. *Current Opinion in Chemical Biology*, *37*, 63–70. https://doi.org/10.1016/j.cbpa.2017.01.013.

Broderick, J. B., Duffus, B. R., Duschene, K. S., & Shepard, E. M. (2014). Radical S-adenosylmethionine enzymes. *Chemical Reviews*, *114*(8), 4229–4317. https://doi.org/10.1021/cr4004709.

Challand, M. R., Driesener, R. C., & Roach, P. L. (2011). Radical S-adenosylmethionine enzymes: Mechanism, control and function. *Natural Product Reports*, *28*(10), 1696–1721. https://doi.org/10.1039/c1np00036e.

Cicchillo, R. M., Iwig, D. F., Jones, A. D., Nesbitt, N. M., Baleanu-Gogonea, C., Souder, M. G., et al. (2004). Lipoyl synthase requires two equivalents of S-adenosyl-L-methionine to synthesize one equivalent of lipoic acid. *Biochemistry*, *43*(21), 6378–6386. https://doi.org/10.1021/bi049528x.

Coulthurst, S. J., Barnard, A. M., & Salmond, G. P. (2005). Regulation and biosynthesis of carbapenem antibiotics in bacteria. *Nature Reviews. Microbiology*, *3*(4), 295–306. https://doi.org/10.1038/nrmicro1128.

Emsley, P., Lohkamp, B., Scott, W. G., & Cowtan, K. (2010). Features and development of coot. *Acta Crystallographica, Section D: Biological Crystallography*, *66*, 486–501. https://doi.org/10.1107/S0907444910007493.

Frenzel, T., Zhou, P., & Floss, H. G. (1990). Formation of 2-methyltryptophan in the biosynthesis of thiostrepton: Isolation of S-adenosylmethionine:Tryptophan 2-methyltransferase. *Archives of Biochemistry and Biophysics*, *278*(1), 35–40.

Frey, P. A., & Booker, S. J. (2001). Radical mechanisms of S-adenosylmethionine-dependent enzymes. *Advances in Protein Chemistry*, *58*(58), 1–45.

Frey, P. A., Hegeman, A. D., & Ruzicka, F. J. (2008). The radical SAM superfamily. *Critical Reviews in Biochemistry and Molecular Biology*, *43*(1), 63–88. https://doi.org/10.1080/10409230701829169.

Harding, M. M., Nowicki, M. W., & Walkinshaw, M. D. (2010). Metals in protein structures: A review of their principal features. *Crystallography Reviews*, *16*(4), 247–302. https://doi.org/10.1080/0889311x.2010.485616.

Holliday, G. L., Akiva, E., Meng, E. C., Brown, S. D., Calhoun, S., Pieper, U., et al. (2018). Atlas of the radical SAM superfamily: Divergent evolution of function using a "plug and play" domain. *Methods in Enzymology*, *606*, 1–71. https://doi.org/10.1016/bs.mie.2018.06.004.

Jin, W. B., Wu, S., Xu, Y. F., Yuan, H., & Tang, G. L. (2019). Recent advances in HemN-like radical S-adenosyl-l-methionine enzyme-catalyzed reactions. *Natural Product Reports*. https://doi.org/10.1039/c9np00032a.

Kabsch, W. (2010a). Integration, scaling, space-group assignment and post-refinement. *Acta Crystallographica. Section D, Biological Crystallography*, *66*(Pt 2), 133–144. https://doi.org/10.1107/S0907444909047374.

Kabsch, W. (2010b). Xds. *Acta Crystallographica, Section D: Biological Crystallography*, *66*(Pt 2), 125–132. https://doi.org/10.1107/S0907444909047337.

Knox, H. L., Chen, P. Y., Blaszczyk, A. J., Mukherjee, A., Grove, T. L., Schwalm, E. L., et al. (2021). Structural basis for non-radical catalysis by TsrM, a radical SAM methylase. *Nature Chemical Biology*. https://doi.org/10.1038/s41589-020-00717-y.

Knox, H. L., Sinner, E. K., Townsend, C. A., Boal, A. K., & Booker, S. J. (2021). Structural basis for carbapenam C-alkylation by TokK, a B12-dependent radical SAM enzyme. *bioRxiv*. 2021.2007.2010.451909. https://doi.org/10.1101/2021.07.10.451909.

Landgraf, B. J., McCarthy, E. L., & Booker, S. J. (2016). Radical S-adenosylmethionine enzymes in human health and disease. *Annual Review of Biochemistry*, *85*(85), 485–514. https://doi.org/10.1146/annurev-biochem-060713-035504.

Lanz, N. D., Blaszczyk, A. J., McCarthy, E. L., Wang, B., Wang, R. X., Jones, B. S., et al. (2018). Enhanced solubilization of class B radical S-adenosylmethionine methylases by improved cobalamin uptake in Escherichia coli. *Biochemistry*, *57*(9), 1475–1490. https://doi.org/10.1021/acs.biochem.7b01205.

Lanz, N. D., Grove, T. L., Gogonea, C. B., Lee, K. H., Krebs, C., & Booker, S. J. (2012). RlmN and AtsB as models for the overproduction and characterization radical SAM proteins. *Natural Product Biosynthesis*, *516*, 125–152. https://doi.org/10.1016/B978-0-12-394291-3.00030-7.

Marblestone, J. G., Edavettal, S. C., Lim, Y., Lim, P., Zuo, X., & Butt, T. R. (2006). Comparison of SUMO fusion technology with traditional gene fusion systems: Enhanced expression and solubility with SUMO. *Protein Science*, *15*(1), 182–189. https://doi.org/10.1110/ps.051812706.

Marous, D. R., Lloyd, E. P., Buller, A. R., Moshos, K. A., Grove, T. L., Blaszczyk, A. J., et al. (2015). Consecutive radical S-adenosylmethionine methylations form the ethyl side chain in thienamycin biosynthesis. *Proceedings of the National Academy of Sciences*, *112*, 10354–10358. https://doi.org/10.1073/pnas.1508615112.

Miller, D. V., Knox, H. L., Wang, B., Schwalm, E. L., Blaszczyk, A. J., & Booker, S. J. (2020). Radical S-adenosylmethionine methylases. In *Vol. 5. Comprehensive natural products III (third edition)* (pp. 24–69). https://doi.org/10.1016/B978-0-12-409547-2.14863-2.

Minor, W., Cymborowski, M., Otwinowski, Z., & Chruszcz, M. (2006). HKL-3000: The integration of data reduction and structure solution - from diffraction images to an initial model in minutes. *Acta Crystallographica, Section D: Biological Crystallography*, *62*, 859–866. https://doi.org/10.1107/S0907444906019949.

Panavas, T., Sanders, C., & Butt, T. R. (2009). SUMO fusion technology for enhanced protein production in prokaryotic and eukaryotic expression systems. *Methods in Molecular Biology*, *497*, 303–317. https://doi.org/10.1007/978-1-59745-566-4_20.

Pierre, S., Guillot, A., Benjdia, A., Sandström, C., Langella, P., & Berteau, O. (2012). Thiostrepton tryptophan methyltransferase expands the chemistry of radical SAM enzymes. *Nature Chemical Biology*, *8*(12), 957–959. https://doi.org/10.1038/nchembio.1091.

Schuttelkopf, A. W., & van Aalten, D. M. (2004). PRODRG: A tool for high-throughput crystallography of protein-ligand complexes. *Acta Crystallographica. Section D, Biological Crystallography*, *60*(Pt 8), 1355–1363. https://doi.org/10.1107/S0907444904011679.

Sevvana, M., Ruf, M., Uson, I., Sheldrick, G. M., & Herbst-Irmer, R. (2019). Non-merohedral twinning: From minerals to proteins. *Acta Crystallographica Section D: Structural Biology*, *75*(Pt 12), 1040–1050. https://doi.org/10.1107/S2059798319010179.

Sinner, E. K., Lichstrahl, M. S., Li, R., Marous, D. R., & Townsend, C. A. (2019). Methylations in complex carbapenem biosynthesis are catalyzed by a single cobalamin-dependent radical S-adenosylmethionine enzyme. *Chemical Communications (Cambridge, England)*, *55*(99), 14934–14937. https://doi.org/10.1039/c9cc07197k.

Sinner, E. K., Marous, D. R., & Townsend, C. A. (2021). Evolution of methods for the study of cobalamin-dependent radical SAM enzymes. *ACS Bio & Med Chem Au. 10.1021/acsbiomedchemau.1c00032(0)*. https://doi.org/10.1021/acsbiomedchemau.1c00032.

Sofia, H. J., Chen, G., Hetzler, B. G., Reyes-Spindola, J. F., & Miller, N. E. (2001). Radical SAM, a novel protein superfamily linking unresolved steps in familiar biosynthetic pathways with radical mechanisms: Functional characterization using new analysis and information visualization methods. *Nucleic Acids Research*, *29*(5), 1097–1106. https://doi.org/10.1093/nar/29.5.1097.

Toh, S. M., Xiong, L., Bae, T., & Mankin, A. S. (2008). The methyltransferase YfgB/RlmN is responsible for modification of adenosine 2503 in 23S rRNA. *RNA*, *14*(1), 98–106. https://doi.org/10.1261/rna.814408.

Zhang, Z., Mahanta, N., Hudson, G. A., Mitchell, D. A., & van der Donk, W. A. (2017). Mechanism of a class C radical S-adenosyl-l-methionine thiazole methyl transferase. *Journal of the American Chemical Society*, *139*(51), 18623–18631. https://doi.org/10.1021/jacs.7b10203.

Zheng, L., Cash, V. L., Flint, D. H., & Dean, D. R. (1998). Assembly of iron-sulfur clusters. Identification of an iscSUA-hscBA-fdx gene cluster from Azotobacter vinelandii. *The Journal of Biological Chemistry*, *273*(21), 13264–13272. https://doi.org/10.1074/jbc.273.21.13264.

Zhou, P., O'Hagan, D., Mocek, U., Zeng, Z., Yuen, L. D., Frenzel, T., et al. (1989). Biosynthesis of the antibiotic thiostrepton. Methylation of tryptophan in the formation of the quinaldic acid moiety by transfer of the methionine methyl group with net retention of configuration. *Journal of the American Chemical Society*, *111*(18), 7274–7276. https://doi.org/10.1021/ja00200a065.

CHAPTER TWO

Purification and characterization of sequential cobalamin-dependent radical SAM methylases ThnK and TokK in carbapenem β-lactam antibiotic biosynthesis

Erica K. Sinner and Craig A. Townsend*
Department of Chemistry, Johns Hopkins University, Baltimore, MD, United States
*Corresponding author: e-mail address: ctownsend@jhu.edu

Contents

Abstract

ThnK and TokK are cobalamin-dependent radical *S*-adenosylmethionine enzymes that catalyze sequential methylations of a common carbapenem biosynthetic intermediate. ThnK was an early characterized member of the subfamily of cobalamin-dependent radical *S*-adenosylmethionine enzymes. Since initial publication of the ThnK function, the field has progressed, and we have made methodological strides in the expression and purification of this enzyme and its ortholog TokK. An optimized protocol for obtaining the enzymes in pure and active form has enabled us to characterize their reactions and gain greater insight into the kinetic behavior of the sequential methylations they catalyze. We share here the methods and strategy that we have developed through our study of these enzymes.

Methods in Enzymology, Volume 669
ISSN 0076-6879
https://doi.org/10.1016/bs.mie.2021.09.014

1. Introduction

Carbapenem antibiotics are a class of β-lactam natural products that have been developed for clinical use in difficult-to-treat infections. Carbapenems are produced by soil-dwelling *Streptomycetes* (Papp-Wallace, Endimiani, Taracila, & Bonomo, 2011; Townsend, 2016), and three cobalamin (Cbl)-dependent radical *S*-adenosylmethionine (rSAM) enzymes are present in all known carbapenem biosynthetic gene clusters. All three have been shown to be essential for carbapenem production (Li, Lloyd, Moshos, & Townsend, 2014; Núñez, Méndez, Braña, Blanco, & Salas, 2003). These enzymes present myriad experimental challenges, including oxygen sensitivity and low solubility upon heterologous expression (Lanz et al., 2018). We have made significant progress in developing experimental methods that allow the study of one of the three enzymes, ThnK, which is involved in thienamycin (**1**) biosynthesis (Marous et al., 2015), and it's ortholog TokK, which plays a similar role in asparenomycin (**2**) biosynthesis (Kawamura, Yasuda, Mayama, & Tanaka, 1982; Sinner, Lichstrahl, Li, Marous, & Townsend, 2019) (Fig. 1). Both enzymes have the distinctive ability to sequentially transfer multiple methyl groups onto their substrates, a trait shared by another recently characterized Cbl-dependent rSAM CysS (Wang, Schnell, Baumann, Müller, & Begley, 2017). PoyC and PoyB, involved in polytheonamide biosynthesis, also appear to catalyze methylations at multiple sites along their polypeptide substrate PoyA (Freeman, Helf, Bhushan, Morinaka, & Piel, 2017; Parent et al., 2016).

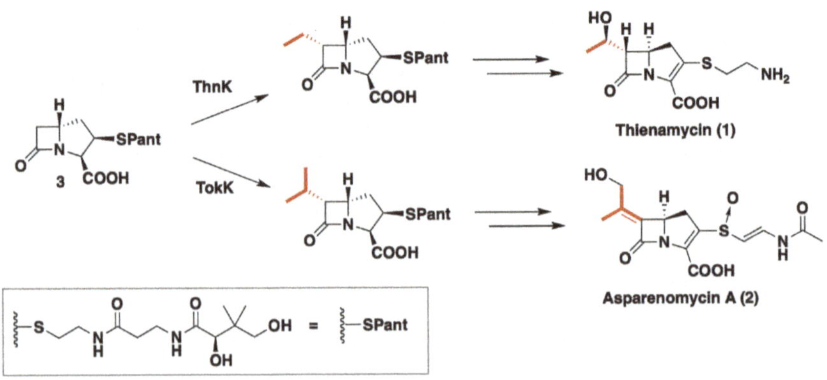

Fig. 1 The roles of ThnK and TokK in the biosynthesis of complex carbapenems thienamycin and asparenomycin, respectively.

ThnK and TokK build the C6 alkyl chain of the carbapenem, which, as a site of structural diversity, contains 1–3 carbon atoms and varies in stereochemistry, oxidation state, and sulfation. Accordingly, ThnK can transfer up to two methyl groups onto the thienamycin precursor **3**, and TokK is able to transfer three methyl groups onto the same substrate **3** *en route* to asparenomycin, as shown in Fig. 1 (Sinner et al., 2019). In both cases, a single Cbl-dependent rSAM enzyme is sufficient to install all of the methionine-derived carbons onto the carbapenem biosynthetic intermediate (Houck, Kobayashi, Williamson, & Floss, 1986; Williamson, 1986).

Both ThnK and TokK are Class B radical SAM methylases, which are classified by the use of a cobalamin cofactor in addition to a [4Fe–4S] cluster and SAM to achieve methyl transfer to unactivated carbon or phosphorus centers during the biosynthesis of a variety of natural products (Bauerle, Schwalm, & Booker, 2015; Wang, 2018; Zhou, Alkhalaf, de los Santos, & Challis, 2016). The majority of Cbl-dependent rSAM methylases, including ThnK and TokK, are believed to operate by the mechanism shown in Fig. 2 wherein one molecule of SAM donates a methyl group to Cob(I)alamin forming methylcobalamin (MeCbl), which acts as an intermediate methyl carrier. A second SAM molecule is reductively cleaved by the [4Fe–4S] cluster to generate a 5′-deoxyadenosyl radical (5′dA•), which abstracts a hydrogen atom from the substrate. The MeCbl species is then attacked by the substrate radical producing a methylated product and Cob(II)alamin, which requires a one-electron reduction to return to its highly nucleophilic Co(I) form for another turnover (Bridwell-Rabb, Grell, & Drennan, 2018; Kim et al., 2013). Both ThnK and TokK produce alkylated product, *S*-adenosylhomocysteine (SAH), and 5′-deoxyadenosine

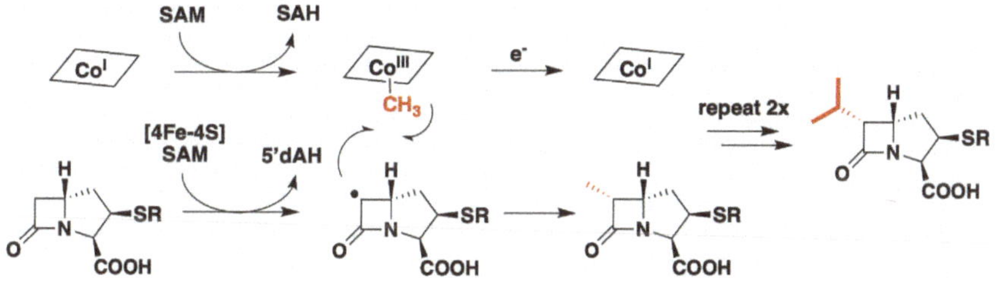

Fig. 2 Proposed mechanism of methyl transfer by Cbl-dependent rSAM enzyme TokK. *Reproduced from Sinner, E. K., Lichstrahl, M. S., Li, R., Marous, D. R., & Townsend, C. A. (2019). Methylations in complex carbapenem biosynthesis are catalyzed by a single cobalamin-dependent radical S-adenosylmethionine enzyme.* Chemical Communications, *55(99), 14934–14937. https://doi.org/10.1039/c9cc07197k with permission from the Royal Society of Chemistry.*

(5′dAH) in an approximately 1:1:1 ratio (Marous et al., 2015; Sinner et al., 2019), consistent with this mechanism. Perhaps because of the complexity of the cycle, catalysis by ThnK and TokK is slow, but we have developed conditions where the enzymes remain active for up to 48h, enabling time-course analysis of the sequential methylation reactions. Because the alkylations are irreversible, the time-course data can be readily fit to a computational model and independent rates can be simultaneously determined for each methyl transfer. This approach has offered powerful kinetic insight into a process that would otherwise be difficult to study. Each methyl transfer occurs at a different rate, and it is advantageous to be able to measure all three rate constants using a single assay that does not require synthesis or isolation of each intermediate.

2. Heterologous expression and purification of ThnK and TokK

Study of the Cbl-dependent rSAM subclass has been impeded by poor solubility of protein expressed in a heterologous host. To maximize overproduction of soluble ThnK or TokK in *Escherichia coli*, co-expression of two plasmids to support cofactor installation are required along with the pET29b plasmid encoding *thnK* or *tokK*. The first accessory plasmid is pDB1282, which harbors the *isc* gene cluster from *Azotobacter vinelandii,* aids in iron-sulfur cluster biogenesis (Lanz et al., 2012; Zheng, Cash, Flint, & Dean, 1998). The second, pBAD42-BtuCEDFB, is a plasmid that encodes genes involved in cobalamin uptake and transport (Lanz et al., 2018). Both accessory plasmids are under the control of an arabinose-inducible promotor, whereas the pET29b plasmid with the rSAM gene of interest is induced with IPTG. This design enables temporal separation of induction so that the proteins that promote proper cofactor installation are present in the cell before induction of the Cbl-dependent rSAM enzyme, which enhances both solubility and cofactor occupancy (Lanz et al., 2018).

To obtain pure and active ThnK or TokK, protein purification must be carried out anaerobically. After a single immobilized-metal affinity chromatography (IMAC) step, the eluted protein contains some impurities (Fig. 3A, lanes 7–8). Therefore, we employ a TEV-protease (Kapust et al., 2001) cleavable *C*-terminal His_6-tag, which allows a second IMAC step in which the desired protein no longer binds the resin, but impurities and the protease remain bound. With this added step, >95% pure protein is obtained (Fig. 3A lanes 10–12). Both ThnK and TokK are nearly 80kDa in size, and the

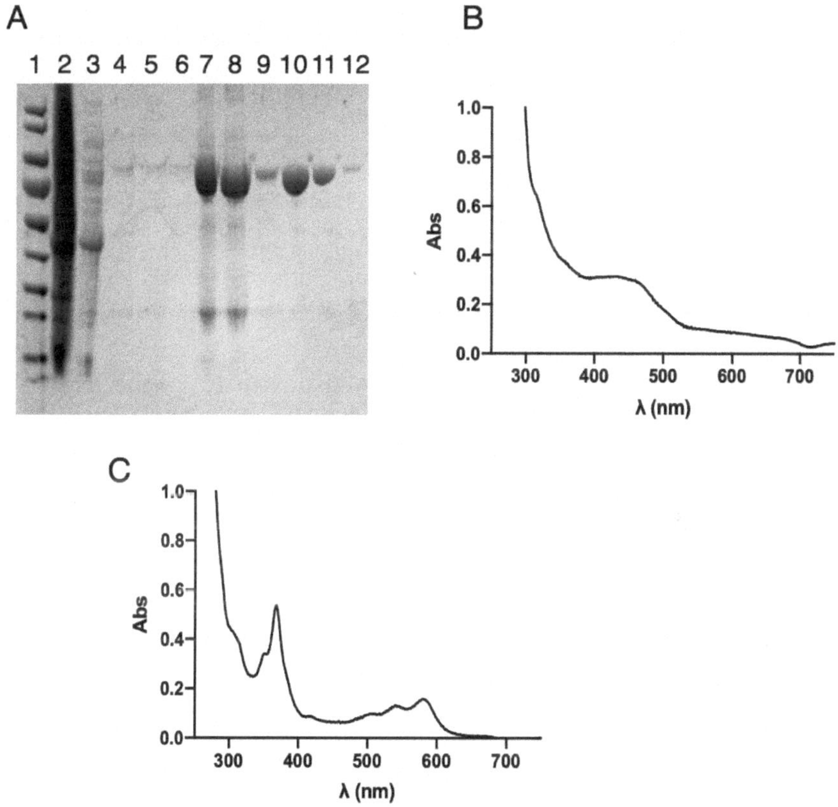

Fig. 3 Characterization of TokK. (A) SDS-PAGE gel of TokK purification, lanes are as follows. 1: PageRuler Protein Ladder (Thermo Scientific, Waltham, MA), 2: flow through, 3–4: wash fractions (lysis buffer), 5–9: elution fractions (1.5 mL each, elution buffer), 10–12: flow through and washes (desalting buffer) from post-TEV cleavage column. (B) UV–visible spectroscopy, and (C) UV–visible spectroscopy of bound cobalamin by treatment with 1 M KCN and heating at 95 °C for 5 min. *Reproduced from Sinner, E. K., Lichstrahl, M. S., Li, R., Marous, D. R., & Townsend, C. A. (2019). Methylations in complex carbapenem biosynthesis are catalyzed by a single cobalamin-dependent radical S-adenosylmethionine enzyme.* Chemical Communications, *55(99), 14934–14937. https://doi.org/10.1039/c9cc07197k with permission from the Royal Society of Chemistry.*

TEV-cleavable His_6-tag enables a flexible purification workflow without greatly increasing the mass of the protein. In a strategy similar to that used for the purification of TsrM (Blaszczyk et al., 2016; Blaszczyk, Wang, & Booker, 2017; Knox et al., 2021), chemical reconstitution with $FeCl_3$, Na_2S, and hydroxycobalamin (OHCbl) can be carried out at the same time as the His_6-tag cleavage reaction, giving maximum cofactor occupancy while minimizing the amount of time from cell lysis to purified protein.

2.1 Overproduction of ThnK/TokK

1. Insert the *thnK* or *tokK* sequence (accession numbers AJ421798 and MN385145, respectively) followed by nucleotides coding for the TEV-protease recognition sequence, ENLYFQ, into a pET29b vector to create a construct with a cleavable *C*-terminal His_6-tag.
2. Transform pDB1282 into BL21(DE3) *E. coli* cells and spread on LB plates containing Ampicillin (100 μg/mL). Grow these cells and prepare them to be electrocompetent using the protocol from New England BioLabs (New England BioLabs, 2012) or similar, and store in flash-frozen aliquots at −80 °C for up to a year.
3. Transform the pET29b plasmid prepared in step 1 and the pBAD42-BtuCEDFB plasmid into pDB1282-electrocompetent cells prepared in step 2. Spread on LB plates containing Kanamycin (50 μg/mL), Spectinomycin (50 μg/mL) and Ampicillin (100 μg/mL).
4. Pick a single colony produced in step 2 and start an overnight culture in LB with the same antibiotics at the concentrations used in step 2.
5. Use 10 mL/L of starter culture to inoculate 15 L LB supplemented with the same antibiotics and 1.3 uM OHCbl. Grow at 37 °C until OD = 0.35, then add solid arabinose (1 g/L), and sterile aqueous solutions of $FeCl_3$ (6.8 mg/L) and L-cysteine (24 mg/L).
6. Continue growth to OD = 0.75, then cold shock on ice ~2 h.
7. Repeat arabinose/$FeCl_3$/Cys induction and add IPTG (1 mM final concentration). Grow ~16 h at 18 °C.
8. Harvest cells by centrifugation at 4000 × *g* (4 °C), and store pellets in liquid N_2 until purification. Typical cell yields are approximately 4 g wet cell mass per liter of culture medium. Cell pellets should be lysed and protein purified as soon as possible following overproduction (within 1–2 weeks), as protein from cells stored for more than 1 month is less active than protein purified from freshly grown cells.

2.2 Anaerobic purification of ThnK/TokK

All steps other than centrifugation should be carried out in an anaerobic chamber with an atmosphere of approximately 98% N_2 and 2% H_2 (±0.5%) and an active palladium catalyst that combines any residual O_2 with H_2 to form H_2O (Coy Laboratory Products). Oxygen is removed from water by boiling outside the chamber until the volume is reduced by 25%. This water is then brought into the anaerobic chamber while still warm, followed by stirring open for approximately 72 h to equilibrate. Solid buffer components are introduced into the chamber in pure form and dissolved in anaerobic water to make stock solutions. A stock solution of 50% glycerol is prepared by introducing 100% glycerol into the chamber and mixing with an equal volume of anaerobic water, followed by stirring open in the chamber for at least 72 h. The presence of β-mercaptoethanol in purification buffers should minimize oxidative damage to the rSAM enzyme from any oxygen that still remains after equilibration of solutions. ThnK and TokK, like many rSAM enzymes, diminish in activity over time. Therefore, the purification should be done in two consecutive days and assays set up immediately upon obtaining pure protein.

1. Prepare buffers for ThnK/TokK purification. All contain 10% glycerol, 300 mM KCl, 50 mM HEPES pH 7.5, and 10 mM β-mercaptoethanol. Lysis buffer contains 5 mM imidazole, elution buffer contains 250 mM imidazole, and desalting buffer does not contain imidazole.
2. On ice, resuspend cell pellet (50–60 g) from overexpression in lysis buffer to 160 mL final volume with 1 mg/mL lysozyme. DNase and PMSF may be added as well, but neither seems to significantly affect protein yield or ease of lysis. After thawing and allowing lysozyme to work for 1 h, sonicate using a Cole-Parmer Ultrasonic Processor at 60%, 9.9 s on/off, for 5 min at a time, stirring to minimize heat buildup in any area. Once the solution is no longer viscous (usually 3–4 rounds of 5 min sonication), place in centrifuge tubes, seal with vinyl tape, and centrifuge at 35,000 × *g* (4 °C) for 35 min.
3. Decant supernatant into a 250 mL media bottle or similar container and add pre-equilibrated Clontech TALON cobalt affinity resin (8 mL 50% suspension). Batch bind on ice for 20 min.
4. Apply mixture to column and collect flow-through. Wash resin bed with 2 × 15 mL lysis buffer. Elute with elution buffer in 1.5 mL fractions. (Note: Additional wash steps with 10–20 mM imidazole buffer can be done before elution, but these steps are unnecessary if a second IMAC column is run following the cleavage of the His_6-tag.)
5. Dark colored elution fractions contain the desired protein. If their volume is greater that 3 mL, concentrate dark colored fractions to 3 mL using an Amicon ultrafiltration device (10 kDa MWCO).
6. Desalt on a Bio-Rad 10-DG column equilibrated with desalting buffer according to the manufacturer's instructions.
7. Set up simultaneous overnight reconstitution and TEV protease cleavage:
   a. Make 1 M aqueous stocks of Na_2S and $FeCl_3$ and weigh out 3 mg OHCbl.
   b. Combine 4.5 mL desalting buffer, 1 mL 1 M HEPES, and 17.5 μL BME.

   c. To the concentrated enzyme solution from step 6, add 50 μL Na_2S stock, mix, then add 50 μL $FeCl_3$ stock.
   d. Incubate on ice 10 min, then check pH to make sure it is near neutral.
   e. Add Cbl, then mix with desalted protein and 400 μL 1 mg/mL TEV protease.
   f. Incubate on ice overnight.
8. Pre-equilibrate 6 mL 50% suspension TALON resin, let settle and decant excess buffer, then batch-bind reconstitution solution with the resin for 20 min.
9. Apply suspension to a column and collect flow-through, then wash with 3–4 mL desalting buffer until eluent runs clear. Resin will remain dark in color from the bound impurities.
10. Combine flow-through and wash fractions, then concentrate in an Amicon ultrafiltration device (10 kDa MWCO) to 3 mL or less.
11. Buffer exchange on a Bio-Rad 10-DG column equilibrated with desalting buffer collecting golden brown colored protein (elutes first) and avoiding the pink Cbl band (elutes second).
12. Concentrate to desired volume and measure protein concentration using the Bradford Assay.
13. Characterize protein by PAGE, UV–visible spectroscopy, and KCN assay to analyze bound Cbl as dicyanocobalamin (Data for TokK shown in Fig. 3).
   a. Run a PAGE gel of fractions taken throughout the purification.
   b. Take a UV–vis spectrum of the protein from 250 to 750 nm and look for the 420 nm shoulder indicative of a bound [4Fe–4S] cluster (Mapolelo, Zhang, Naik, Huynh, & Johnson, 2012).
   c. Incubate protein (~10–20 μM) with 100 mM KCN at 95 °C for 5 min to release bound Cbl as dicyanocobalamin (λ_{max} = 367, ε_{367} = 30,800 $M^{-1} cm^{-1}$) (Ljungdahl, LeGall, & Lee, 1973). Use a benchtop centrifuge at maximum speed for 5 min to pellet the precipitated protein, then record a UV–visible spectrum of supernatant.

3. Time-course assays of ThnK and TokK

Once the protein is purified, time-course assays can be performed. Because the enzymes only remain active for approximately 48 h and turnover is slow, the relative concentration of substrate to enzyme is a critical factor in being able to observe all sequential methylations. If substrate concentration is too high to ever be significantly depleted during the timeframe of the assay, the first methylation reaction will dominate, and subsequent methylations will be difficult to detect. However, when substrate and enzyme are present in equimolar amounts, the initial substrate is depleted, intermediates will accumulate, and the entire reaction course can be observed. For ThnK, the reaction goes to completion in 48 h, indicating each ThnK molecule does 2 methylations on average. The TokK reaction does not quite reach completion under the conditions used, but by the end of 48 h the trimethylated product is the major species present. Therefore, each TokK molecule catalyzes between 2 and 3 methylations on average. It should be noted that the maximum velocity of these enzymes is not reached under these conditions, as substrate concentration is limited in order to observe all methylation events. For the purpose of fitting the time-course data to our computational model, we assume that the radical-mediated carbon-carbon bond formation by ThnK and TokK is irreversible.

3.1 Assay setup

1. Immediately after protein purification, combine enzyme (100 μM) with HEPES (100 mM, pH 7.5), KCl (200 mM), SAM (1 mM), methyl viologen (1 mM), NADPH (1 mM), and substrate (100 μM). MeCbl (0.5 mM) can be added, but it is not absolutely required for reaction to occur.
2. Incubate at room temperature in the dark under anaerobic conditions for the duration of the time course. Due to the relatively long incubation time, if the assays are run exposed to light, any added MeCbl will photolyze over time. Therefore, to maximize consistency throughout the time course, reactions are run in the dark.
3. At each time point, take a 20–30 μL aliquot from the reaction mixture, dilute 5 × with a solution of 125 μM *L*-phenylalanine (used as an internal standard with a final concentration of 100 μM) in H_2O. Filter the diluted aliquot through a 0.5 mL Amicon ultrafiltration device (10 kDa MWCO) and analyze the filtrate by UPLC-HRMS.
4. Generate extracted ion chromatograms of each species of interest using the masses shown in Table 1. Under the approximation that each species ionizes in a consistent fashion as they only differ by the addition of hydrocarbon, peak areas can be used to determine the relative concentration of each species.

3.2 Using COPASI to estimate rate constants

1. Download COPASI (Hoops et al., 2006) from copasi.org according to your system's specifications. If you are a first-time COPASI user, there are a number of valuable resources, including video tutorials, a manual, and a user forum at copasi.org/Support.

Table 1 Monoisotopic masses for species monitored in ThnK and TokK time courses.

Species	Mass
Substrate (**3**)	432.18 ± 0.01
Methyl (**4**)	446.20 ± 0.01
Ethyl (**5**)	460.21 ± 0.01
Isopropyl (**6**)	474.23 ± 0.01

2. Open COPASI, name your model, and enter preferred units for each dimension under the **Details** tab. Choose "Concentrations" from the dropdown menu.
3. Build a COPASI model for the reaction you are monitoring:
   a. Under **Model** > **Biochemical** > **Compartments**, define a compartment by naming it, selecting "fixed" from the **type** dropdown menu, and entering its initial volume. For the TokK reaction, we have a single compartment which we named "Eppendorf," which has an initial volume of 200 μL.
   b. Under **Model** > **Biochemical** > **Species**, define each species by naming them, and putting them in your previously defined compartment. Under **type**, select "reactions," and enter initial concentrations. With TokK, we are monitoring four species, named "Substrate," "Methyl," "Ethyl," and "Isopropyl." "Substrate" is at an initial concentration of 100 μM, and all others have an initial concentration of 0 μM.
   c. Define each reaction under **Model** > **Biochemical** > **Reaction**s. For TokK we name three reactions "M1," "M2," and "M3," for each respective methylation event.
   d. Once you have named a reaction, a corresponding entry will appear on the left sidebar under **Reactions**. Select a reaction on that sidebar to assign characteristics, as shown in Fig. 4. Use the species names that have previously been defined. For example, the first methylation event for TokK, which we have named "M1" is defined as the reaction Substrate ->Methyl. The ->symbol denotes an irreversible reaction. Select "mass action (irreversible)" from the **rate law** dropdown to give the reaction a single definable first-order rate constant in units of 1/h.
   e. Repeat steps c. and d. for "M2" and "M3" Once you have defined each reaction, your model is complete.
4. After you collect your time-course data, save it as a tab separated values (tsv) file with five columns in the following order: (1) Time, (2) Substrate, (3) Methyl, (4) Ethyl, (5) Isopropyl.

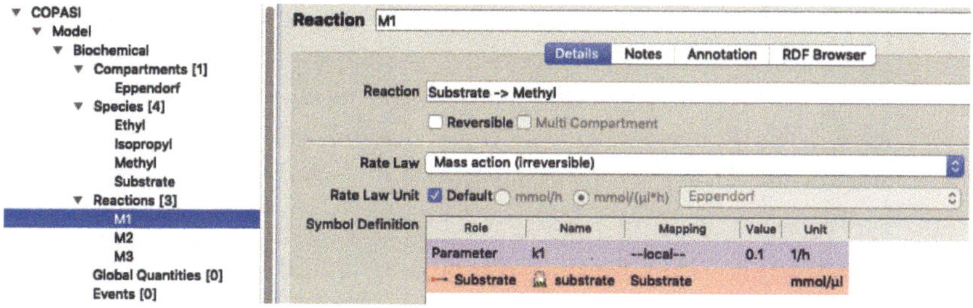

Fig. 4 Defining reactions in COPASI.

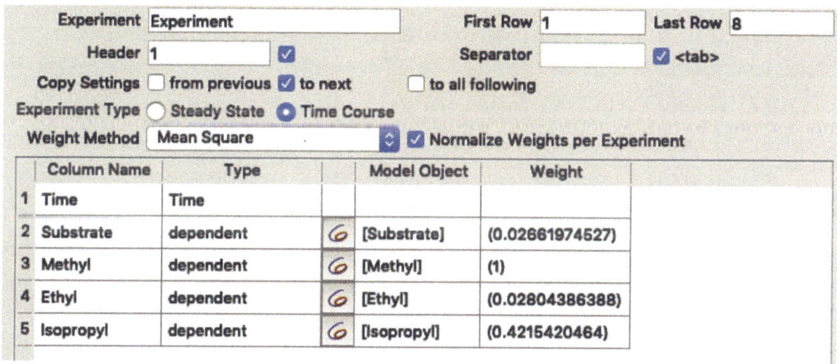

Fig. 5 Entering time-course data into COPASI for parameter estimation.

5. Under **Tasks > Parameter Estimation**, click the "Experimental Data" button. In the pop-up screen, click the green "+" symbol to upload your tsv file. For **Experiment Type**, select "Time Course".
6. Define each column from your tsv in the table as shown in Fig. 5, by selecting "dependent" under **Type**, and choosing "transient concentrations [Species]" for each in the **Model Object** column. Once you have done this, click "Run" at the bottom of the screen. This action will create a report in the **Parameter Estimation > Result** tab on the left sidebar with estimated rate constants listed for each reaction.

3.3 Using VCell to simulate reaction curves

1. Download VCell (Schaff, Fink, Slepchenko, Carson, & Loew, 1997) from vcell.org.
2. Build a model that mirrors the one you built in COPASI. Hovering over the icons causes named labels to show up, which we have used for clarity in the following steps.

   a. Using the reaction diagram tool, place 4 green spheres using the species tool into the compartment, then connect them with arrows using the RX connection tool to depict the reaction pathway as shown in Fig. 6. You can then name each species ("Substrate," "Methyl," "Ethyl," and "Isopropyl").
   b. Go to the Reaction tab on the left side of the screen. Name each reaction ("M1," "M2," "M3") and define reaction properties. Under Kinetic Type, select "Mass Action," and make sure that the reversible box is unchecked (Fig. 6).
   c. For each reaction, enter the rate constant from the COPASI parameter estimation in the table as Kf as shown in Fig. 7.
3. On the left, under the **Applications** tab, select **Specifications**, and go to the "species" tab to set the initial condition as it is in the assay. For TokK, there is 100 μM initial substrate concentration and 0 μM for all others. Then select **Simulations** and set the end time as desired (48 h). Then click the blue play button to run the simulation, and a screen will show a graphical representation of the time course. Clicking the table symbol on the bottom right gives the data numerically, which can be copied into another program to be plotted. Time courses of ThnK and TokK along with simulated curves are shown in Fig. 8.

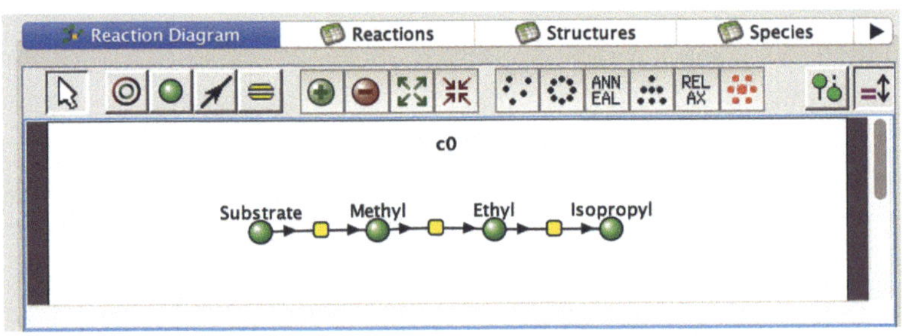

Fig. 6 Building a model using the Reaction Diagram tool in VCell.

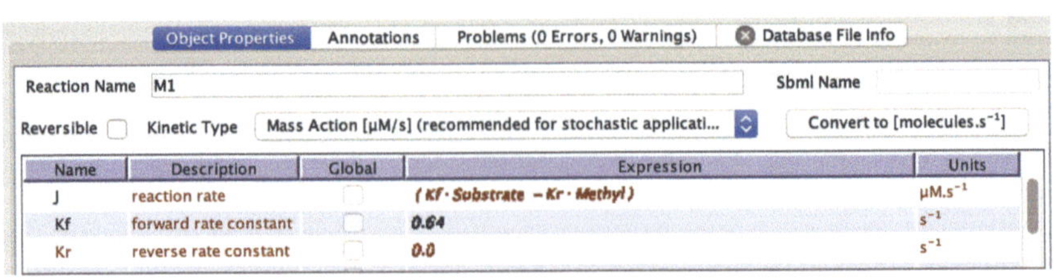

Fig. 7 Entering rates from COPASI parameter estimation into VCell for simulation.

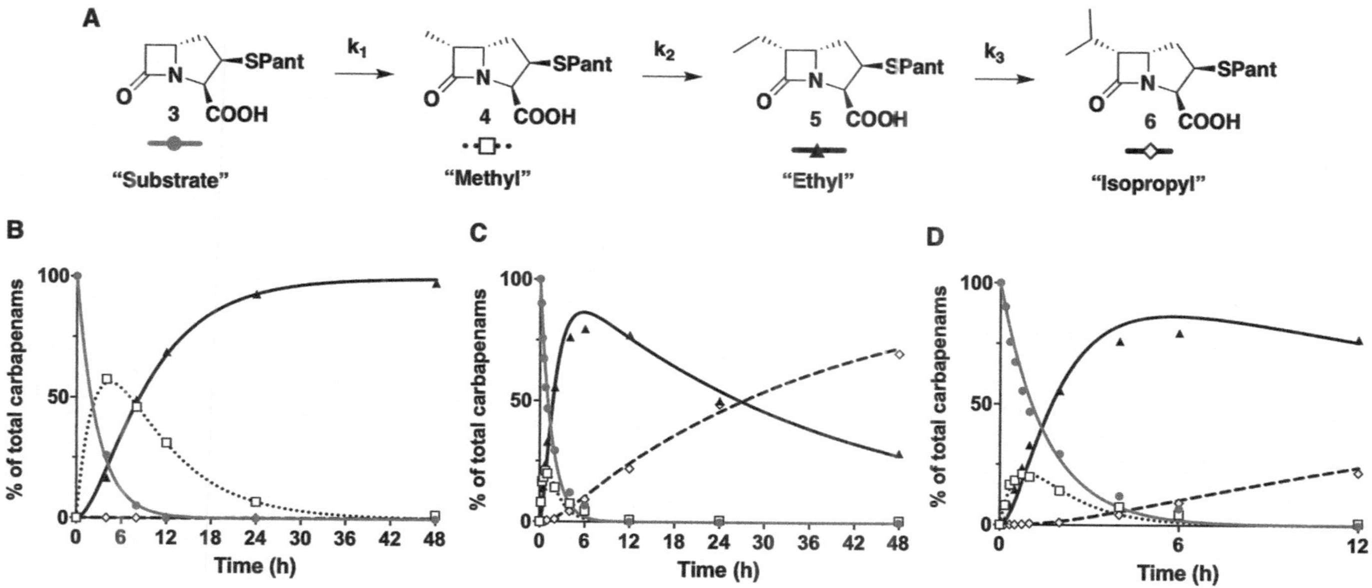

Fig. 8 Time courses of ThnK and TokK. (A) Reaction scheme of sequential methylations. (B) Time course of ThnK. C. Time course of TokK. D. Expansion of the first 12 h of the TokK Time Course. For (B)–(D), symbols represent experimental data at each time point, and curves represent the simulation from VCell using parameter estimation from COPASI. *Adapted from Sinner, E. K., Lichstrahl, M. S., Li, R., Marous, D. R., & Townsend, C. A. (2019). Methylations in complex carbapenem biosynthesis are catalyzed by a single cobalamin-dependent radical S-adenosylmethionine enzyme.* Chemical Communications, *55(99), 14934–14937. https://doi.org/10.1039/c9cc07197k with permission from the Royal Society of Chemistry.*

3.4 Using the model fit to understand the mechanism of sequential methylation

The fact that each intermediate can be observed by mass spectrometry indicates that the reaction likely occurs in a non-processive manner. Each time a substrate binds and is methylated, it must be released from the enzyme along with the SAM coproducts. The cofactors are then re-primed for catalysis, and a new substrate or intermediate can bind and be methylated. The close fit of the irreversible, stepwise kinetic model described above to simulate reaction curves also supports this exclusively non-processive mechanistic hypothesis. It appears that this non-processive model is also operational in the sequential methylase CysS (Wang & Begley, 2020).

4. Conclusion

Co-expression of *thnK* and *tokK* with plasmids that support iron-sulfur cluster biogenesis and cobalamin uptake allows for the preparation of soluble, pure, and active enzymes. ThnK and TokK are distinctive in their ability to perform consecutive methylations on their substrate, and careful consideration of assay conditions has allowed simultaneous observation of each methylation event as well as an understanding of their relative rates. These rates do not directly correspond to the predicted radical stability of each position, indicating that the 5′-deoxyadenosyl radical is sufficiently reactive to overcome even the low stability of forming a primary methyl radical. Since a single enzyme active site must be flexible enough to accommodate two to three competent substrates, it seems that factors other than radical stability are governing the relative rates of each methylation event.

References

Bauerle, M. R., Schwalm, E. L., & Booker, S. J. (2015). Mechanistic diversity of radical *S*-adenosylmethionine (SAM)-dependent methylation. *The Journal of Biological Chemistry*, *290*(7), 3995–4002. https://doi.org/10.1074/jbc.R114.607044.

Blaszczyk, A. J., Silakov, A., Zhang, B., Maiocco, S. J., Lanz, N. D., Kelly, W. L., et al. (2016). Spectroscopic and electrochemical characterization of the iron–sulfur and cobalamin cofactors of TsrM, an unusual radical *S*-adenosylmethionine methylase. *Journal of the American Chemical Society*, *138*(10), 3416–3426. https://doi.org/10.1021/jacs.5b12592.

Blaszczyk, A. J., Wang, R. X., & Booker, S. J. (2017). TsrM as a model for purifying and characterizing cobalamin-dependent radical *S*-adenosylmethionine methylases. *Methods in Enzymology*, *595*, 303–329. https://doi.org/10.1016/bs.mie.2017.07.007.

Bridwell-Rabb, J., Grell, T. A. J., & Drennan, C. L. (2018). A rich man, poor man story of *S*-adenosylmethionine and cobalamin revisited. *Annual Review of Biochemistry*, *87*(1), 555–584. https://doi.org/10.1146/annurev-biochem-062917-012500.

Freeman, M. F., Helf, M. J., Bhushan, A., Morinaka, B. I., & Piel, J. (2017). Seven enzymes create extraordinary molecular complexity in an uncultivated bacterium. *Nature Chemistry*, *9*(4), 387–395. https://doi.org/10.1038/nchem.2666.

Hoops, S., Gauges, R., Lee, C., Pahle, J., Simus, N., Singhal, M., et al. (2006). COPASI—A COmplex PAthway SImulator. *Bioinformatics*, *22*(24), 3067–3074. https://doi.org/10.1093/bioinformatics/btl485.

Houck, D. R., Kobayashi, K., Williamson, J. M., & Floss, H. G. (1986). Stereochemistry of methylation in thienamycin biosynthesis: Example of a methyl transfer from methionine with retention of configuration. *Journal of the American Chemical Society*, *108*(17), 5365–5366. https://doi.org/10.1021/ja00277a063.

Kapust, R. B., Tözsér, J., Fox, J. D., Anderson, D. E., Cherry, S., Copeland, T. D., et al. (2001). Tobacco etch virus protease: Mechanism of autolysis and rational design of stable mutants with wild-type catalytic proficiency. *Protein Engineering, Design and Selection*, *14*(12), 993–1000. https://doi.org/10.1093/protein/14.12.993.

Kawamura, Y., Yasuda, Y., Mayama, M., & Tanaka, K. (1982). Asparenomycins A, B and C, new carbapenem antibiotics. I. Taxonomic studies on the producing microorganisms. *The Journal of Antibiotics*, *35*(1), 10–14. https://doi.org/10.7164/antibiotics.35.10.

Kim, H. J., McCarty, R. M., Ogasawara, Y., Liu, Y. N., Mansoorabadi, S. O., LeVieux, J., et al. (2013). GenK-catalyzed C-6′ methylation in the biosynthesis of gentamicin: Isolation and characterization of a cobalamin-dependent radical SAM enzyme. *Journal of the American Chemical Society*, *135*(22), 8093–8096. https://doi.org/10.1021/ja312641f.

Knox, H. L., Chen, P. Y.-T., Blaszczyk, A. J., Mukherjee, A., Grove, T. L., Schwalm, E. L., et al. (2021). Structural basis for non-radical catalysis by TsrM, a radical SAM methylase. *Nature Chemical Biology*, *17*(4), 485–491. https://doi.org/10.1038/s41589-020-00717-y.

Lanz, N. D., Blaszczyk, A. J., McCarthy, E. L., Wang, B., Wang, R. X., Jones, B. S., et al. (2018). Enhanced solubilization of class B radical *S*-adenosylmethionine methylases by improved cobalamin uptake in *Escherichia coli*. *Biochemistry*, *57*(9), 1475–1490. https://doi.org/10.1021/acs.biochem.7b01205.

Lanz, N. D., Grove, T. L., Gogonea, C. B., Lee, K.-H., Krebs, C., & Booker, S. J. (2012). RlmN and AtsB as models for the overproduction and characterization of radical SAM proteins. *Methods in Enzymology*, *516*, 125–152. https://doi.org/10.1016/B978-0-12-394291-3.00030-7.

Li, R., Lloyd, E. P., Moshos, K. A., & Townsend, C. A. (2014). Identification and characterization of the carbapenem MM 4550 and its gene cluster in *Streptomyces argenteolus* ATCC 11009. *Chembiochem*, *15*(2), 320–331. https://doi.org/10.1002/cbic.201300319.

Ljungdahl, L. G., LeGall, J., & Lee, J.-P. (1973). Isolation of a protein containing tightly bound 5-methoxybenzimidazolylcobamide (factor IIIm) from *Clostridium thermoaceticum*. *Biochemistry*, *12*(9), 1802–1808. https://doi.org/10.1021/bi00733a022.

Mapolelo, D. T., Zhang, B., Naik, S. G., Huynh, B. H., & Johnson, M. K. (2012). Spectroscopic and functional characterization of iron–sulfur cluster-bound forms of *Azotobacter vinelandii* ^{Nif}IscA. *Biochemistry*, *51*(41), 8071–8084. https://doi.org/10.1021/bi3006658.

Marous, D. R., Lloyd, E. P., Buller, A. R., Moshos, K. A., Grove, T. L., Blaszczyk, A. J., et al. (2015). Consecutive radical *S*-adenosylmethionine methylations form the ethyl side chain in thienamycin biosynthesis. *Proceedings of the National Academy of Sciences of the United States of America*, *112*(33), 10354–10358. https://doi.org/10.1073/pnas.1508615112.

New England BioLabs. (2012). *Making your own electrocompetent cells*. NEB. https://www.neb.com/protocols/2012/06/21/making-your-own-electrocompetent-cells.

Núñez, L. E., Méndez, C., Braña, A. F., Blanco, G., & Salas, J. A. (2003). The biosynthetic gene cluster for the β-lactam carbapenem thienamycin in *Streptomyces cattleya*. *Chemistry and Biology*, *10*(4), 301–311. https://doi.org/10.1016/S1074-5521(03)00069-3.

Papp-Wallace, K. M., Endimiani, A., Taracila, M. A., & Bonomo, R. A. (2011). Carbapenems: Past, present, and future. *Antimicrobial Agents and Chemotherapy*, *55*(11), 4943–4960. https://doi.org/10.1128/AAC.00296-11.

Parent, A., Guillot, A., Benjdia, A., Chartier, G., Leprince, J., & Berteau, O. (2016). The B_{12}-radical SAM enzyme PoyC catalyzes valine C_{β}-methylation during polytheonamide biosynthesis. *Journal of the American Chemical Society*, *138*(48), 15515–15518. https://doi.org/10.1021/jacs.6b06697.

Schaff, J., Fink, C. C., Slepchenko, B., Carson, J. H., & Loew, L. M. (1997). A general computational framework for modeling cellular structure and function. *Biophysical Journal*, *73*(3), 1135–1146. https://doi.org/10.1016/S0006-3495(97)78146-3.

Sinner, E. K., Lichstrahl, M. S., Li, R., Marous, D. R., & Townsend, C. A. (2019). Methylations in complex carbapenem biosynthesis are catalyzed by a single cobalamin-dependent radical *S*-adenosylmethionine enzyme. *Chemical Communications*, *55*(99), 14934–14937. https://doi.org/10.1039/c9cc07197k.

Townsend, C. A. (2016). Convergent biosynthetic pathways to β-lactam antibiotics. *Current Opinion in Chemical Biology*, *35*, 97–108. https://doi.org/10.1016/j.cbpa.2016.09.013.

Wang, S. C. (2018). Cobalamin-dependent radical *S*-adenosyl-L-methionine enzymes in natural product biosynthesis. *Natural Product Reports*, *35*(8), 707–720. https://doi.org/10.1039/C7NP00059F.

Wang, Y., & Begley, T. P. (2020). Mechanistic studies on CysS – A vitamin B_{12}-dependent radical SAM methyltransferase involved in the biosynthesis of the *tert*-butyl group of cystobactamid. *Journal of the American Chemical Society*, *142*(22), 9944–9954. https://doi.org/10.1021/jacs.9b06454.

Wang, Y., Schnell, B., Baumann, S., Müller, R., & Begley, T. P. (2017). Biosynthesis of branched alkoxy groups: Iterative methyl group alkylation by a cobalamin-dependent radical SAM enzyme. *Journal of the American Chemical Society*, *139*(5), 1742–1745. https://doi.org/10.1021/jacs.6b10901.

Williamson, J. M. (1986). The biosynthesis of thienamycin and related carbapenems. *Critical Reviews in Biotechnology*, *4*(1), 111–131. https://doi.org/10.3109/07388558609150792.

Zheng, L., Cash, V. L., Flint, D. H., & Dean, D. R. (1998). Assembly of iron-sulfur clusters. Identification of an iscSUA-hscBA-fdx gene cluster from *Azotobacter vinelandii*. *Journal of Biological Chemistry*, *273*(21), 13264–13272. https://doi.org/10.1074/jbc.273.21.13264.

Zhou, S., Alkhalaf, L. M., de los Santos, E. L., & Challis, G. L. (2016). Mechanistic insights into class B radical-*S*-adenosylmethionine methylases: Ubiquitous tailoring enzymes in natural product biosynthesis. *Current Opinion in Chemical Biology*, *35*, 73–79. https://doi.org/10.1016/j.cbpa.2016.08.021.

CHAPTER THREE

Characterization of the cobalamin-dependent radical *S*-adenosyl-L-methionine enzyme *C*-methyltransferase Fom3 in fosfomycin biosynthesis

Shusuke Sato, Fumitaka Kudo, and Tadashi Eguchi*
Department of Chemistry, Tokyo Institute of Technology, Tokyo, Japan
*Corresponding author: e-mail address: eguchi@chem.titech.ac.jp

Contents

Abstract

Fosfomycin is a clinically used broad-spectrum antibiotic that has the structure of an oxirane ring with a phosphonic acid substituent and a methyl substituent. In nature,

Methods in Enzymology, Volume 669
ISSN 0076-6879
https://doi.org/10.1016/bs.mie.2021.11.025

fosfomycin is produced by *Streptomyces* spp. and *Pseudomonas* sp., but biosynthesis of fosfomycin significantly differs between the two bacteria, especially in the incorporation mechanism of the methyl group. It has been proposed that the cobalamin-dependent radical *S*-adenosyl-L-methionine (SAM) enzyme Fom3 is responsible for the methyl-transfer reaction in *Streptomyces* fosfomycin biosynthesis. In this chapter, we describe the experimental methods to characterize Fom3. We performed the methylation reaction with the purified recombinant Fom3, revealing that Fom3 recognizes a cytidylylated 2-hydroxyethylphosphonate as a substrate and catalyzes stereoselective methylation of the sp^3 carbon at the C2 position to afford cytidylylated (*S*)-2-hydroxypropylphosphonate. Reaction analysis using deuterium-labeled substrates showed that the 5′-deoxyadenosyl radical generated by reductive cleavage of SAM stereoselectively abstracts the *pro-R* hydrogen atom of the C—H bond at the C2 position of cytidylylated 2-hydroxyethylphosphonate. Therefore, the *C*-methylation reaction catalyzed by Fom3 proceeds with inversion of the configuration at the C2 position. Experimental methods to elucidate the chemical structures of the substrate and products and the stereochemical course in the Fom3-catalyzed reaction could give information to progress investigation of cobalamin-dependent radical SAM *C*-methyltransferases.

Abbreviations

3-OBPn	3-oxo-4-phosphonobutanoate
5′-dA·	5′-deoxyadenosyl radical
5′-dA	5′-deoxyadenosine
(*S*)-HPP-CMP	cytidylyl (*S*)-2-hydroxypropylphosphonate
CyTase	cytidylyltransferase
DTT	dithiothreitol
HEP	2-hydroxyethylphosphonate
HEP-CMP	cytidylyl 2-hydroxyethylphosphonate
HPLC	high-performance liquid chromatography
HPP	2-hydroxypropylphosphonate
IPTG	isopropyl-β-D-thiogalactopyranoside
KIE	kinetic isotope effect
LB	Luria–Bertani
LC–ESI–MS	liquid chromatography–electrospray ionization–mass spectrometry
MeCbl	methylcobalamin
MV	methylviologen
OD_{600}	optical density at 600 nm
OPP	2-oxopropylphosphonate
PEP	phosphoenolpyruvate
PnMM	2-phosphonomethymalate
PnPy	phosphonopyruvate
SAH	*S*-adenosylhomocysteine
SAM	*S*-adenosyl-L-methionine
SDS-PAGE	sodium dodecyl sulfate-polyacrylamide gel electrophoresis
SUF	sulfur utilization factors

1. Introduction

Fosfomycin is a broad-spectrum antibiotic discovered in the cultures of *Streptomyces* spp. (Christensen et al., 1969; Hendlin et al., 1969) and *Pseudomonas* sp. (Katayama, Tsubotani, Nozaki, Harada, & Ono, 1990; Shoji et al., 1986). Fosfomycin reacts with a cysteine residue in the active site of UDP-*N*-acetylglucosamine enolpyruvyltransferase (MurA) to form an irreversible covalent bond, causing inactivation of MurA (Kahan, Kahan, Cassidy, & Kropp, 1974; Skarzynski et al., 1996). Inhibition of MurA-targeted peptidoglycan biosynthesis by fosfomycin ensures its antibiotic effect, and, consequently, fosfomycin has been included in the World Health Organization model list of essential medicines (21st list, 2019). Although fosfomycin is a relatively small molecule among natural products, it has a unique structure in which a phosphonic acid group and a methyl group are directly bound to an oxirane ring (Fig. 1). Genome sequencing analysis has revealed fosfomycin gene clusters, and the biosynthetic mechanisms have been found to be different between *Streptomyces* spp. (Hidaka et al., 1995; Woodyer et al., 2006) and *Pseudomonas* sp. (Kim et al., 2012).

The processes common to the two bacteria are the carbon–phosphorus bond formation step and oxirane ring formation step. In the first step of

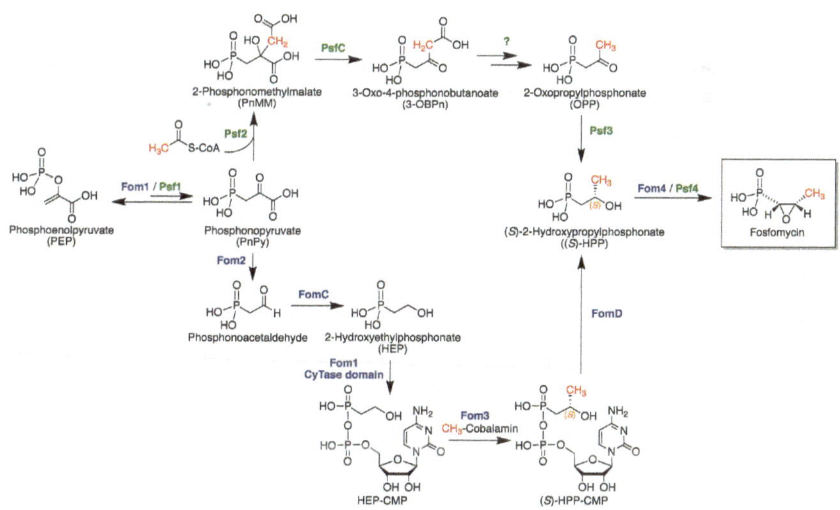

Fig. 1 Fosfomycin biosynthetic pathway.

fosfomycin biosynthesis, the phosphomutase Fom1/Psf1 converts phosphoenolpyruvate (PEP) to phosphonopyruvate (PnPy) (Fig. 1) (Kim et al., 2012; Seidel, Freeman, Seto, & Knowles, 1988; Seto & Kuzuyama, 1999). The C—P bond formation reaction catalyzed by Fom1/Psf1 is unfavorable because the dissociation energy of the O—P bond of PEP is 17–24 kcal mol^{-1} greater than that of the C—P bond of PnPy (Bowman, McQueney, Barry, & Dunaway-Mariano, 1988). Therefore, it is believed that PnPy must be consumed by a subsequent irreversible reaction to afford the C—P bond (Horsman & Zechel, 2017). In the final step of fosfomycin biosynthesis, (*S*)-2-hydroxypropylphosphonate (HPP) is converted to fosfomycin (Fig. 1) (Liu et al., 2001; Munos et al., 2008; Seto et al., 1991). This stereospecific radical epoxidation reaction is catalyzed by the non-heme iron-dependent epoxidase Fom4/Psf4 (HppE) using hydroperoxide as an oxidant (Wang et al., 2013).

The process that differs between the two bacteria is the introduction mechanism of the methyl group. In *Pseudomonas* fosfomycin biosynthesis, PnPy produced by Psf1 reacts with acetyl-CoA, and then the CoA thioester is hydrolyzed to form 2-phosphonomethymalate (PnMM) (Fig. 1, *Pseudomonas* pathway). The aldol-type condensation reaction and subsequent hydrolysis are catalyzed by the citrate synthase-like protein Psf2 to yield PnMM (Kim et al., 2012). PnMM is then oxidatively decarboxylated by the diiron metalloenzyme PsfC (Psf13) to give 3-oxo-4-phosphonobutanoate (3-OBPn) (Simon, Ongpipattanakul, Nair, & van der Donk, 2021). Although the biosynthetic enzyme that converts 3-OBPn to 2-oxopropylphosphonate (OPP) has not been identified, the NADP-dependent dehydrogenase Psf3 stereoselectively reduces the ketone group of OPP to afford (*S*)-HPP (Olivares, Ulrich, Chekan, van der Donk, & Nair, 2017). Hence, in the case of *Pseudomonas* sp., the methyl group of fosfomycin is derived from acetyl-CoA.

Streptomyces wedmorensis introduces the methyl group on cobalamin into fosfomycin (Kuzuyama, Hidaka, Kamigiri, Imai, & Seto, 1992). In *Streptomyces* fosfomycin biosynthesis, the phosphonic acid moiety of 2-hydroxyethylphosphonate (HEP), which is formed from PnPy by the PnPy decarboxylase Fom2 (Hidaka, Iwakura, Imai, & Seto, 1992; Nakashita, Watanabe, Hara, Hidaka, & Seto, 1997) and the iron-containing NADP-dependent alcohol dehydrogenase FomC (Shao et al., 2008; Woodyer, Li, Zhao, & van der Donk, 2007), is cytidylylated by the cytidylyltransferase (CyTase) domain on the N-terminus of Fom1 to give cytidylylated 2-hydroxyethylphosphonate (HEP-CMP) (Fig. 1, *Streptomyces* pathway) (Cho et al., 2017). The cobalamin-dependent radical *S*-adenosyl-L-methionine (SAM) methyltransferase Fom3 then recognizes HEP-CMP as a substrate and stereoselectively transfers the methyl group of methylcobalamin (MeCbl) at the C2 position, affording cytidylyl (*S*)-2-hydroxypropylphosphonate ((*S*)-HPP-CMP) (Lanz et al., 2018; McLaughlin & van der Donk, 2018; Sato, Kudo, Kim, Kuzuyama, & Eguchi, 2017; Sato, Kudo, Kuzuyama, Hammerschmidt, & Eguchi, 2018; Wang et al., 2018). Finally, the binuclear metallohydrolase FomD catalyzes hydrolysis to remove the cytidylyl group (McLaughlin & van der Donk, 2018; Sato, Miyanaga, et al., 2018; Wang et al., 2018). Steady-state kinetic analysis of Fom1 and FomD suggests that the cytidylylation reaction catalyzed by the CyTase domain of Fom1 prefers HEP rather than (*S*)-HPP as the substrate, and the hydrolysis reaction catalyzed by FomD prefers (*S*)-HPP-CMP rather than HEP-CMP (Sato, Miyanaga, et al., 2018).

Radical SAM enzymes bind a [4Fe—4S] cluster in the active site through three cysteine residues, which form a characteristic CxxxCxxC motif in the primary sequence (Broderick, Duffus, Duschene, & Shepard, 2014). Radical SAM enzymes can catalyze diverse radical reactions using the 5′-deoxyadenosyl radical (5′-dA•) as a radical initiator, which is generated by reductive cleavage of SAM coordinated to the $[4Fe—4S]^+$ cluster. More than 600,000 sequences of proteins belonging to the radical SAM enzyme superfamily are currently known (RadicalSAM.org, https://radicalsam.org/index.php). The radical SAM methyltransferases are classified into four subgroups according to the differences in the cofactors and the domains that make up the proteins (Bauerle, Schwalm, & Booker, 2015). Fom3 belongs to class B because it has a cobalamin-binding domain on the N-terminus and a radical SAM domain on the C-terminus. Because it is often difficult to obtain class B enzymes as soluble proteins when preparing the enzymes by heterologous expression, their function has been estimated by disrupting biosynthetic genes and isolating the unmethylated products or biosynthetic intermediates. Fom3 is no exception to this trend. Booker's group found that cobalamin-dependent radical SAM enzymes can be obtained as soluble proteins by overexpressing a set of proteins (i.e., the Btu cluster) for cobalamin uptake in the cell membrane of *Escherichia coli* to increase the intracellular cobalamin concentration (Lanz et al., 2018).

In this chapter, we describe the experimental methods used to elucidate the true substrate of Fom3 and to investigate the stereochemistry of the *C*-methylation reaction. HEP-CMP, synthesized enzymatically using the CyTase domain of Fom1, reacted with purified Fom3 in the presence of SAM, MeCbl, dithiothreitol (DTT), methylviologen (MV) and NADH. As a result, HPP-CMP, 5′-deoxyadenosine (5′-dA), and *S*-adenosylhomocysteine (SAH) were produced in a 1:1:1 ratio. Production of 5′-dA could not be

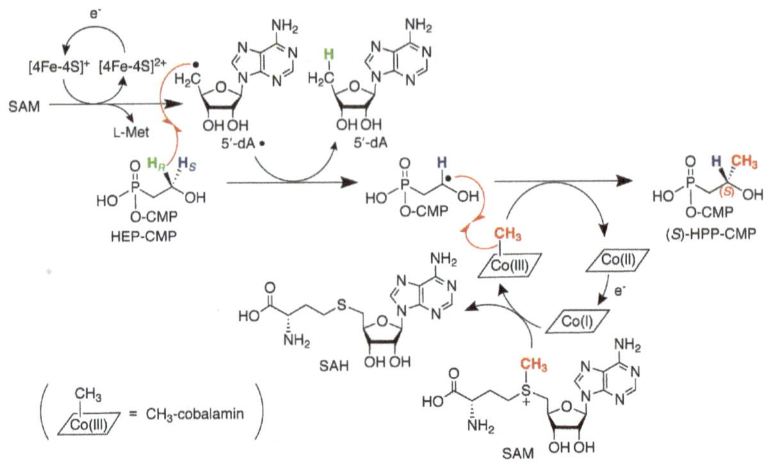

Fig. 2 Proposed mechanism of the *C*-methylation reaction catalyzed by Fom3.

detected when HEP was used instead of HEP-CMP. This result indicated that the CMP moiety is important for substrate recognition of Fom3 (Sato et al., 2017). The stereochemistry of the C2 position of the methylation product was determined to be the *S*-configuration by high-performance liquid chromatography (HPLC) analysis using a chiral ligand-exchange column. Moreover, we performed the Fom3 reaction with deuterium-labeled substrates, which revealed that 5′-dA• selectively abstracted the *pro-R* hydrogen atom from the two diastereotopic hydrogen atoms in the C2 position of HEP-CMP to afford 5′-dA and the radical intermediate. We thus propose that *C*-methylation catalyzed by Fom3 proceeds with inversion of the configuration at the C2 position owing to methyl transfer to the radical intermediate from the opposite side of the hydrogen atom abstracted by 5′-dA• (Fig. 2) (Sato, Kudo, et al., 2018).

2. Heterologous expression and purification of Fom3

Coexpression of iron–sulfur cluster biosynthetic machinery with radical SAM enzymes often enhances the cluster content and improves the yield of radical SAM enzymes (Bodea & Balskus, 2018; Cicchillo et al., 2004; Lewis, Bruender, & Bandarian, 2018; Pang & Yokoyama, 2018). We previously prepared the radical SAM enzymes NeoN and AprD4 by coexpression of *E. coli*-derived sulfur utilization factors (SUF) (Kudo, Hoshi, Kawashima, Kamachi, & Eguchi, 2014; Kudo, Tokumitsu, & Eguchi, 2017). For overexpression of SUF in *E. coli*, the plasmid

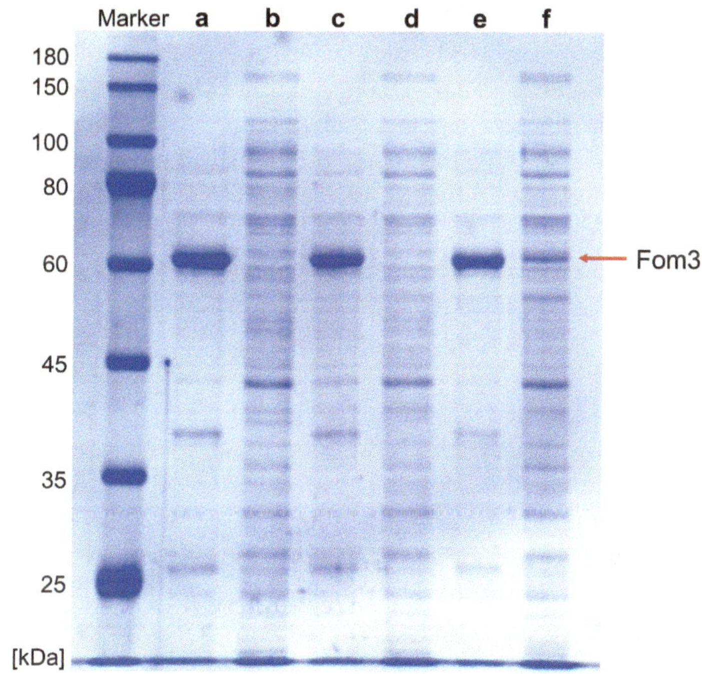

Fig. 3 SDS-PAGE (10% acrylamide gel) analysis of heterologous expression of Fom3. (A) Insoluble and (B) soluble fractions of the Fom3 expression system. (C) Insoluble and (D) soluble fractions of the Fom3–SUF expression system. (E) Insoluble and (F) soluble fractions of the Fom3–SUF–Btu expression system. The theoretical molecular weight of Fom3 fused with a His_6-tag at the N-terminus is 62.5 kDa. We constructed the plasmid pBAD24_BtuCEDFB (Maruyama et al., 2020), which carries the cobalamin uptake genes, and coexpressed Fom3 with the Btu cluster according to the method described by Booker's group (Lanz et al., 2018).

pRKSUF017 was used, in which the SUF is inserted under the control of the lactose-promoter of the vector pRK415 (Takahashi & Tokumoto, 2002). The small amount of Fom3 obtained as a soluble protein by coexpression with SUF was used for analysis, although coexpression barely improved the yield of soluble Fom3 (Fig. 3). It should be noted that the technique of coexpressing the Btu cluster developed by Booker's group dramatically increased the yield of soluble Fom3 (Lanz et al., 2018).

2.1 Protocol for expression and purification of Fom3

1. *E. coli* BL21(DE3) was transformed with the plasmid pET28-fom3 and pRKSUF017. The *fom3* gene was amplified by polymerase chain reaction using the cosmid pFBG51, which contains part of the fosfomycin

biosynthesis gene cluster from *S. wedmorensis* 144-91 (Hidaka et al., 1995) as a template, and inserted into the *Nde*I-*Hind*III site of the expression vector pET28a.

2. The transformant in 5 mL of Luria–Bertani (LB) medium containing 50 μg/mL kanamycin and 10 μg/mL tetracycline was incubated at 37 °C and shaken at 250 rpm overnight.
3. Portions of the resulting pre-culture (1 mL) were added to 200 mL of LB medium containing 50 μg/mL kanamycin and 10 μg/mL tetracycline in 500 mL baffled flasks, followed by incubation at 37 °C and shaking at 200 rpm until the optical density at 600 nm (OD_{600}) was 0.8–1.0.
4. The flasks were then cooled on ice for 30 min.
5. $Fe(NH_4)_2(SO_4)_2$, L-cysteine, and isopropyl-β-D-thiogalactopyranoside (IPTG) were added to each final concentration of 0.2 mM.
6. The cultures were allowed to incubate at 15 °C for 20–24 h with 80 rpm agitation. Before harvesting the cells by centrifugation, 3 mL of the culture medium was removed from each flask and immediately prepared for analysis by sodium dodecyl sulfate-polyacrylamide gel electrophoresis (SDS-PAGE).
7. The bacteria were harvested by centrifugation at 10,000 × *g* and 4 °C for 10 min, and the resulting cell paste (ca. 3 g/L) was stored at −30 °C until use.
8. Buffer A [50 mM HEPES-NaOH (pH 8.0), 300 mM NaCl, 10% glycerol] was degassed by a freeze–thaw procedure under reduced pressure, and then bubbled with a mixture of argon and hydrogen gas purified through a reduced Cu catalyst. The degassed buffer was stored in a glove box ($[O_2] \leq 8$ ppm, UNICO, Moriya, Japan).
9. The frozen Fom3 cell paste (7 g) was transferred to a glove box, and it was then suspended in 40 mL of degassed buffer A.
10. The cell suspension was disrupted by sonication with sonication bursts of 5 s with a 5 s interval (total of 4 min) on cooled aluminum beads.
11. The tubes were tightly capped and the edges were sealed with vinyl tape before removing them from the glove box.
12. The lysate was centrifuged at 14,000 × *g* and 4 °C for 20 min.
13. The tubes were transferred back to the glove box and the supernatant was carefully decanted into a column loaded with TALON resin (Clontech, Mountain View, CA) pre-equilibrated with degassed buffer A.

14. The resin adsorbed by Fom3 was washed with degassed buffer A containing 15 mM imidazole.
15. Fom3 was eluted with degassed buffer A containing 200 mM imidazole.
16. Imidazole was removed from the Fom3 solution by using a PD-10 desalting column (GE Healthcare, Buckinghamshire, UK).
17. The ultraviolet (UV) absorption of the protein solution was measured at 280 nm by spectrophotometry (NanoDrop Lite, Thermo Fisher Scientific K.K., Tokyo, Japan) and the concentration of Fom3 was determined using the extinction coefficient ($\varepsilon_{280nm} = 81.5\ mM^{-1}\ cm^{-1}$) determined by the Edelhoch method (Edelhoch, 1967).

2.2 Chemical reconstitution of the iron–sulfur cluster bound to Fom3

Generally, the [4Fe—4S] clusters bound to radical SAM enzymes are highly unstable to oxygen, and thus reconstitution of the [4Fe—4S] clusters is required to detect the activity. To reconstitute the [4Fe—4S] cluster lost in the purification process, we reacted purified Fom3 with DTT, sulfide ions, and iron ions. We measured the UV–visible spectrum of the resulting enzyme solution to detect formation of a $[4Fe—4S]^{2+}$ cluster, which has a characteristic maximal absorption at 420 nm (Fig. 4). It should be noted that Wang's group obtained Fom3 by refolding and then detected a $[4Fe—4S]^{+}$ cluster bound to Fom3 by electron paramagnetic resonance (EPR) spectroscopy after chemical reconstitution and reduction with dithionite (Allen & Wang, 2014). Booker's group purified soluble Fom3 from their coexpression system, and they also detected a [4Fe—4S] cluster bound to Fom3 by EPR and Mössbauer spectroscopy without chemical reconstitution (Wang et al., 2018).

The protocol for chemical reconstitution of the iron–sulfur cluster bound to Fom3 was as follows:

1. The purified Fom3 was treated with 5 mM DTT for 15 min at room temperature under anaerobic conditions.
2. $Fe(NH_4)_2(SO_4)_2$, $FeCl_3$, and Na_2S were added to final concentrations of 0.25, 0.25, and 0.5 mM, respectively.
3. The mixture was allowed to incubate at room temperature for 1 h.
4. Excess iron sulfide was removed from the reconstituted Fom3 solution by using a PD-10 desalting column.
5. The UV–visible absorption spectrum of the reconstituted Fom3 solution in a quartz cell with a screw cap was measured.

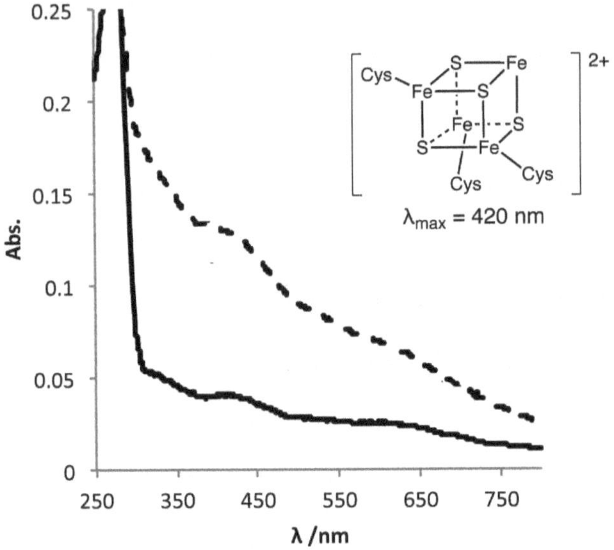

Fig. 4 UV–visible spectra of the purified (solid line) and reconstituted (dashed line) Fom3 solution.

3. Enzymatic synthesis of HEP-CMP and HPP-CMP using the cytidylyltransferase domain of Fom1

3.1 Protocol for expression and purification of the cytidylyltransferase domain of Fom1

The phosphomutase Fom1 from *S. wedmorensis* is composed of an N-terminal CyTase domain (1–139 aa) and a C-terminal phosphomutase domain (140–435 aa) (Cho et al., 2017). In this study, only the CyTase domain was expressed in *E. coli*, and it was purified by affinity column chromatography. We confirmed that Fom1 with a truncated C-terminal domain and full-length Fom1 had the same level of CyTase activity (data not shown).

The protocol for expression and purification of the cytidylyltransferase domain of Fom1 was as follows:

1. *E. coli* BL21(DE3) was transformed with the plasmid pHis8_NtFom1CyTase (Cho et al., 2017).
2. The transformant was incubated in 5 mL of LB medium containing 50 μg/mL kanamycin at 37 °C and shaken at 250 rpm overnight.
3. Portions of the resulting pre-culture (1 mL) were added to 200 mL of LB medium containing 50 μg/mL kanamycin in 500 mL baffled flasks,

followed by incubation at 37 °C and shaking at 200 rpm until $OD_{600}=0.6$.

4. IPTG was added to a final concentration of 0.2 mM.
5. The cultures were allowed to incubate at 15 °C for 20–24 h with 200 rpm agitation.
6. The bacteria were harvested by centrifugation at 10,000 × *g* and 4 °C for 10 min, and the resulting cell paste (ca. 3 g/L) was stored at −30 °C until use.
7. A sample of the frozen Fom1-CyTase cell paste (3 g) was suspended in 30 mL of buffer B [50 mM HEPES-NaOH (pH 8.0)].
8. The cell suspension was disrupted by sonication with sonication bursts of 5 s with a 5 s interval (total of 2 min) on ice.
9. The lysate was centrifuged at 14,000 × *g* and 4 °C for 10 min.
10. The supernatant was decanted into a column loaded with Ni Superflow Resin (Clontech) pre-equilibrated with buffer B.
11. The resin adsorbed to Fom1-CyTase was washed with buffer B containing 5 mM imidazole.
12. Fom1-CyTase was eluted with buffer B containing 200 mM imidazole.
13. Imidazole was removed from the Fom1-CyTase solution by using a PD-10 desalting column.
14. The obtained Fom1-CyTase was concentrated with an Amicon Ultra 3 K centrifugal filter (Merck KGaA).
15. The UV absorption of the protein solution at 280 nm was measured by spectrophotometry (NanoDrop Lite) and the concentration of Fom1-CyTase was determined using the extinction coefficient ($\varepsilon_{280nm}=17.8\,mM^{-1}\,cm^{-1}$) determined by the Edelhoch method.
16. The purified Fom1-CyTase was stored at −30 °C until use.

3.2 Protocol for enzymatic synthesis of HEP-CMP and HPP-CMP

We performed cytidylylation reactions of HEP, (*S*)-HPP, or (*R*)-HPP using the purified Fom1-CyTase, and then purified the cytidylylated compounds by anion-exchange chromatography and gel-filtration chromatography. HEP was kindly donated by Professor Tomohisa Kuzuyama (Graduate School of Agricultural and Life Science, The University of Tokyo). (*S*)- and (*R*)-HPP were synthesized from diethyl (2-oxopropyl)phosphonate according to a published procedure (Hammerschmidt, 1991; Woschek, Lindner, & Hammerschmidt, 2003). The catalytic efficiency of the cytidylylation activity of Fom1 against (*S*)- and (*R*)-HPP has been estimated to be one-tenth and one-half of that against HEP, respectively (Sato, Miyanaga, et al., 2018). HEP (1 mM) was fully cytidylylated using 10 μM of Fom1-CyTase at 28 °C for 4 h. In the case of (*S*)-HPP, the reaction progress was improved by overnight reaction with a high concentration of Fom1-CyTase. Cytidylylation of (*R*)-HPP proceeded sufficiently under the same conditions as in the case of HEP.

The protocol for enzymatic synthesis of HEP-CMP and HPP-CMP was as follows

1. HEP or (*R*)-HPP (1 mM), CTP (3 mM), and $MgCl_2$ (1 mM) were added to Fom1-CyTase (10 μM) solution in a 1.5 mL Eppendorf tube. The volume of the reaction solution was 500 μL × 24. In the case of (*S*)-HPP, the concentration of Fom1-CyTase was 43 μM
2. Incubation was performed at 28 °C for 4 h. In the case of (*S*)-HPP, the solution was allowed to react for at least 20 h.
3. The reaction mixture was loaded onto a DOWEX AG1-X8 column ($HCOO^-$ form, 15 mm × 40 mm, Bio-Rad Laboratories, Hercules, CA), and then the column was washed with water.
4. A total of 500 mL of aqueous ammonium formate solution (pH 7.5 adjusted with aqueous ammonia) was eluted in a linear gradient of the concentration from 0 to 1 M of ammonium formate and fractionated into 5 mL fractions. The UV absorption of each fraction was measured at 280 nm by spectrophotometry (NanoDrop Lite) to confirm that the cytidine derivatives were obtained. Typically, HEP-CMP and HPP-CMP were eluted prior to the nucleotides.
5. The fractions containing the cytidylylated compound were lyophilized.
6. The resulting powder was resuspended in 500 μL of H_2O, and it was then loaded on a Sephadex G-10 column (10 mm × 880 mm, GL Healthcare).
7. Elution was performed with water and fractionated into 1 mL fractions. The UV absorption of each fraction was measured at 280 nm by spectrophotometry (NanoDrop Lite) to confirm that the cytidylylated compound was obtained.
8. The fractions containing the cytidylylated compound were lyophilized, resuspended in D_2O, and analyzed by NMR.

3.3 Separation of (*S*)- and (*R*)-HPP-CMP in HPLC analysis using a chiral column

We hypothesized that chelation of HPP-CMP to metal ions with the 2-OH group and phosphonate group would make a difference in the

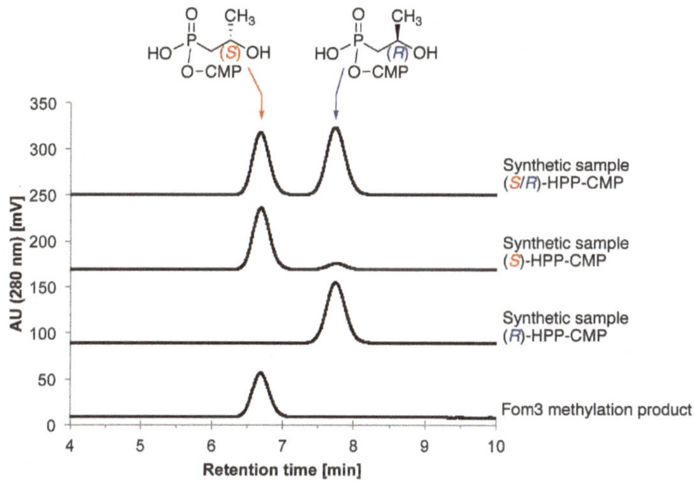

Fig. 5 Chiral ligand-exchange chromatography to separate (*S*)- and (*R*)-HPP-CMP. The synthetic (*S*)-HPP-CMP contained a small amount of (*R*)-HPP-CMP.

conformations of the two diastereomers, leading to good separation. We thus attempted chiral ligand-exchange chromatography with an *N*,*S*-dioctyl-D-penicillamine octadecyl-coated silica gel column using a copper(II)-containing solution. The HPP-CMP solutions were injected into a HPLC system equipped with a SUMICHIRAL OA-5000 column (5 μm, 4.6 mm × 150 mm, Sumika Chemical Analysis Service Ltd.) equilibrated with solvent C [99% buffer containing Cu^{2+} (0.5 mM cupric acetate, 100 mM ammonium acetate, pH 5.0), 1% 2-propanol]. Elution was performed with an isocratic solvent C at a flow rate of 1.0 mL/min at 40 °C. Chromatography was monitored at 280 nm. (*S*)- and (*R*)-HPP-CMP were clearly separated (Fig. 5).

4. Elucidation of the true substrate and product of Fom3

4.1 Protocol for enzymatic assay of Fom3

Using the reconstituted Fom3 and synthesized HEP-CMP, we attempted to assess the methylation activity by HPLC analysis. The chemically reconstituted 4Fe—4S cluster bound to Fom3 was in the oxidative form based on the UV–visible spectrum. It was therefore necessary to perform activation to reduce the 4Fe—4S cluster with suitable reduction reagents. We used an artificial electron-transfer system composed of MV and NADH for reduction of the $[4Fe—4S]^{2+}$ cluster, which was used to activate

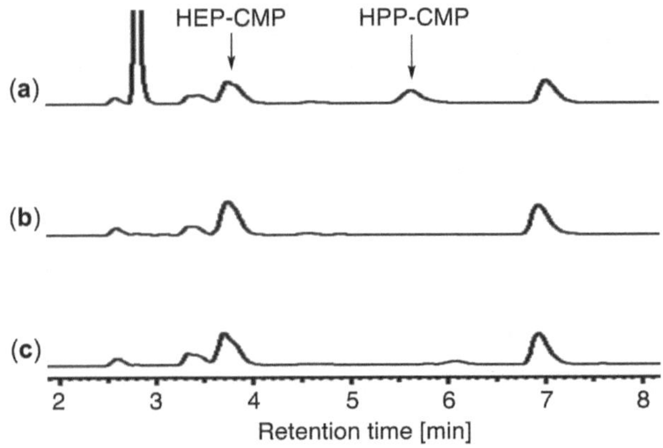

Fig. 6 HPLC analysis of the Fom3-catalyzed reaction using (A) 1 mM MV and 4 mM NADH, (B) 5 mM sodium dithionite, or (C) 25 μM flavodoxin, 25 μM flavodoxin reductase and 4 mM NADPH as reduction reagents for the iron–sulfur cluster.

the cobalamin-dependent radical SAM methyltransferase GenK (Kim et al., 2013). As a result, HPP-CMP was formed (Fig. 6). 5′-dA and SAH were also detected in HPLC analysis of the Fom3-catalyzed methylation reaction of HEP-CMP. When HEP was used as the substrate instead of HEP-CMP, 5′-dA was not formed. These results indicate that HEP-CMP is the true substrate of the Fom3-catalyzed reaction.

The protocol for enzymatic assay of Fom3 was as follows:

1. NADH (4 mM), MV (1 mM), and MeCbl (0.1 mM) were added to Fom3 (2 μM) solution containing 10 mM DTT in a 1.5 mL brown tube under anaerobic conditions. The volume of the solution was 100 μL.
2. HEP-CMP (0.5 mM) and SAM (4 mM) were added to the reaction mixture, followed by incubation at 28 °C overnight under anaerobic conditions. SAM *p*-toluenesulfonate purchased from Sigma-Aldrich (St. Louis, MO) was used for enzymatic assay of Fom3.
3. The reaction was then quenched by removing the reaction mixture from the glove box.
4. A 10 μL sample of the reaction solution was injected into a HPLC system equipped with an InertSustain AQ-C18 column (5 μm, 4.6 mm × 250 mm) equilibrated with solvent A [20 mM phosphate buffer, pH 2.3]. Elution was performed at a flow rate of 1.0 mL/min at 40 °C using solvent A and solvent B [80% methanol, 20% 20 mM phosphate buffer, pH 2.3]: 0% B for 15 min, 0–20% B for 15–30 min, and 20% B for 30–35 min. Chromatography was monitored at 254 nm (for 5′-dA and SAH) and 280 nm (for HPP-CMP). Note that (*S*)- and (*R*)-HPP-CMP could not be separated under the above conditions.

4.2 Stereochemistry at the C2 position of HPP-CMP produced by Fom3

To determine the stereochemistry at the C2 position of HPP-CMP produced by Fom3, we isolated the methylation product and then analyzed it by chiral ligand-exchange chromatography (see Section 3.3). As a result, the retention time of the methylation product was identified by that of (*S*)-HPP-CMP (Fig. 5). Hence, the stereochemistry at the C2 position of the methylation product was the *S*-configuration.

The protocol for isolation of the methylation product was as follows:

1. A total of 3 mL of the resulting reaction mixture was loaded onto a DOWEX AG1-X8 column ($HCOO^-$ form, 15 mm × 40 mm), and then the column was washed with water.
2. A total of 500 mL of aqueous ammonium formate solution (pH 7.5 adjusted with aqueous ammonia) was eluted in a linear gradient of the concentration from 0.0 to 1.0 M of ammonium formate, and fractionated into 5 mL fractions. Each fraction was analyzed by HPLC to confirm that the methylation product was obtained.
3. The fractions containing the methylation product were lyophilized.
4. The resulting powder was suspended in 200 μL of H_2O.
5. The crude solution was injected into a HPLC system equipped with an InertSustain AQ-C18 column (5 μm, 4.6 mm × 250 mm). Elution was then performed with isocratic 20 mM phosphate buffer (pH 2.3) at a flow rate of 1.0 mL/min at 40 °C. The peak of the methylation product with retention time of 6 min was fractionated.
6. The collecting solution was neutralized with aqueous NaOH solution and then lyophilized to obtain the phosphate salt containing the methylation product.
7. The white solid was resuspended in 400 μL of H_2O.
8. A portion of the resulting sample (5 μL) was injected into a liquid chromatography–electrospray ionization–mass spectrometry (LC–ESI–MS) instrument equipped with a TSK-GEL Amide-80 column (3 μm, 2.0 mm × 150 mm, TOSOH) pre-equilibrated with 80% acetonitrile and 20% H_2O containing 10 mM ammonium formate (pH 3.2). Elution was then performed at a flow rate of 0.3 mL/min at 40 °C using ammonium formate buffer and acetonitrile: 80% MeCN for 5 min, 80–50% MeCN for 5–15 min, 50–10% MeCN for 15–20 min, and 10% MeCN for 20–25 min. The peak of HPP-CMP was eluted at retention time of 13 min. The MS spectrum of the isolated methylation product showed $m/z = 444$ in the negative scan, corresponding to $[M-H]^-$ of HPP-CMP.

9. The resulting sample (10 µL) was analyzed by chiral ligand-exchange chromatography (see Section 3.3).

4.3 Stoichiometry of the Fom3 reaction products

We quantified production of (*S*)-HPP-CMP, 5′-dA, and SAH in the Fom3-catalyzed reaction by HPLC analysis (Section 4.1, step 4). The concentration of authentic CMP was correlated with the peak area in the chromatogram monitored at 280 nm to prepare a standard curve, which was used to quantify production of (*S*)-HPP-CMP. The concentration of authentic 5′-dA was correlated with the peak area in the chromatogram monitored at 254 nm to prepare a standard curve, which was used to quantify production of 5′-dA and SAH. SAH was also formed non-enzymatically, probably by nucleophilic attack of DTT or Co(I)balamin at the methyl group of SAM. The amount of SAH produced in the control reaction without Fom3 was quantified and subtracted from that produced in the reaction with Fom3 at the same reaction time. Production of each product was plotted against the reaction time, revealing that the production rates of (*S*)-HPP-CMP, 5′-dA, and SAH were almost the same (Fig. 7).

4.4 Steady-state kinetic analysis

For steady-state kinetic analysis, 50, 100, 200, 500, 1000, and 2000 µM of HEP-CMP were incubated in the presence of 4 mM SAM, 0.1 mM MeCbl, 1 mM MV, 4 mM NADH, and 10 mM DTT with 2 µM Fom3 in 1.5 mL brown microtubes at 28 °C for 1.5, 6, or 16 h. Production of

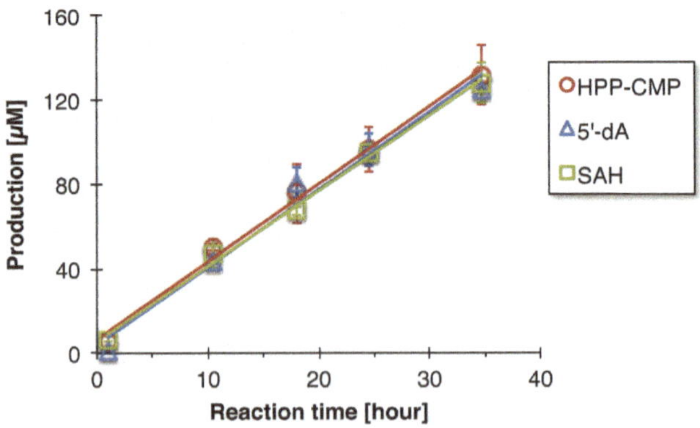

Fig. 7 Production rates of (*S*)-HPP-CMP (red), 5′-dA (blue), and SAH (green) in the Fom3-catalyzed reaction. Each plot represents the average of two replicates, and the error bars are the differences between the two measurements.

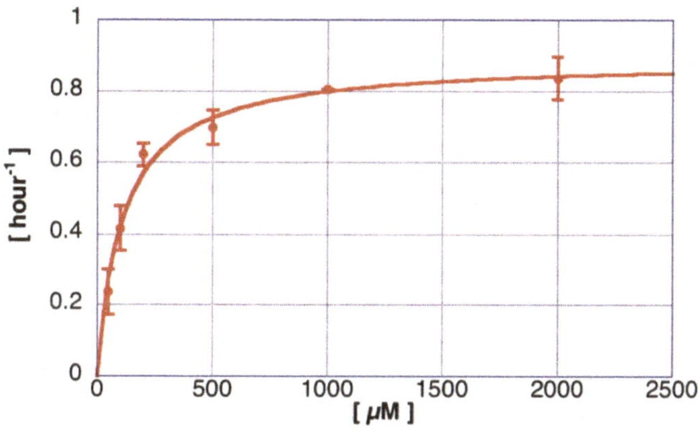

Fig. 8 Michaelis–Menten plot of the Fom3-catalyzed reaction for HEP-CMP. Each plot represents the average of two replicates, and the error bars are the differences between the two measurements.

5′-dA in each analysis was measured by HPLC and then the initial velocity was calculated for each substrate concentration. Non-linear fits to the Michaelis–Menten equation were performed using KaleidaGraph software (Synergy Software, Reading, PA). As a result, the values of K_M and k_{cat} for HEP-CMP were estimated to be $113 \pm 15\,\mu M$ and $0.89 \pm 0.03\,h^{-1}$, respectively (Fig. 8).

5. Enzyme reaction of Fom3 using deuterium-labeled substrates

5.1 Regioselectivity of hydrogen abstraction by the 5′-deoxyadenosyl radical

One equivalent of 5′-dA was formed in the Fom3-catalyzed reaction, suggesting that 5′-dA• generated by reductive cleavage of SAM abstracts a hydrogen atom from HEP-CMP. We hypothesized that 5′-dA• abstracts a hydrogen atom from the C—H bond of the carbon atom to be methylated and performed the Fom3-catalyzed reaction using [2,2-$^2H_2$]HEP-CMP as the substrate. [2,2-$^2H_2$]HEP-CMP was synthesized following the protocol described in Section 3.3 using the [2,2-$^2H_2$]HEP cyclohexylamine salt (Hammerschmidt & Kählig, 1991) donated by Professor Friedrich Hammerschmidt (Institute of Organic Chemistry, University of Vienna). The Fom3 reaction using [2,2-$^2H_2$]HEP-CMP instead of HEP-CMP was performed following the protocol described at Section 4.1. As a result, the MS spectrum of 5′-dA in the positive scan showed $m/z = 253$, which

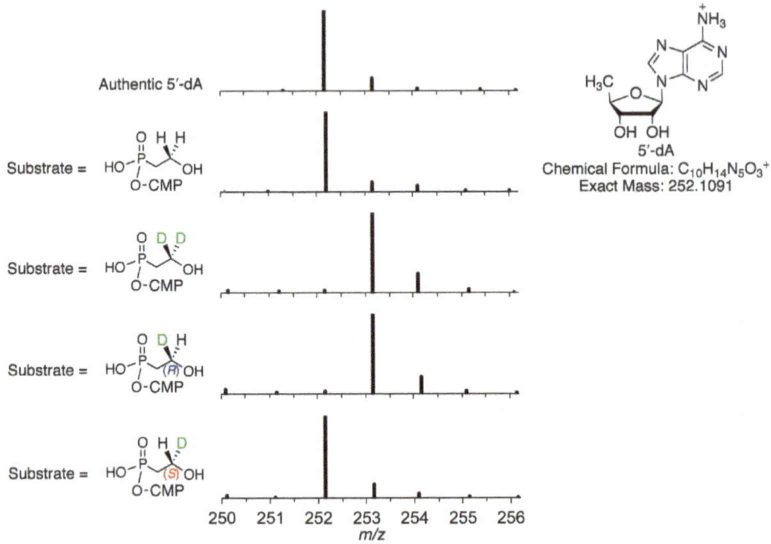

Fig. 9 MS spectra of 5′-dA formed by the Fom3-catalyzed reaction.

is 1 Da larger than the pseudo-molecular ion peak $[M+H]^+$ of 5′-dA (Fig. 9). The formed 5′-dA was isolated and analyzed by ^{1}H and ^{2}H NMR spectroscopy. In the ^{1}H NMR spectrum, the integrated value of the doublet signal of the proton at C5′ corresponded to two protons, and a deuteron signal was observed at the C5′ position in the ^{2}H NMR spectrum. These results indicated that a H (D) atom at the C2 position of HEP-CMP transfers to the C5′ position of 5′-dA during the Fom3-catalyzed reaction. Therefore, 5′-dA• seems to regioselectively abstract a hydrogen atom from the C—H bond at the C2 position of HEP-CMP.

The protocol for measurement of the MS spectrum of 5′-dA formed by the Fom3-catalyzed reaction and isolation was as follows:

1. The Fom3 reaction was performed using [2,2-$^2H_2$]HEP-CMP instead of HEP-CMP following the protocol described at Section 4.1.
2. The reaction mixture was deproteinized using an Amicon Ultra 10 K centrifugal filter
3. A portion of the resulting sample (5 μL) was injected into a LC–ESI–MS instrument equipped with a Kinetex 2.6u C18 100A column (2.1 mm × 150 mm, Phenomenex) pre-equilibrated with 80% H_2O and 20% MeOH. Elution was then performed at a flow rate of 0.1 mL/min at 40 °C using water and methanol: 20% MeOH for 20 min, 20–90% MeOH for 20–21 min, and 90% MeOH for 21–26 min. 5′-dA was eluted at retention time of 12 min.

4. To isolate deuterium-incorporated 5′-dA, a total of 7.5 mL of the reaction mixture was loaded onto a DOWEX AG1-X8 column ($HCOO^-$ form, 15 mm × 40 mm).
5. The flow-through fraction and H_2O-wash fraction were collected.
6. The collected fractions were concentrated by evaporation.
7. The resulting mixture was injected into a HPLC system equipped with a Triart C18 column (5 μm, 10 mm × 250 mm, YMC). Elution was then performed at a flow rate of 4.0 mL/min at 40 °C using water and methanol: 10% MeOH for 8 min, 10–25% MeOH for 8–9 min, 25% MeOH for 9–20 min, and 90% MeOH for 20–25 min. The peak of 5′-dA at retention time of 16 min was fractionated.
8. The solvent was removed under reduced pressure.
9. The resulting solid was resuspended in 400 μL of D_2O for ^1H NMR analysis.
10. The solvent was evaporated and the solid was resuspended in deuterium-depleted H_2O (≤1 ppm deuterium oxide, ISOTEC) for ^2H NMR analysis.

5.2 Stereoselectivity of hydrogen abstraction by the 5′-deoxyadenosyl radical

We investigated which of the two diastereotopic hydrogen atoms at the C2 position that 5′-dA• would abstract. According to feeding experiments by Hammerschmidt in which (*S*)- or (*R*)-[2−$^2H_1$]HEP was given to the fosfomycin-producing bacterium *Streptomyces fradiae*, the *pro-R* hydrogen atom at the C2 position of HEP is lost during fosfomycin biosynthesis (Hammerschmidt, 1992). To clarify the stereoselectivity of hydrogen abstraction by 5′-dA•, we performed the Fom3-catalyzed reaction with (*S*)- or (*R*)-[2-$^2H_1$]HEP-CMP as the substrate. (*S*)- and (*R*)-[2-$^2H_1$]HEP-CMP were synthesized following the protocol described at Section 3.3 using (*S*)- and (*R*)-[2-$^2H_1$]HEP cyclohexylamine salt (Hammerschmidt, 1992) donated by Professor Hammerschmidt. The Fom3 reaction using (*S*)- or (*R*)-[2-$^2H_1$]HEP-CMP instead of HEP-CMP was performed following the protocol described at Section 4.1. The MS spectrum of the 5′-dA formed was measured following the protocol described at Section 5.1 (steps 2 and 3). As a result, $m/z = 252$ corresponding to the pseudo-molecular ion $[M+H]^+$ of 5′-dA was observed when (*S*)-[2-$^2H_1$]HEP-CMP was reacted, whereas $m/z = 253$ was observed when (*R*)-[2-$^2H_1$]HEP-CMP was reacted (Fig. 9). Moreover, the MS spectrum of the methylation product formed was measured applying the condition described at Section 4.2 (step 8).

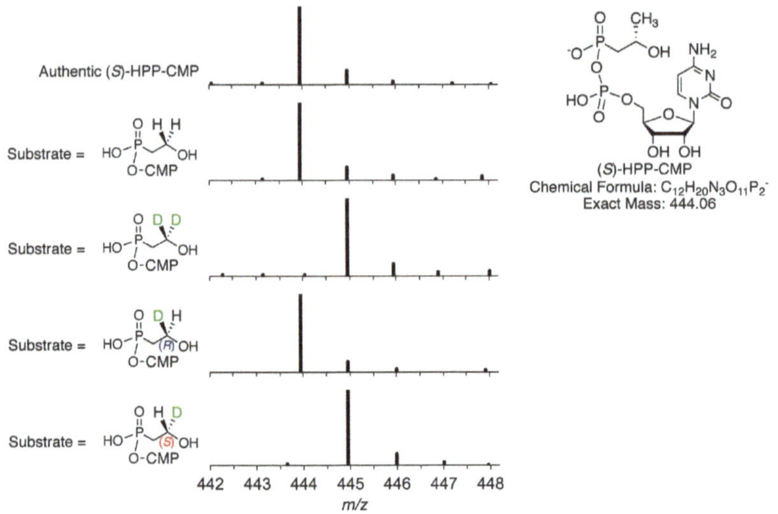

Fig. 10 MS spectra of (*S*)-HPP-CMP formed by the Fom3-catalyzed reaction.

When (*R*)-[2-$^2H_1$]HEP-CMP was used as the substrate, $m/z=444$ corresponding to the pseudo-molecular ion $[M-H]^-$ of HPP-CMP was observed (Fig. 10). These results showed that the deuterium atom of (*R*)-[2-$^2H_1$]HEP-CMP transfers to 5′-dA and no deuterium atoms remain in the methylation product. Hence, 5′-dA• stereoselectively abstracts the *pro-R* hydrogen atom at the C2 position of HEP-CMP.

5.3 Kinetic isotope effect

The Fom3-catalyzed reaction with 1 mM HEP-CMP, [2,2-$^2H_2$]HEP-CMP, (*S*)-[2-$^2H_1$]HEP-CMP, or (*R*)-[2-$^2H_1$]HEP-CMP was performed according to Section 4.1. The amounts of 5′-dA produced in the reaction at 1.5, 7, 10, and 14 h were quantified by HPLC analysis to estimate the reaction rate of Fom3 for each substrate. As a result, the reaction rates for [2,2-$^2H_2$]HEP-CMP and (*R*)-[2-$^2H_1$]HEP-CMP were slower than those for unlabeled HEP-CMP and (*S*)-[2-$^2H_1$]HEP-CMP (Fig. 11). The reaction rate for unlabeled HEP-CMP was divided by that for [2,2-$^2H_2$]HEP-CMP to obtain the kinetic isotope effect (KIE) of 1.18 (Table 1). Because the KIE was observed, the hydrogen atom abstraction step is one of the steps involved in determining the reaction rate. However, because the observed KIE was relatively small, it is presumed that the slow step of the reaction occurs later than the hydrogen atom abstraction step.

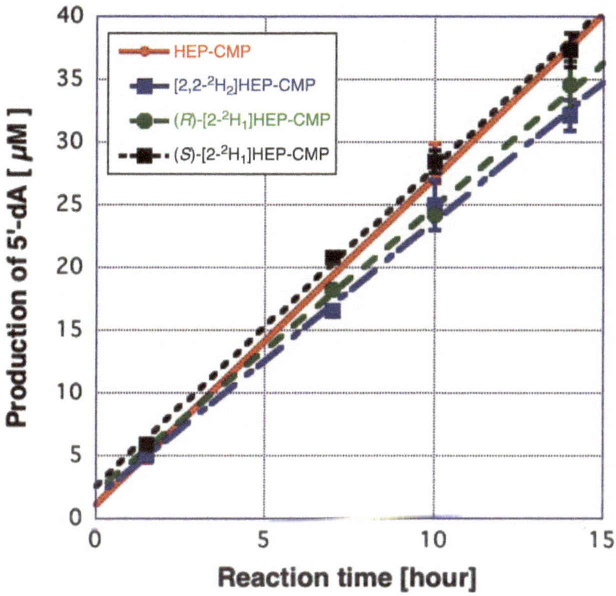

Fig. 11 Production rate of 5′-dA in the Fom3-catalyzed reaction with HEP-CMP (red solid line), [2,2-$^2H_2$]HEP-CMP (blue dashed line), (*R*)-[2-$^2H_1$]HEP-CMP (green dashed line), and (*S*)-[2-$^2H_1$]HEP-CMP (black dashed line).

Table 1 Kinetic isotope effect.

Substrate	V_{Max} **[μM/h]**	$V_{Max,HEP\text{-}CMP}/V_{Max,deuteriated\ HEP\text{-}CMP}$
HEP-CMP	2.60 ± 0.13	1.00
[2,2-$^2H_2$]HEP-CMP	2.21 ± 0.11	1.18
(*R*)-[2-$^2H_1$]HEP-CMP	2.28 ± 0.07	1.14
(*S*)-[2-$^2H_1$]HEP-CMP	2.53 ± 0.09	1.03

6. Perspective

In this chapter, we have described in vitro analysis of the cobalamin-dependent radical SAM *C*-methyltransferase Fom3. Reports on functional analysis of Fom3 have been published by Wang's group (Allen & Wang, 2014), van der Donk's group (McLaughlin & van der Donk, 2018), and Booker's group (Lanz et al., 2018; Wang et al., 2018), in addition to our group. In the study by Wang's group, HEP was used as the substrate. They reported that a small amount of the generated HPP was detected by ^{31}P NMR, but they did not describe formation of 5′-dA (Allen & Wang, 2014). 5′-dA was not formed when HEP was used for the Fom3-catalyzed reaction in this study. Formation of 5′-dA was observed for the first time when HEP-CMP was used as the substrate, suggesting that the true substrate of Fom3 is cytidylylated HEP. In the case of Wang's group, the methylation reaction was performed using a very high concentration of Fom3, which may cause the reaction to slightly proceed even when using non-cytidylylated HEP.

van der Donk's group and Booker's group successfully obtained Fom3 as a soluble enzyme by coexpression with the Btu cluster. In the kinetic analysis performed by Booker's group, the k_{cat} value of Fom3 obtained with MeCbl for HEP-CMP was estimated to be $1.81\,h^{-1}$ in the presence of MV and NADH (Lanz et al., 2018). This value is approximately twice the value estimated in this study. Therefore, it is likely that binding of MeCbl to Fom3 during the heterologous expression process in *E. coli* results in correct folding and enhances the activity. Booker's group reported that when the reducing agent for the iron–sulfur cluster was changed from MV/NADH to Ti(III) citrate, the activity of Fom3 was further enhanced by about 26-fold ($k_{cat}=0.78\,min^{-1}$) (Wang et al., 2018). In characterization of cobalamin-dependent radical SAM *C*-methyltransferases, such as Fom3, binding of cobalamin to the enzyme and suitable reducing agents are the key factors.

In this study, chiral HPLC analysis revealed that the stereochemistry at the C2 position of the methylation product is the *S*-configuration, and reaction analysis using deuterium-labeled substrates showed that 5′-dA• stereoselectively abstracts the *pro-R* hydrogen atom at the C2 position during the Fom3-catalyzed reaction. Therefore, the *C*-methylation reaction catalyzed by Fom3 proceeds with inversion of the configuration at the C2 position. Conversely, cobalamin-dependent radical SAM *C*-methyltransferase GenK in gentamicin biosynthesis catalyzes the *C*-methylation reaction at the C6′ position of gentamicin X2 with retention of the configuration (Kim, Liu, McCarty, & Liu, 2017). We are thus led to the hypothesis that *C*-methylation reactions catalyzed by cobalamin-dependent radical SAM enzymes can be classified into those that proceed with inversion of the configuration at the reaction site (Fom3 type) and those that proceed with retention of the configuration at the reaction site (GenK type). The difference in the stereochemical course of methyl transfer is probably due to the locations of the iron–sulfur cluster, MeCbl, and the methyl acceptor in the active site.

Acknowledgments

This work was supported in part by Grants-in-Aid for Scientific Research (B) from the Japan Society for the Promotion of Science (15H03834 and 18H02095 to T.E.), Grants-in-Aid for Scientific Research in Innovative Areas from the Ministry of Education, Culture, Sports, Science, and Technology (16H06451 to T.E.), and the Japan Society for the Promotion of Science A3 Foresight Program.

References

Allen, K. D., & Wang, S. C. (2014). Initial characterization of Fom3 from *Streptomyces wedmorensis*: The methyltransferase in fosfomycin biosynthesis. *Archives of Biochemistry and Biophysics*, *543*, 67–73.

Bauerle, M. R., Schwalm, E. L., & Booker, S. J. (2015). Mechanistic diversity of radical *S*-adenosylmethionine (SAM)-dependent methylation. *Journal of Biological Chemistry*, *290*, 3995–4002.

Bodea, S., & Balskus, E. P. (2018). Purification and characterization of the choline trimethylamine-lyase (CutC)-activating protein CutD. *Methods in Enzymology*, *606*, 73–94.

Bowman, E., McQueney, M., Barry, R. J., & Dunaway-Mariano, D. (1988). Catalysis and thermodynamics of the phosphoenolpyruvate/phosphonopyruvate rearrangement. Entry into the phosphonate class of naturally occurring organophosphorus compounds. *Journal of the American Chemical Society*, *110*, 5575–5576.

Broderick, J. B., Duffus, B. R., Duschene, K. S., & Shepard, E. M. (2014). Radical *S*-adenosylmethionine enzymes. *Chemical Reviews*, *114*, 4229–4317.

Cho, S.-H., Kim, S.-Y., Tomita, T., Shiraishi, T., Park, J.-S., Sato, S., et al. (2017). Fosfomycin biosynthesis *via* transient cytidylylation of 2-hydroxyethylphosphonate by the bifunctional Fom1 enzyme. *ACS Chemical Biology*, *12*, 2209–2215.

Christensen, B. G., Leanza, W. J., Beattie, T. R., Patchett, A. A., Arison, B. H., Ormond, R. E., et al. (1969). Phosphonomycin: Structure and synthesis. *Science*, *166*, 123–125.

Cicchillo, R. M., Iwig, D. F., Jones, A. D., Nesbitt, N. M., Baleanu-Gogonea, C., Souder, M. G., et al. (2004). Lipoyl synthase requires two equivalents of *S*-adenosyl-L-methionine to synthesize one equivalent of lipoic acid. *Biochemistry*, *43*, 6378–6386.

Edelhoch, H. (1967). Spectroscopic determination of tryptophan and tyrosine in proteins. *Biochemistry*, *6*, 1948–1954.

Hammerschmidt, F. (1991). Markierte vertreter eines möglichen zwischenprodukts der biosynthese von fosfomycin in *Streptomyces fradiae*: darstellung von (*R*,*S*)-(2-Hydroxypropyl)-, (*R*,*S*)-, (*R*)-, (*S*)-(2-Hydroxy-[1,1-$^{2}H_{2}$]propyl)-und (*R*,*S*)-(2-[^{18}O] Hydroxypropyl)phosphonsäure. *Monatshefte für Chemie Chemical Monthly*, *122*, 389–398.

Hammerschmidt, F. (1992). Biosynthesis of natural products with a phosphorus-carbon bond. IX. Synthesis and incorporation of (*S*)- and (*R*)-2-hydroxy-[2-$^{2}H_{1}$]ethylphosphonic acid into fosfomycin by *Streptomyces fradiae*. *Liebigs Annalen der Chemie*, *1992*, 553–557.

Hammerschmidt, F., & Kählig, H. (1991). Biosynthesis of natural products with a phosphorus-carbon bond. 7. *The Journal of Organic Chemistry*, *56*, 2364–2370.

Hendlin, D., Stapley, E. O., Jackson, M., Wallick, H., Miller, A. K., Wolf, F. J., et al. (1969). Phosphonomycin, a new antibiotic produced by strains of *Streptomyces*. *Science*, *166*, 122–123.

Hidaka, T., Goda, M., Kuzuyama, T., Takei, N., Hidaka, M., & Seto, H. (1995). Cloning and nucleotide sequence of fosfomycin biosynthetic genes of *Streptomyces wedmorensis*. *Molecular & General Genetics*, *249*, 274–280.

Hidaka, T., Iwakura, H., Imai, S., & Seto, H. (1992). Studies on the biosynthesis of fosfomycin. 3. Detection of phosphoenol-pyruvate phosphomutase activity in a fosfomycin high-producing strain of *Streptomyces wedmorensis* and characterization of its blocked mutant NP-7. *Journal of Antibiotics*, *45*, 1008–1010.

Horsman, G. P., & Zechel, D. L. (2017). Phosphonate biochemistry. *Chemical Reviews*, *117*, 5704–5783.

Kahan, F. M., Kahan, J. S., Cassidy, P. J., & Kropp, H. (1974). Mechanism of action of fosfomycin (phosphonomycin). *Annals of the New York Academy of Sciences*, *235*, 364–386.

Katayama, N., Tsubotani, S., Nozaki, Y., Harada, S., & Ono, H. (1990). Fosfadecin and fosfocytocin, new nucleotide antibiotics produced by bacteria. *The Journal of Antibiotics*, *43*, 238–246.

Kim, S. Y., Ju, K., Metcalf, W. W., Evans, B. S., Kuzuyama, T., & van der Donk, W. A. (2012). Different biosynthetic pathways to fosfomycin in *Pseudomonas syringae* and *Streptomyces* species. *Antimicrobial Agents and Chemotherapy*, *56*, 4175–4183.

Kim, H. J., Liu, Y., McCarty, R. M., & Liu, H.-w. (2017). Reaction catalyzed by GenK, a cobalamin-dependent radical *S*-adenosyl-L-methionine methyltransferase in the biosynthetic pathway of gentamicin, proceeds with retention of configuration. *Journal of the American Chemical Society*, *13*, 16084–16087.

Kim, H. J., McCarty, R. M., Ogasawara, Y., Liu, Y., Mansoorabadi, S. O., LeVieux, J., et al. (2013). GenK-catalyzed C-6′ methylation in the biosynthesis of gentamicin: Isolation and characterization of a cobalamin-dependent radical SAM enzyme. *Journal of the American Chemical Society*, *135*, 8093–8096.

Kudo, F., Hoshi, S., Kawashima, T., Kamachi, T., & Eguchi, T. (2014). Characterization of a radical *S*-adenosyl-L-methionine epimerase, NeoN, in the last step of neomycin B biosynthesis. *Journal of the American Chemical Society*, *136*, 13909–13915.

Kudo, F., Tokumitsu, T., & Eguchi, T. (2017). Substrate specificity of radical *S*-adenosyl-L-methionine dehydratase AprD4 and its partner reductase AprD3 in the C3′-deoxygenation of aminoglycoside antibiotics. *The Journal of Antibiotics*, *70*, 423–428.

Kuzuyama, T., Hidaka, T., Kamigiri, K., Imai, S., & Seto, H. (1992). Studies on the biosynthesis of fosfomycin. 4. The biosynthetic origin of the methyl group of fosfomycin. *The Journal of Antibiotics*, *45*, 1812–1814.

Lanz, N. D., Blaszczyk, A. J., McCarthy, E. L., Wang, B., Wang, R. X., Jones, B. S., et al. (2018). Enhanced solubilization of class B radical *S*-adenosylmethionine methylases by improved cobalamin uptake in *Escherichia coli*. *Biochemistry*, *57*, 1475–1490.

Lewis, J. K., Bruender, N. A., & Bandarian, V. (2018). QueE: A radical SAM enzyme involved in the biosynthesis of 7-deazapurine containing natural products. *Methods in Enzymology*, *606*, 95–118.

Liu, P., Murakami, K., Seki, T., He, X., Yeung, S.-M., Kuzuyama, T., et al. (2001). Protein purification and function assignment of the epoxidase catalyzing the formation of fosfomycin. *Journal of the American Chemical Society*, *123*, 4619–4620.

Maruyama, C., Chinone, Y., Sato, S., Kudo, F., Ohsawa, K., Kubota, J., et al. (2020). *C*-Methylation of *S*-adenosyl-L-methionine occurs prior to cyclopropanation in the biosynthesis of 1-amino-2-methylcyclopropanecarboxylic acid (norcoronamic acid) in a bacterium. *Biomolecules*, *10*, 775.

McLaughlin, M. I., & van der Donk, W. A. (2018). Stereospecific radical-mediated B_{12}-dependent methyl transfer by the fosfomycin biosynthesis enzyme Fom3. *Biochemistry*, *57*, 4967–4971.

Munos, J. W., Moon, S.-J., Mansoorabadi, S. O., Chang, W., Hong, L., Yan, F., et al. (2008). Purification and characterization of the epoxidase catalyzing the formation of fosfomycin from *Pseudomonas syringae*. *Biochemistry*, *47*, 8726–8735.

Nakashita, H., Watanabe, K., Hara, O., Hidaka, T., & Seto, H. (1997). Studies on the biosynthesis of bialaphos. Biochemical mechanism of C-P bond formation: Discovery of phosphonopyruvate decarboxylase which catalyzes the formation of phosphonoacetaldehyde from phosphonopyruvate. *The Journal of Antibiotics*, *50*, 212–219.

Olivares, P., Ulrich, E. C., Chekan, J. R., van der Donk, W. A., & Nair, S. K. (2017). Characterization of two late-stage enzymes involved in fosfomycin biosynthesis in *Pseudomonads*. *ACS Chemical Biology*, *12*, 456–463.

Pang, H., & Yokoyama, K. (2018). Lessons from the studies of a C-C bond forming radical SAM enzyme in molybdenum cofactor biosynthesis. *Methods in Enzymology*, *606*, 485–522.

Sato, S., Kudo, F., Kim, S.-Y., Kuzuyama, T., & Eguchi, T. (2017). Methylcobalamin-dependent radical SAM *C*-methyltransferase Fom3 recognizes cytidylyl-2-hydroxyethylphosphonate and catalyzes the nonstereoselective *C*-methylation in fosfomycin biosynthesis. *Biochemistry*, *56*, 3519–3522.

Sato, S., Kudo, F., Kuzuyama, T., Hammerschmidt, F., & Eguchi, T. (2018). *C*-Methylation catalyzed by Fom3, a cobalamin-dependent radical *S*-adenosyl-L-methionine enzyme in fosfomycin biosynthesis, proceeds with inversion of configuration. *Biochemistry*, *57*, 4963–4966.

Sato, S., Miyanaga, A., Kim, S. Y., Kuzuyama, T., Kudo, F., & Eguchi, T. (2018). Biochemical and structural analysis of FomD that catalyzes the hydrolysis of cytidylyl (*S*)-2-Hydroxypropylphosphonate in fosfomycin biosynthesis. *Biochemistry*, *57*, 4858–4866.

Seidel, H. M., Freeman, S., Seto, H., & Knowles, J. R. (1988). Phosphonate biosynthesis: Isolation of the enzyme responsible for the formation of a carbon–phosphorus bond. *Nature*, *335*, 457–458.

Seto, H., Hidaka, T., Kuzuyama, T., Shibata, S., Usui, T., Sakanaka, O., et al. (1991). Studies on the biosynthesis of fosfomycin. 2. Conversion of 2-hydoxypropylphosphonic acid to fosfomycin by blocked mutants of *Streptomyces wedmorensis*. *The Journal of Antibiotics*, *44*, 1286–1288.

Seto, H., & Kuzuyama, T. (1999). Bioactive natural products with carbon–phosphorus bonds and their biosynthesis. *Natural Product Reports*, *16*, 589–596.

Shao, Z., Blodgett, J. A. V., Circello, B. T., Eliot, A. C., Woodyer, R., Li, G., et al. (2008). Biosynthesis of 2-hydroxyethylphosphonate, an unexpected intermediate common to multiple phosphonate biosynthetic pathways. *Journal of Biological Chemistry*, *283*, 23161–23168.

Shoji, J., Kato, T., Hinoo, H., Hattori, T., Hirooka, K., Matsumoto, K., et al. (1986). Production of fosfomycin (phosphonomycin) by *Pseudomonas syringae*. *The Journal of Antibiotics*, *39*, 1011–1012.

Simon, M. A., Ongpipattanakul, C., Nair, S. K., & van der Donk, W. A. (2021). Biosynthesis of fosfomycin in pseudomonads reveals an unexpected enzymatic activity in the metallohydrolase superfamily. *Proceedings of the National Academy of Sciences of the United States of America*, *118*, e2019863118.

Skarzynski, T., Mistry, A., Wonacott, A., Hutchinson, S. E., Kelly, V. A., & Duncan, K. (1996). Structure of UDP-*N*-acetylglucosamine enolpyruvyl transferase, an enzyme essential for the synthesis of bacterial peptidoglycan, complexed with substrate UDP-*N*-acetylglucosamine and the drug fosfomycin. *Structure*, *4*, 1465–1474.

Takahashi, Y., & Tokumoto, U. (2002). A third bacterial system for the assembly of iron-sulfur clusters with homologs in Archaea and plastids. *Journal of Biological Chemistry*, *277*, 28380–28383.

Wang, B., Blaszczyk, A. J., Knox, H. L., Zhou, S., Blaesi, E. J., Krebs, C., et al. (2018). Stereochemical and mechanistic investigation of the reaction catalyzed by Fom3 from *Streptomyces fradiae*, a cobalamin-dependent radical *S*-adenosylmethionine methylase. *Biochemistry*, *57*, 4972–4984.

Wang, C., Chang, W.-c., Guo, Y., Huang, H., Peck, S. C., Pandelia, M. E., et al. (2013). Evidence that the fosfomycin-producing epoxidase, HppE, is a non-heme-iron peroxidase. *Science*, *342*, 991–995.

Woodyer, R. D., Li, G., Zhao, H., & van der Donk, W. A. (2007). New insight into the mechanism of methyl transfer during the biosynthesis of fosfomycin. *Chemical Communications*, *4*, 359–361.

Woodyer, R. D., Shao, Z., Thomas, P. M., Kelleher, N. L., Blodgett, J. A. V., Metcalf, W. W., et al. (2006). Heterologous production of fosfomycin and identification of the minimal biosynthetic gene cluster. *Chemistry & Biology*, *13*, 1171–1182.

Woschek, A., Lindner, W., & Hammerschmidt, F. (2003). Enzymes in organic chemistry, 11: Hydrolase-catalyzed resolution of α- and β-hydroxyphosphonates and synthesis of chiral, non-racemic β-aminophosphonic acids. *Advanced Synthesis & Catalysis*, *345*, 1287–1298.

CHAPTER FOUR

Studies of GenK and OxsB, two B_{12}-dependent radical SAM enzymes involved in natural product biosynthesis

Yu-Hsuan Lee[a] and Hung-wen Liu[a,b,*]
[a]Department of Chemistry, University of Texas at Austin, Austin, TX, United States
[b]Division of Chemical Biology and Medicinal Chemistry, College of Pharmacy, University of Texas at Austin, Austin, TX, United States
*Corresponding author: e-mail address: h.w.liu@mail.utexas.edu

Contents

Abstract

The B_{12}-dependent radical SAM enzymes are an emerging subgroup of biological catalysts that bind a cobalamin cofactor in addition to the canonical $[Fe_4S_4]$ cluster characteristic of radical SAM enzymes. Most of the B_{12}-dependent radical SAM enzymes that have been characterized mediated methyltransfer reactions; however, a small number are known to catalyze more diverse reactions such as ring contractions. Thus, Genk is a methyltransferase from the gentamicin C biosynthetic pathway, whereas OxsB catalyzes the oxidative ring contraction of 2′-deoxyadenosine 5′-phosphates to generate an oxetane aldehyde during the biosynthesis of oxetanocin A. The preparation and *in vitro* characterization of such enzymes is complicated by the presence of two redox sensitive cofactors in addition to challenges in obtaining soluble protein for study.

Methods in Enzymology, Volume 669
ISSN 0076-6879
https://doi.org/10.1016/bs.mie.2021.12.014

This chapter describes expression, purification and assay methodologies for GenK and OxsB highlighting the use of denaturation/refolding protocols for solubilizing inclusion bodies as well as the use of cluster assembly and cobalamin uptake machinery during *in vivo* expression.

1. Introduction

Radical *S*-adenosyl-L-methionine (SAM) enzymes catalyze transformations involving radical intermediates. They are characterized by the presence of a low potential $[Fe_4S_4]^{1+}$ cluster typically bound to the highly conserved CX_3CX_2C motif in the active site (Broderick, Duffus, Duschene, & Shepard, 2014; Frey, Hegeman, & Ruzicka, 2008; Sofia, Chen, Hetzler, Reyes-Spindola, & Miller, 2001). This cluster serves to induce reductive homolysis of the C5′–S bond of SAM (**1**). The resulting 5′-deoxyadenosyl radical (**2**) or its equivalent then abstracts a hydrogen atom from the substrate (**3**) to form 5′-deoxyadenosine (**4**), methionine (**5**), and a substrate radical (**6**), which can undergo a large number of possible transformations leading to a wide array of potential products depending on the constraints of the active site (see Fig. 1).

A number of radical SAM enzymes are also known to bind cobalamin (coenzyme B_{12}) in their active sites and are referred to as B_{12}-dependent radical SAM enzymes (Wang, 2018). Cobalamin (Cbl) is a cobalt-containing tetrapyrrole and can participate in both one- and two-electron processes since the cobalt center can accommodate oxidation states from 1+ to 3+.

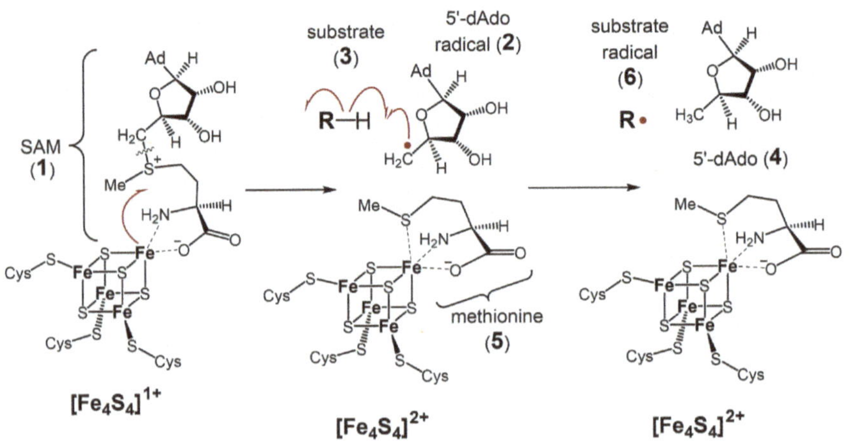

Fig. 1 The initial steps of reactions catalyzed by radical SAM enzymes.

While more than 9700 UniRef50 sequence clusters have been annotated as B_{12}-dependent radical SAM enzymes (Oberg, Precord, Mitchell, & Gerlt, 2021), only a handful have been characterized (Allen & Wang, 2014a; Bridwell-Rabb, Zhong, Sun, Drennan, & Liu, 2017; Huang et al., 2015; Kim, Liu, McCarty, & Liu, 2017; Kim et al., 2013; Marous et al., 2015; Maruyama et al., 2020; Parent et al., 2016; Pierre et al., 2012; Radle, Miller, Laremore, & Booker, 2019; Sinner, Lichstrahl, Li, Marous, & Townsend, 2019; Wang, Schnell, Baumann, Müller, & Begley, 2017; Werner et al., 2011). Moreover, only three B_{12}-dependent radical SAM enzymes, OxsB (Bridwell-Rabb et al., 2017), TsrM (Knox et al., 2021) and TokK (Knox, Sinner, Townsend, Boal, & Booker, 2021), have been structurally characterized to date.

The majority of B_{12}-dependent radical SAM enzymes that have been characterized *in vitro* are limited to catalyzing methylation reactions with methylcobalamin (**7**) serving as the immediate methyl donor (McLaughlin & van der Donk, 2018; Sato, Kudo, Kim, Kuzuyama, & Eguchi, 2017; Werner et al., 2011). GenK represents one such example that catalyzes C6′ methylation of gentamicin X_2 (**9**, GenX_2) to G418 (**11**) during the biosynthesis of gentamicin C_1 (**13**) (Kim et al., 2017, 2013; Unwin et al., 2004). GenK catalysis proceeds via abstraction of the C6′ *pro-R* H-atom of **9** by the 5′-deoxyadenosyl radical equivalent (**2**) to give the substrate radical intermediate **10** (Fig. 2). Transfer of a methyl radical from Cbl(III)-Me (**7**) generates G418 (**11**) and Cbl(II) (**8**). Two additional reducing equivalents are then required to both reduce the $[Fe_4S_4]^{2+}$ cluster back to the $[Fe_4S_4]^{1+}$ state and regenerate Me-Cbl(III) from Cbl(II) and an additional molecule of SAM yielding SAH (**12**) as a byproduct.

OxsB is the only B_{12}-dependent radical SAM enzyme characterized *in vitro* to catalyze an oxidative ring contraction, which takes place during

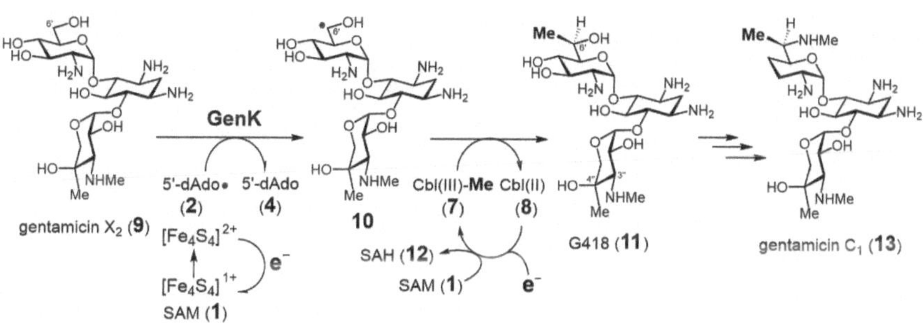

Fig. 2 Reaction catalyzed by GenK.

the biosynthesis of oxetanocin A (OXT-A, **18**) (Bridwell-Rabb et al., 2017). Oxetanocin A is a nucleoside analog having potent antiviral, antitumor, and antimicrobial activities due to its ability to inhibit DNA polymerases (Shimada et al., 1986; Ueda, Tsurimoto, Nagahata, Chisaka, & Matsubara, 1989). Its biosynthesis has been shown to be catalyzed by the cobalamin-dependent radical SAM enzyme OxsB (Bridwell-Rabb et al., 2017). The transformation catalyzed by OxsB involves a sequence of **14** → **15** → **16**; however, the functional form of the cobalamin cofactor as well as its catalytic role remain unclear (Fig. 3).

A major hurdle in the study of B_{12}-dependent radical SAM enzymes is acquiring sufficient protein in soluble form. Manipulation of buffer salt concentration can be utilized to solubilize inclusion bodies and then refold the denatured proteins (Allen & Wang, 2014a, 2014b; Kim et al., 2013). Another approach is co-expression of the radical SAM enzyme with $[Fe_4S_4]$ cluster biogenesis enzymes in order to facilitate *in vivo* reconstitution of the clusters (Anton et al., 2008; Chatterjee et al., 2008; Grove, Lee, St Clair, Krebs, & Booker, 2008; Hänzelmann et al., 2004; Kriek, Peters, Takahashi, & Roach, 2003; Wecksler et al., 2009). This can lead to improved solubility and two plasmid systems have been frequently utilized. These include pDB1282, which encodes the *isc* operon from *Azotobacter vinelandii* (Zheng, Cash, Flint, & Dean, 1998), and pRKSUF017, which encodes the *suf* operon from *Escherichia coli* (Takahashi & Tokumoto, 2002). Moreover, Lanz et al. have also reported that co-expression of B_{12}-dependent radical SAM enzymes with the pBAD42-BtuCEDFB plasmid, which encodes the Btu system for cobalamin uptake in *E. coli*, can also improve the yield of soluble protein (Lanz et al., 2018).

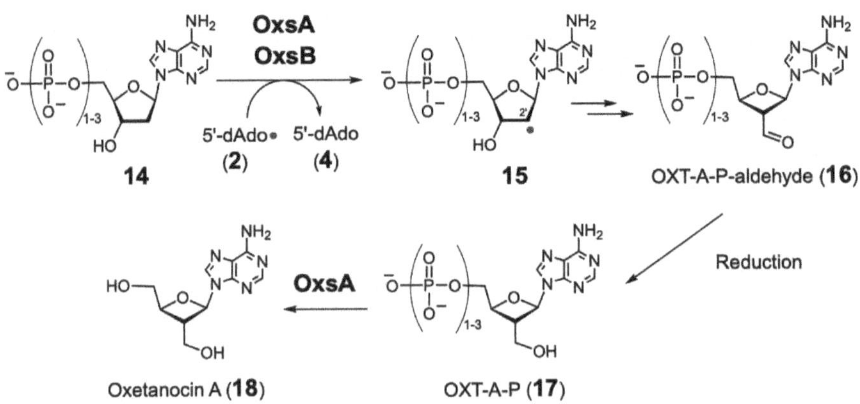

Fig. 3 Reaction catalyzed by OxsB/OxsA.

The following sections provide methodologies currently in practice for the preparation and assay of GenK and OxsB. GenK typically expresses as inclusion bodies necessitating a refolding protocol in order to obtain soluble protein. In contrast, OxsB can be isolated as a soluble protein without the need for refolding. However, the preparation and purification of OxsB can be improved by codon-optimization and co-expression with $[Fe_4S_4]$ cluster biogenesis enzymes. Therefore, this chapter provides details for each of these two different approaches to the *in vitro* study of B_{12}-dependent radical SAM enzymes.

2. GenK catalyzes 6′-C methylation in gentamicin biosynthesis

2.1 Cloning, overexpression, purification and chemical reconstitution of GenK

Either plasmid pDB1282 or pRKSUF017 can be used for co-expression of $[Fe_4S_4]$ biogenesis machinery (Takahashi & Tokumoto, 2002; Zheng et al., 1998), whereas pBAD42-BtuCEDFB encodes proteins for the cobalamin uptake system (Lanz et al., 2018). In the absence of these plasmids, expression of GenK lacking an affinity tag is primarily in the form of inclusion bodies. While coexpression of native GenK with pDB1282 or pRKSUF017/pBAD42-BtuCEDFB does appear to improve the yield of soluble enzyme it may also make purification of native protein more challenging. Consequently, this approach may be inferior to the expression of GenK alone followed by denaturation and refolding.

2.1.1 Cloning

The gene encoding GenK is polymerase chain reaction (PCR)-amplified from the genomic DNA of the gentamicin producing strain *Micromonospora echinospora* (NRRL 2953) using KOD polymerase (Novagen). The forward primer (5′-CAACATATGAACGCGCTGGTGGCAGC-3′) and reverse primer (5′-TAACGAATTCAGTGGGAAACCGCCTCGG-3′) for the PCR reaction are engineered with an *Nde*I and an *Eco*RI digestion site, respectively underlined. The resulting DNA fragment is then subcloned into a pCR:blunt vector (Invitrogen) followed by excision with *Nde*I and *Eco*RI (New England Biolabs, NEB) and ligated into pET-24b + (Novagen). The resulting plasmid is designated GenK:pET-24 and encodes a construct of native GenK that lacks an affinity tag.

2.1.2 Overexpression

GenK is susceptible to proteolytic degradation especially when expressed at low temperatures (e.g., 18 °C) where inclusion bodies are less stable (De Groot & Ventura, 2006). Yields appear to be improved, however, when expression is carried out at a more elevated temperature (e.g., 25 °C) with proteinase inhibitors included in the lysis buffer.

1. Competent *E. coli* BL21 star (DE3) cells are transformed with the GenK: pET-24 plasmid and selected by plating on Luria-Bertani (LB) agar plates with 50 μg/mL kanamycin at 37 °C for 16 h
2. A single colony is used to inoculate 10 mL LB broth containing 50 μg/mL kanamycin. This starter culture is incubated at 37 °C with 200 rpm shaking for 12 h
3. The starter culture is added to 1 L LB broth containing 50 μg/mL kanamycin. The resulting culture is incubated at 37 °C with 200 rpm shaking until the optical density at 600 nm (OD_{600}) reaches 0.6
4. The culture is then cooled to 25 °C, and 40 mg ammonium iron(II) sulfate is added along with isopropyl-β-D-thiogalactopyranoside (IPTG) to a final concentration of 0.1 mM
5. The culture is incubated at 25 °C for another 20 h with 120 rpm shaking
6. Cells are harvested by centrifugation at 4000 *g* and 4 °C for 12 min, and the cell pellet can be stored at −80 °C

2.1.3 Purification

GenK is isolated in soluble form via a denaturation/refolding process. The entire purification process may take several days to complete and can be performed aerobically. The purity of isolated GenK is examined by SDS-PAGE. Refolded GenK can then be flash frozen and stored at −80 °C. Prior to assay, GenK is thawed, and the $[Fe_4S_4]$ clusters are reconstituted under anaerobic conditions. The following purification procedure is designed for a 3 L expression culture as described above. All steps are carried out at 4 °C under aerobic conditions.

Buffers. The purification and reconstitution are carried out using the following buffer solutions. All buffers are chilled at 4 °C and β-mercaptoethanol is freshly added before use. Phenylmethylsulfonyl fluoride (PMSF) can be stored in isopropanol as a 100 mM stock solution and is freshly added before use.

Lysis buffer: 50 mM Tris·HCl (pH 8.0), 1 mM EDTA, 1 mM PMSF, 1 mM β-mercaptoethanol and 10% (*v*/*v*) glycerol.

Denaturation buffer: 50 mM Tris·HCl (pH 8.0), 8 M urea, 1 mM β-mercaptoethanol and 10% (*v*/*v*) glycerol.

Refolding buffer: 50 mM Tris·HCl (pH 8.0), 1 mM $MgSO_4$, 400 mM L-arginine, 5 mM glutathione (reduced), 0.5 mM glutathione (oxidized), 5 mM DTT and 10% (*v*/*v*) glycerol.

Reconstitution buffer: 50 mM Tris·HCl (pH 8.0), 0.15 M NaCl and 20% (*v*/*v*) glycerol.

1. The cell pellet from the 3 L culture is suspended in 75 mL lysis buffer
2. The suspended cell pellet is sonicated with a Fisher Scientific 550 Dismembrator at power 9 for 300 s total on-time with each 10 s burst followed by a 25 s pause
3. The cell lysate is pelleted by centrifugation at 18000 *g* for 25 min at 4 °C
4. The lysate pellet is suspended in 75 mL chilled lysis buffer, centrifuged at 18000 *g* for 25 min at 4 °C and the supernatant is discarded. This step is repeated at least twice
5. The lysate pellet is suspended in 15 mL denaturation buffer using sonication for 40 s total on-time with each 5 s burst followed by a 20 s pause
6. The suspension from step 5 is centrifuged at 18000 *g* for 25 min. The dark yellow supernatant (ca. 15 mL) can be either flash frozen with liquid N_2 for storage at −80 °C or immediately refolded
7. The urea buffer containing denatured GenK from step 6 (10 mL) is combined with 200 mL refolding buffer slowly over 2 days with gentle stirring. This is done by adding ca. 1 mL of the denatured GenK dropwise over 1 min to the refolding buffer approximately every 3 h
8. The refolded GenK is concentrated to ca. 5 mg/mL protein, aliquoted, flash frozen and stored at −80 °C or immediately reconstituted. The purity of GenK can be examined by SDS-PAGE (Fig. 4).

2.1.4 Reconstitution of the [Fe_4S_4]

Reconstitution is carried out in an anaerobic chamber with buffers that have been equilibrated with the anaerobic atmosphere.

1. An aliquot of 2.85 mL containing ca. 5 mg refolded GenK (diluted with reconstitution buffer) is equilibrated with the atmosphere of the anaerobic chamber by stirring gently for 2.5 h in an uncapped vial at 4 °C
2. To the reconstitution solution is added 30 μL of 1 M dithiothreitol (DTT), and the mixture is gently stirred for another 30 min
3. A total of 60 μL 100 mM $Fe(NH_4)_2(SO_4)_2$ in water is added to the reconstitution mixture over 10 min and then stirred for another 10 min

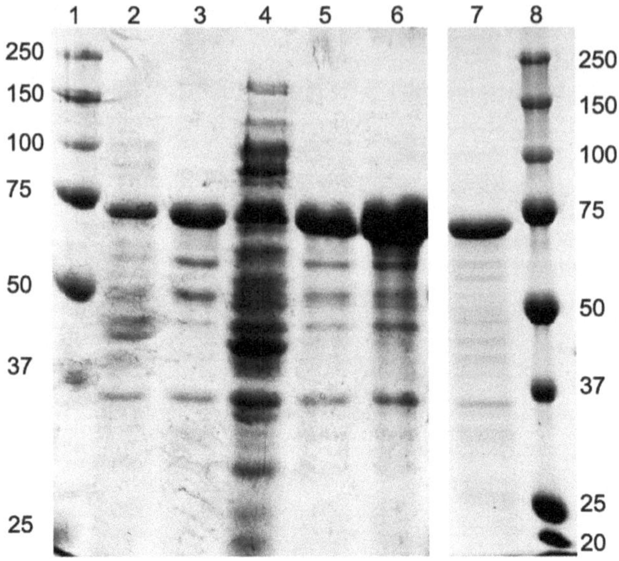

Fig. 4 SDS-PAGE analysis of GenK (69 kDa). Lanes 1 and 8 are molecular weight markers; lanes 2–4 are whole cell lysate, pellet and supernatant; lanes 5–6 are pellet and supernatant in denaturation buffer and lane 7 is GenK from refolding buffer.

4. A total of 60 μL 100 mM Na_2S in reconstitution buffer is added to the reconstitution mixture over 20 min and stirred for another 3 h. The resulting dark brown mixture contains ca. 70 μM GenK, 5 mM DTT, 2 mM iron and 2 mM sulfide in a volume of ca. 3 mL
5. The reconstitution mixture is desalted using an Econo-Pac 10DG desalting column from Bio-Rad Laboratories pre-equilibrated with anaerobic reconstitution buffer. The enzyme elutes in olive-brown colored aliquots, which are combined and concentrated by centrifugal filtration to yield ca. 70 μM GenK monomer

2.2 *In vitro* functional characterization

GenK activity can be assessed using the $GenX_2$ substrate (**9**). Under anaerobic conditions, a 50 μL solution containing 5 μM reconstituted GenK, 10 mM DTT, 1 mM methyl viologen, 4 mM NADPH, 4 mM SAM (**1**) and 1 mM methylcobalamin (Me-Cbl, **7**) is prepared in 50 mM Tris·HCl buffer (pH 8.0). The reaction is initiated with the addition of 1 mM $GenX_2$ followed by incubation at room temperature over 20 h and halted by filtration to remove the protein. The deproteinized solution can then be analyzed by electrospray ionization mass spectrometry (ESI-MS).

Under positive ion mode, the $[M+H]^+$ ions of substrate $GenX_2$ (**9**) and product G418 (**11**) are detected at m/z 483.27 and 497.28, respectively.

Reduction and demethylation of SAM can also be followed by HPLC without the need for mass spectrometry. Thus, a 27 μL assay solution reaction can be quenched by adding 3 μL 30% (*v/v*) aqueous trichloroacetic acid and centrifuging to remove the precipitated protein. The resulting supernatant can then be analyzed by HPLC using UV absorbance detection at 260 nm to follow the formation and consumption of the adenine nucleosides SAM (**1**), 5′-deoxyadenosine (**4**) and SAH (**12**). Chromatographic separation is achieved with a Varian Microsorb-MV 100–5 C18 (4.6 mm × 250 mm) column. A two-solvent system is used with 0.1% trifluoroacetic acid (TFA) in ddH_2O as mobile phase A and 0.1% TFA in acetonitrile as mobile phase B. The column is eluted at a flow rate of 1 mL/min with a linear gradient from 0% to 20% B over 30 min.

Despite the lack of a chromophore in $GenX_2$ (**9**) and G418 (**11**), these aminoglycosides can also be detected by UV–vis absorption following derivatization with 1-fluoro-2,4-dinitrobenzene (**19**, DNFB) (Ryan, 1984). Mass spectrometry demonstrates complete modification of all three primary amino groups in **9** or **11** with the 2,4-dinitrophenyl (DNP) moiety (Fig. 5). Thus, a 30 μL aliquot of the GenK reaction solution described above is combined with 120 μL of a mixture prepared from 112 μL MeOH, 1.8 μL of 8.4 M DNFB and 6 μL of 0.5 M sodium hydroxide. The resulting solution is heated at 80 °C for 5 min and then centrifuged at 15000*g* for 5 min to pellet any particulates and protein precipitate.

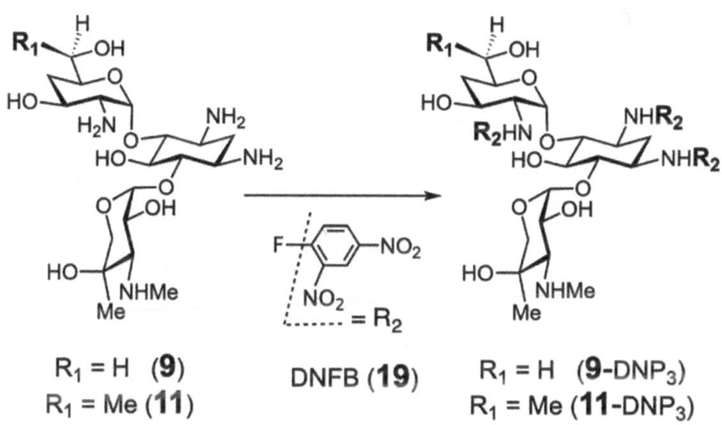

Fig. 5 DNFB derivatization of substrate $GenX_2$ (**9**) and product G418 (**11**). The derivatization installs three DNP moieties and yields **9**-DNP_3 and **11**-DNP_3.

The supernatant can be analyzed by HPLC with UV–vis absorbance detection at 340 nm using a Varian Microsorb-MV 100–5 C18 (4.6 mm × 250 mm) column. The column is eluted with ddH_2O as mobile phase A and acetonitrile as mobile phase B at a flow rate of 1 mL/min and a linear gradient from 40% to 50% B over 20 min.

3. OxsB and OxsA catalyzes ring contraction in oxetanocin A biosynthesis

The oxetanocin A (OXT-A, **18**) biosynthetic gene cluster along with potential resistance determinants is plasmid borne (pOXT1) in the producing strain *Bacillus megaterium* strain *NK84-0128* and encoded on a single 6.8 kb fragment (*Bgl*II-D) obtained via the enzymatic digestion of pOXT1 with the restriction enzyme *Bgl*II (Morita et al., 1999). Only two genes encoded by this cluster are strictly required for OXT-A biosynthesis, namely *oxsA* and *oxsB* (Bridwell-Rabb et al., 2017; Zhong, Lee, Liu, & Liu, 2021). OxsB is the B_{12}-dependent radical SAM enzyme directly responsible for the oxidative ring contraction **14** → **16**. OxsA is an HD (His-Asp) domain phosphohydrolase capable of dephosphorylating oxetanocin A 5′-triphosphate (**17**) to OXT-A (Bridwell-Rabb, Kang, Zhong, Liu, & Drennan, 2016) (Fig. 3). However, *in vivo* and *in vitro* assays have shown that the presence of OxsA is required for OxsB activity suggesting a direct interaction between the two proteins (Bridwell-Rabb et al., 2017; Zhong et al., 2021).

3.1 Cloning, overexpression, purification and chemical reconstitution of OxsB

3.1.1 Cloning

The *oxsB* gene in the original studies (Bridwell-Rabb et al., 2017) was synthesized by the Mr. Gene service according to the published primary sequence of *oxsB* (NCBI, accession No. AB005787) following codon optimization. Once obtained, the *oxsB* gene can be PCR-amplified and subcloned into the *Nde*I and *Hind*III sites of pET-28b. This modifies *oxsB* to encode the *N*-His_6 tagged OxsB protein, and the resulting plasmid is designated OxsB:pET-28b.

An expression construct can also be generated for the isolation of untagged OxsB. This is accomplished by subcloning the PCR-amplified *oxsB* gene fragment into a MalE-pET vector at the *Nde*I and *Hind*III cut sites. This plasmid is designated as OxsB:MalE-pET. The MalE-pET vector allows for expression of OxsB as a fusion protein with *C*-His_{10} tagged maltose binding protein (MBP) linked to the *N*-terminus of OxsB via a 7-residue (ENLYFQG) linker that can be cleaved by the tobacco etch virus (TEV) protease (Kapust et al., 2001).

3.1.2 Overexpression

Two *E. coli* strains have been investigated for the expression of both OxsB constructs. The first is *E. coli* BL21 star (DE3), which harbors the pDB1282 plasmid for *in vivo* reconstitution of [Fe_4S_4] clusters (Zheng et al., 1998). The second is *E. coli* BL21 star (DE3) containing both the pDB1282 and pBAD42-BtuCEDFB plasmids to allow for reconstitution of both the [Fe_4S_4] cluster as well as the cobalamin cofactor (Lanz et al., 2018).

Overexpression from *E. coli* BL21 star (DE3)/pDB1282. This protocol is applicable to both the OxsB:pET-28b and OxsB:MalE-pET plasmids. The culture broth will assume a pale pink color upon addition of cyanocobalamin; however, the harvested cell pellet generally does not have a pink color.

1. Competent *E. coli* BL21 star (DE3) cells are sequentially transformed with pDB1282, which confers ampicillin resistance, and the OxsB containing plasmid, which confers kanamycin resistance. The transformants containing both plasmids are selected by plating on LB agar containing 50 μg/mL kanamycin and 100 μg/mL ampicillin at 37 °C for 16 h
2. A single colony is selected and used to inoculate 10 mL LB broth containing 50 μg/mL kanamycin and 100 μg/mL ampicillin. The resulting starter culture is incubated at 37 °C with 200 rpm shaking for 12 h
3. The 10 mL starter culture is then used to inoculate 1 L LB broth containing 30 μg/mL kanamycin, 50 μg/m L ampicillin, 0.1% (*v*/*v*) ethanolamine and 3 mg/L cyanocobalamin (CN-Cbl). The inoculated culture is incubated at 37 °C with 200 rpm shaking until the OD_{600} reaches 0.2
4. Once the OD_{600} reaches 0.2, expression of the pDB1282 plasmid is induced by the addition of 0.5 g L-arabinose along with 40 mg $Fe(NH_4)_2(SO_4)_2$ and 20 mg L-cysteine. Incubation is continued at 37 °C and 200 rpm shaking
5. Once the OD_{600} reaches 0.6, the culture is chilled on ice for 10 min, and overexpression of the OxsB construct is induced by adding IPTG to a final concentration of 0.1 mM. The resulting culture is further incubated at 18 °C for 20 h with 150 rpm shaking
6. The cell pellet is then harvested by centrifugation at 4000 *g* and 4 °C for 12 min and can be immediately purified or stored at −80 °C

Overexpression from BL21 star (DE3)/pDB1282/pBAD42-BtuCEDFB. This protocol is applicable to both the OxsB:pET-28b and OxsB:MalE-pET plasmids. The procedure is the same as that from *E. coli* BL21 star (DE3)/pDB1282; however, 100 μg/mL ampicillin, 50 μg/mL kanamycin and 50 μg/mL streptomycin are added to the selection agar, starter culture and expression culture following cotransformation of the *E. coli* BL21 star (DE3) cells with the pDB1282, pBAD42-BtuCEDFB and OxsB containing plasmids. In contrast to coexpression with the pDB1282 plasmid alone, the harvested cell pellet is also pink in color consistent with coexpression of the cobalamin uptake system.

3.1.3 Purification

Purifications are carried out using the following buffers, which are prepared with fresh β-mercaptoethanol. All procedures are carried out aerobically at 4 °C.

Lysis buffer: 50 mM 4-(2-hydroxyethyl)-1-piperazineethanesulfonate (HEPES, pH 8.0), 300 mM NaCl, 10 mM imidazole, 2 mM β-mercaptoethanol and 10% (*v/v*) glycerol.

Wash buffer: 50 mM HEPES (pH 8.0), 500 mM NaCl, 25 mM imidazole, 2 mM β-mercaptoethanol and 10% (*v/v*) glycerol.

Elution buffer: 50 mM HEPES (pH 8.0), 300 mM NaCl, 200 mM imidazole, 2 mM β-mercaptoethanol and 10% (*v/v*) glycerol.

Dialysis buffer: 50 mM HEPES (pH 8.0), 100 mM NaCl, 2 mM β-mercaptoethanol and 10% (*v/v*) glycerol.

Reconstitution buffer: 25 mM HEPES (pH 7.5).

Purification of *N*-His$_6$-tagged OxsB from plasmid OxsB:pET-28b. The quantities are quoted for a 6 L culture. The typical yield is 120–200 mg protein per 6 L culture.

1. The harvested cell pellet is suspended in 180 mL lysis buffer on ice
2. The suspended cell pellet is sonicated for 180 s total on-time with each 10 s burst followed by a 30 s pause
3. The cell debris is pelleted by centrifugation at 18000*g* and 4 °C for 25 min. The supernatant is then decanted onto a chilled column prepacked with 10 mL Ni-NTA resin pre-equilibrated with lysis buffer. The initial flow through is poured back through the column at least two times
4. The column is washed with wash buffer until Bradford assay shows no further elution of non-specific binding proteins (ca. 500 mL).

5. *N*-His_6-tagged OxsB is then eluted with the elution buffer, and the pooled fractions are dialyzed (12–14 kDa MWCO) three times against 1 L dialysis buffer for 1 h each at 4 °C. The dialyzed protein is then concentrated to ca. 0.1 mM, at which point it can be flash frozen and stored at −80 °C. The final protein solution from overexpression in the presence of pBAD42-BtuCEDFB is typically clear and bright pink in color. The overall cobalamin occupancy is estimated to be ca. 32% (Zhong et al., 2021).

Purification of untagged OxsB from plasmid OxsB:MalE-pET. Steps 1–5 are the same as in the protocol for the purification of *N*-His_6-tagged OxsB. The typical yield is 120–200 mg protein per 6 L culture. A His_6-tagged TEV protease with the S219V point mutation is used in this experiment which has been reported to show improved efficiency in this methodology (Kapust et al., 2001).

6. Approximately 1 mg TEV protease per 10 mg MBP-OxsB is added to the OxsB preparation, and the mixture is incubated at 4 °C for 12 h. If a significant level of MBP-tagged OxsB (131 kDa) remains, additional TEV protease can be added and the incubation extended for another 12 h
7. The MBP tag (44 kDa) and TEV protease (29 kDa) are then bound by adding 10 mL Ni-NTA resin. The mixture is then gently agitated at 4 °C for 1 h
8. The slurry is then poured into an empty column, and the flow through is poured back to the column at least three times to maximize binding between Ni-NTA and the tagged impurities
9. After the flow-through is collected, the Ni-NTA column is washed with another 10 mL dialysis buffer to elute any bound OxsB. The purity of OxsB (87 kDa) can be checked by SDS-PAGE (Fig. 6).
10. The purified protein is concentrated to ca. 0.1 mM OxsB monomer and can be aliquoted, flash frozen and stored at −80 °C or immediately reconstituted

3.1.4 Reconstitution of OxsB

Iron sulfur reconstitution of OxsB follows essentially the same procedure as that for GenK (see Section 2.1.4).

3.1.5 Cobalamin reconstitution of OxsB

This step is unnecessary if cobalamin is to be added to the assays. If performed, however, it must follow iron-sulfur reconstitution.

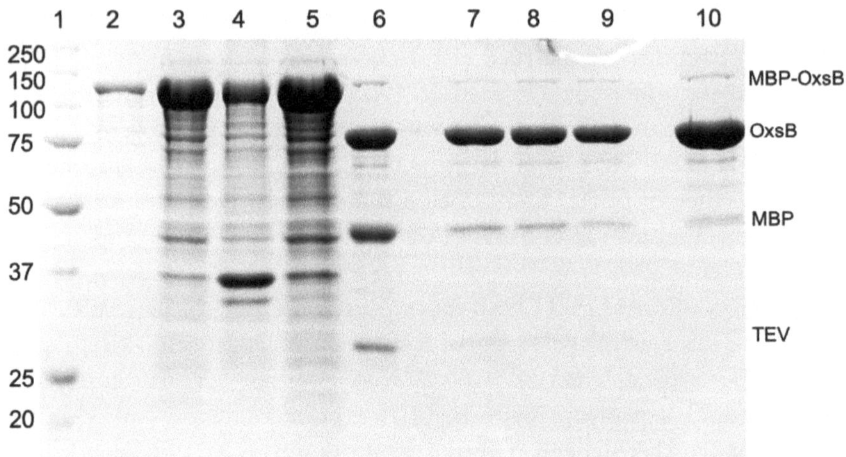

Fig. 6 SDS-PAGE analysis of MBP-OxsB (131 kDa), OxsB (87 kDa), MBP (44 kDa) and TEV (29 kDa). Lane 1 is molecular weight marker; lanes 2–4 are whole cell lysate, pellet and supernatant; lane 5 is the flow through; lane 6 is after incubation with TEV for 24 h; lanes 7–10 are sequential flow through showing increasing purity of the untagged OxsB.

1. To a 1.2 mL solution of 30 μM [Fe_4S_4]-reconstituted OxsB is added 36 μL of 10 mM hydroxocobalamin (HO-Cbl) prepared in reconstitution buffer
2. The mixture is stirred at 4 °C ice for at least 3 h. Although HO-Cbl is not as light sensitive as the organocobalamin cofactors (i.e., Adenosylcobalamin & Me-Cbl), the mixture is nevertheless covered with foil to minimize exposure to ambient light
3. Centrifugal filtration (MWCO 10 kDa) is used to remove the unbound cobalamin via buffer exchange
4. The reconstituted enzyme can be stored anaerobically in an amber vial at ca. 10 °C

3.2 Cloning, overexpression and purification of OxsA

3.2.1 Cloning

The OxsA expression construct in the original studies was synthesized according to the published primary sequence using the Mr. Gene service after codon optimization (Bridwell-Rabb et al., 2016). The *oxsA* gene can then be PCR-amplified and subcloned into the *Nde*I and *Hind*III sites of the pET-28b and pET-24 plasmids, which are designated OxsA:pET-28b and OxsA:pET-24, respectively. The former construct encodes the OxsA protein with an *N*-His_6 affinity tag. The latter construct yields the OxsA protein in its native form with no modifications.

3.2.2 Overexpression

Overexpression of OxsA from *E. coli* BL21 star (DE3). The following protocol is applicable to both the OxsA:pET-24 and OxsA:pET-28b plasmids.

1. Competent *E. coli* BL21 star (DE3) cells are transformed with the relevant OxsA plasmid. The correct transformant can be selected after plating on LB agar containing 50 μg/mL kanamycin at 37 °C for 16 h
2. A single colony is selected to inoculate 10 mL LB broth containing 50 μg/mL kanamycin. The starter culture is incubated at 37 °C with 200 rpm shaking for 12 h
3. The starter culture is used to inoculate 1 L LB broth containing 30 μg/mL kanamycin. The resulting culture is incubated at 37 °C with 200 rpm shaking until the OD_{600} reaches 0.6
4. The turbid culture is chilled on ice for 10 min, and OxsA overexpression is induced by the addition of IPTG to a final concentration of 0.1 mM
5. The culture is incubated at 18 °C for another 20 h with 150 rpm shaking
6. The cell pellet is harvested at 4000 *g* and 4 °C for 12 min and can be immediately purified or stored at −80 °C

3.2.3 Purification

Purification of *N*-His_6 tagged OxsA follows the standard procedure for His_6-tagged proteins. Due to the high yield of OxsA expression, ammonium sulfate precipitation of the untagged OxsA followed by anion exchange chromatography yields OxsA as the major component on SDS-PAGE analysis. However, OxsB catalysis is sluggish under the assay conditions currently identified such that trace enzyme contaminants can lead to side reactions when the OxsB/OxsA assay is conducted over long time intervals (Zhong et al., 2021). Therefore, size-exclusion chromatography can be used to further purify the untagged OxsA.

Purification of *N*-His_6-tagged OxsA (from OxsA:pET-28b). The purification of *N*-His_6-tagged OxsA follows the same procedure as that described for *N*-His_6-tagged OxsB (Section 3.1.3). The typical yield of *N*-His_6-tagged OxsA from a 2 L culture is 10 mg.

Purification of untagged OxsA (from OxsA:pET-24). The purification of untagged OxsA uses Tris·HCl buffers during ammonium sulfate precipitation and (diethylaminoethyl-cellulose) DEAE column chromatography. The purified protein is then dialyzed against HEPES buffer for storage.

Lysis buffer: 50 mM Tris·HCl (pH 8.0), 300 mM NaCl and 10% (*v*/*v*) glycerol.

Dialysis buffer: 50 mM Tris·HCl (pH 8.0) and 10% (*v*/*v*) glycerol.

Elution buffer: 50 mM Tris·HCl (pH 8.0), 1 M NaCl and 10% (*v*/*v*) glycerol.

Storage buffer: 50 mM HEPES (pH 8.0), 100 mM NaCl and 10% (*v*/*v*) glycerol.

1. The OxsA cell pellet from a 1 L culture is suspended in 35 mL lysis buffer on ice
2. The suspended cell pellet is sonicated for 100 s total on-time with each 10 s burst followed by a 30 s pause
3. The cell debris is pelleted by centrifugation at 18000*g* and 4 °C for 25 min
4. The isolated supernatant is stirred slowly at 4 °C, and solid ammonium sulfate (47.2 g/100 mL supernatant) is added over 4 h. The solution will become opaque
5. The precipitated OxsA solution is centrifuged at 18000*g* and 4 °C for 25 min. The precipitated OxsA pellet is then dissolved in a minimum volume (ca. 2 mL) of the dialysis buffer and dialyzed (3.5 kDa MWCO membrane) against 1 L dialysis buffer three times over 3 h
6. A column packed with 250 mL DEAE cellulose is pre-equilibrated with dialysis buffer
7. A two-chamber gradient mixer is prepared with 300 mL dialysis buffer in the first chamber and 300 mL elution buffer in the second. The dialyzed OxsA isolate is applied to the DEAE column. The first chamber is then connected to the column and 50 mL of dialysis buffer is run through the column before opening the second chamber to begin the gradient elution
8. Eluted fractions containing OxsA by SDS-PAGE are pooled and concentrated to 30 mL before dialysis
9. The concentrated OxsA is dialyzed for 1 h against 1 L storage buffer three times
10. The resulting OxsA can be aliquoted, flash frozen and stored at −80 °C
11. If further purification is necessary, FPLC size exclusion chromatography using a Superdex 75 10/300 GL column (GE healthcare) can be used. A 0.8 mL sample containing around 150 μM OxsA is eluted isocratically with storage buffer at a flow rate of 0.5 mL/min

3.3 *In vitro* functional characterization

Activity of reconstituted OxsB can be assayed anaerobically using 2′-dAMP, 2′-dADP or 2′-dATP as the substrate (Bridwell-Rabb et al., 2017). A 50 μL solution containing 1 mM substrate, 1 mM SAM, 1 mM DTT, 10 mM $MgCl_2$, 0.2 mM HO-Cbl, 1 mM NADPH, 0.25 mM methyl viologen is prepared in 25 mM HEPES buffer (pH 7.5). The reaction is initiated by adding OxsA and reconstituted OxsB each to a final concentration of 20 μM, and the reaction is allowed to incubate for 12 h. The reaction is then deproteinized by centrifugal filtration and can be analyzed by HPLC using a Dionex column (CarboPac PA1, Thermo Scientific, 250 mm × 4 mm). The elution is monitored by UV absorbance detection at 260 nm. Separation can be achieved by running ddH_2O as mobile phase A and 1.5 M NH_4OAc as mobile phase B at a flow rate of 1 mL/min with the following linear gradient: 0–5 min 10% B, 5–34 min 10–30% B, 34–35 min 30% B, 35–37 min 30–10% B, 37–41 min 10% B.

Improved separation of the reaction products using reversed phase HPLC can be obtained following dephosphorylation and reduction of the product aldehyde using $NaBH_4$ to yield OXT-A (**18**). A 50 μL aliquot of the reaction mixture is incubated with 5 μL CutSmart buffer (NEB) and 0.5 μL calf intestinal phosphatase (NEB) for 30 min at 37 °C before adding 55 μL acetonitrile and centrifuging to remove the precipitates. To 90 μL of the supernatant is then added 9 μL of 100 mM aqueous $NaBH_4$ prepared fresh. The reaction is incubated at room temperature for 30 min before adding 1 μL glacial acetic acid. The quenched reaction is then evaporated and dissolved with ddH_2O before HPLC analysis using a C18 column (Agilent Microsorb-MV 100–5 C18 250 × 4.6 mm) with UV absorbance at 260 nm. Separation can be achieved by running 1% (*w*/*v*) NH_4OAc as mobile phase A and acetonitrile as mobile phase B at a flow rate of 1 mL/min with the following linear gradient: 0–5 min 2% B, 5–23 min 2–10% B, 23–32 min 10–35% B, 32–35 min 35% B, 35–37 min 35–2% B, 37–47 min 2% B. OXT-A (**18**) elutes at ca. 18 min.

4. Conclusion

As more and more radical SAM enzymes are discovered and investigated, significant new insights are made into the chemistry of these metalloenzymes and the roles played by radicals in enzymological systems. At the same time, however, just as many new questions and unusual mechanisms of catalysis are becoming apparent and in need of further study.

This is especially true of the B_{12}-dependent radical SAM subgroup, which often poses challenges with respect to both obtaining soluble protein for study and the identification of suitable assay conditions for *in vitro* characterization. While much work is left to be done, progress is nevertheless being made. The methodologies provided in this chapter thus afford a starting point for the enzymologist interested in the study of these complex enzyme systems.

Acknowledgment

We are grateful to all coworkers who have made contributions to our studies of GenK and OxsB over the years. We would also like to thank Dr. Mark Ruszczycky for his valuable comments of this paper. The plasmid pDB1282 was a generous gift from Professor Dennis Dean at the Virginia Polytechnic Institute and State University. The plasmid pBAD42-BtuCEDFB was a kind gift from Professor Squire Booker at Pennsylvania State University. This work is supported by grants from the National Institutes of Health (GM035906 and GM040541) and the Welch Foundation (F-1511).

References

Allen, K. D., & Wang, S. C. (2014a). Initial characterization of Fom3 from *Streptomyces wedmorensis*: The methyltransferase in fosfomycin biosynthesis. *Archives of Biochemistry and Biophysics*, *543*, 67–73.

Allen, K. D., & Wang, S. C. (2014b). Spectroscopic characterization and mechanistic investigation of P-methyl transfer by a radical SAM enzyme from the marine bacterium *Shewanella denitrificans* OS217. *Biochimica et Biophysica Acta*, *1844*(12), 2135–2144.

Anton, B. P., Saleh, L., Benner, J. S., Raleigh, E. A., Kasif, S., & Roberts, R. J. (2008). RimO, a MiaB-like enzyme, methylthiolates the universally conserved Asp88 residue of ribosomal protein S12 in *Escherichia coli*. *Proceedings of the National Academy of Sciences of the United States of America*, *105*(6), 1826–1831.

Bridwell-Rabb, J., Kang, G., Zhong, A., Liu, H.-w., & Drennan, C. L. (2016). An HD domain phosphohydrolase active site tailored for oxetanocin-A biosynthesis. *Proceedings of the National Academy of Sciences of the United States of America*, *113*(48), 13750–13755.

Bridwell-Rabb, J., Zhong, A., Sun, H. G., Drennan, C. L., & Liu, H.-w. (2017). A B_{12}-dependent radical SAM enzyme involved in oxetanocin A biosynthesis. *Nature*, *544*(7650), 322–326.

Broderick, J. B., Duffus, B. R., Duschene, K. S., & Shepard, E. M. (2014). Radical S-adenosylmethionine enzymes. *Chemical Reviews*, *114*(8), 4229–4317.

Chatterjee, A., Li, Y., Zhang, Y., Grove, T. L., Lee, M., Krebs, C., et al. (2008). Reconstitution of ThiC in thiamine pyrimidine biosynthesis expands the radical SAM superfamily. *Nature Chemical Biology*, *4*(12), 758–765.

De Groot, N. S., & Ventura, S. (2006). Effect of temperature on protein quality in bacterial inclusion bodies. *FEBS Letters*, *580*(27), 6471–6476.

Frey, P. A., Hegeman, A. D., & Ruzicka, F. J. (2008). The radical SAM superfamily. *Critical Reviews in Biochemistry and Molecular Biology*, *43*(1), 63–88.

Grove, T. L., Lee, K. H., St Clair, J., Krebs, C., & Booker, S. J. (2008). *In vitro* characterization of AtsB, a radical SAM formylglycine-generating enzyme that contains three [4Fe-4S] clusters. *Biochemistry*, *47*(28), 7523–7538.

Hänzelmann, P., Hernández, H. L., Menzel, C., García-Serres, R., Huynh, B. H., Johnson, M. K., et al. (2004). Characterization of MOCS1A, an oxygen-sensitive iron-sulfur protein involved in human molybdenum cofactor biosynthesis. *The Journal of Biological Chemistry*, *279*(33), 34721–34732.

Huang, C., Huang, F., Moison, E., Guo, J., Jian, X., Duan, X., et al. (2015). Delineating the biosynthesis of gentamicin X2, the common precursor of the gentamicin C antibiotic complex. *Chemistry & Biology*, *22*(2), 251–261.

Kapust, R. B., Tözsér, J., Fox, J. D., Anderson, D. E., Cherry, S., Copeland, T. D., et al. (2001). Tobacco etch virus protease: Mechanism of autolysis and rational design of stable mutants with wild-type catalytic proficiency. *Protein Engineering*, *14*(12), 993–1000.

Kim, H. J., Liu, Y. N., McCarty, R. M., & Liu, H.-w. (2017). Reaction catalyzed by GenK, a cobalamin-dependent radical *S*-adenosyl-L-methionine methyltransferase in the biosynthetic pathway of gentamicin, proceeds with retention of configuration. *Journal of the American Chemical Society*, *139*(45), 16084–16087.

Kim, H. J., McCarty, R. M., Ogasawara, Y., Liu, Y. N., Mansoorabadi, S. O., LeVieux, J., et al. (2013). GenK-catalyzed C-6′ methylation in the biosynthesis of gentamicin: Isolation and characterization of a cobalamin-dependent radical SAM enzyme. *Journal of the American Chemical Society*, *135*(22), 8093–8096.

Knox, H. L., Chen, P. Y., Blaszczyk, A. J., Mukherjee, A., Grove, T. L., Schwalm, E. L., et al. (2021). Structural basis for non-radical catalysis by TsrM, a radical SAM methylase. *Nature Chemical Biology*, *17*(4), 485–491.

Knox, H. L., Sinner, E. K., Townsend, C. A., Boal, A. K., & Booker, S. J. (2021). Structural basis for carbapenam C-alkylation by TokK, a B_{12}-dependent radical SAM enzyme. *BioRxiv*. https://doi.org/10.1101/2021.07.10.451909. 2021.07.10.451909.

Kriek, M., Peters, L., Takahashi, Y., & Roach, P. L. (2003). Effect of iron-sulfur cluster assembly proteins on the expression of *Escherichia coli* lipoic acid synthase. *Protein Expression and Purification*, *28*(2), 241–245.

Lanz, N. D., Blaszczyk, A. J., McCarthy, E. L., Wang, B., Wang, R. X., Jones, B. S., et al. (2018). Enhanced solubilization of class B radical *S*-adenosylmethionine methylases by improved cobalamin uptake in *Escherichia coli*. *Biochemistry*, *57*(9), 1475–1490.

Marous, D. R., Lloyd, E. P., Buller, A. R., Moshos, K. A., Grove, T. L., Blaszczyk, A. J., et al. (2015). Consecutive radical *S*-adenosylmethionine methylations form the ethyl side chain in thienamycin biosynthesis. *Proceedings of the National Academy of Sciences of the United States of America*, *112*(33), 10354–10358.

Maruyama, C., Chinone, Y., Sato, S., Kudo, F., Ohsawa, K., Kubota, J., et al. (2020). *C*-methylation of *S*-adenosyl-L-methionine occurs prior to cyclopropanation in the biosynthesis of 1-amino-2-methylcyclopropanecarboxylic acid (Norcoronamic acid) in a bacterium. *Biomolecules*, *10*(5), 775.

McLaughlin, M. I., & van der Donk, W. A. (2018). Stereospecific radical-mediated B_{12}-dependent methyl transfer by the fosfomycin biosynthesis enzyme Fom3. *Biochemistry*, *57*(33), 4967–4971.

Morita, M., Tomita, K., Ishizawa, M., Takagi, K., Kawamura, F., Takahashi, H., et al. (1999). Cloning of oxetanocin A biosynthetic and resistance genes that reside on a plasmid of *Bacillus megaterium* strain NK84-0128. *Bioscience, Biotechnology, and Biochemistry*, *63*(3), 563–566.

Oberg, N., Precord, T. W., Mitchell, D. A., & Gerlt, J. A. (2021). RadicalSAM.org: A resource to interpret sequence-function space and discover new radical SAM enzyme chemistry. *ACS Bio & Med Chem Au*.

Parent, A., Guillot, A., Benjdia, A., Chartier, G., Leprince, J., & Berteau, O. (2016). The B_{12}-rdical SAM enzyme PoyC catalyzes valine C_{β}-methylation during polytheonamide biosynthesis. *Journal of the American Chemical Society*, *138*(48), 15515–15518.

Pierre, S., Guillot, A., Benjdia, A., Sandström, C., Langella, P., & Berteau, O. (2012). Thiostrepton tryptophan methyltransferase expands the chemistry of radical SAM enzymes. *Nature Chemical Biology*, *8*(12), 957–959.

Radle, M. I., Miller, D. V., Laremore, T. N., & Booker, S. J. (2019). Methanogenesis marker protein 10 (Mmp10) from *Methanosarcina acetivorans* is a radical *S*-adenosylmethionine methylase that unexpectedly requires cobalamin. *The Journal of Biological Chemistry*, *294*(31), 11712–11725.

Ryan, J. A. (1984). Colorimetric determination of gentamicin, kanamycin, tobramycin, and amikacin aminoglycosides with 2,4-dinitrofluorobenzene. *Journal of Pharmaceutical Sciences*, *73*(9), 1301–1302.

Sato, S., Kudo, F., Kim, S. Y., Kuzuyama, T., & Eguchi, T. (2017). Methylcobalamin-dependent radical SAM C-methyltransferase Fom3 recognizes cytidylyl-2-hydroxyethylphosphonate and catalyzes the nonstereoselective C-methylation in fosfomycin biosynthesis. *Biochemistry*, *56*(28), 3519–3522.

Shimada, N., Hasegawa, S., Harada, T., Tomisawa, T., Fujii, A., & Takita, T. (1986). Oxetanocin, a novel nucleoside from bacteria. *The Journal of Antibiotics*, *39*(11), 1623–1625.

Sinner, E. K., Lichstrahl, M. S., Li, R., Marous, D. R., & Townsend, C. A. (2019). Methylations in complex carbapenem biosynthesis are catalyzed by a single cobalamin-dependent radical *S*-adenosylmethionine enzyme. *Chemical Communications (Cambridge, England)*, *55*(99), 14934–14937.

Sofia, H. J., Chen, G., Hetzler, B. G., Reyes-Spindola, J. F., & Miller, N. E. (2001). Radical SAM, a novel protein superfamily linking unresolved steps in familiar biosynthetic pathways with radical mechanisms: Functional characterization using new analysis and information visualization methods. *Nucleic Acids Research*, *29*(5), 1097–1106.

Takahashi, Y., & Tokumoto, U. (2002). A third bacterial system for the assembly of iron-sulfur clusters with homologs in archaea and plastids. *The Journal of Biological Chemistry*, *277*(32), 28380–28383.

Ueda, K., Tsurimoto, T., Nagahata, T., Chisaka, O., & Matsubara, K. (1989). An *in vitro* system for screening anti-hepatitis B virus drugs. *Virology*, *169*(1), 213–216.

Unwin, J., Standage, S., Alexander, D., Hosted, T., Jr., Horan, A. C., & Wellington, E. M. (2004). Gene cluster in *Micromonospora echinospora* ATCC15835 for the biosynthesis of the gentamicin C complex. *The Journal of Antibiotics*, *57*(7), 436–445.

Wang, S. C. (2018). Cobalamin-dependent radical *S*-adenosyl-L-methionine enzymes in natural product biosynthesis. *Natural Product Reports*, *35*(8), 707–720.

Wang, Y., Schnell, B., Baumann, S., Müller, R., & Begley, T. P. (2017). Biosynthesis of branched alkoxy groups: Iterative methyl group alkylation by a cobalamin-dependent radical SAM enzyme. *Journal of the American Chemical Society*, *139*(5), 1742–1745.

Wecksler, S. R., Stoll, S., Tran, H., Magnusson, O. T., Wu, S. P., King, D., et al. (2009). Pyrroloquinoline quinone biogenesis: Demonstration that PqqE from *Klebsiella pneumoniae* is a radical *S*-adenosyl-L-methionine enzyme. *Biochemistry*, *48*(42), 10151–10161.

Werner, W. J., Allen, K. D., Hu, K., Helms, G. L., Chen, B. S., & Wang, S. C. (2011). *In vitro* phosphinate methylation by PhpK from *Kitasatospora phosalacinea*. *Biochemistry*, *50*(42), 8986–8988.

Zheng, L., Cash, V. L., Flint, D. H., & Dean, D. R. (1998). Assembly of iron-sulfur clusters. Identification of an iscSUA-hscBA-fdx gene cluster from *Azotobacter vinelandii*. *The Journal of Biological Chemistry*, *273*(21), 13264–13272.

Zhong, A., Lee, Y. H., Liu, Y. N., & Liu, H.-w. (2021). Biosynthesis of oxetanocin-a includes a B_{12}-dependent radical SAM enzyme that can catalyze both oxidative ring contraction and the demethylation of SAM. *Biochemistry*, *60*(7), 537–546.

CHAPTER FIVE

Purification and structural elucidation of a cobalamin-dependent radical SAM enzyme

Zerick Dill[a,b], Bin Li[a], and Jennifer Bridwell-Rabb[a,b,*]
[a]Department of Chemistry, University of Michigan, Ann Arbor, MI, United States
[b]Program in Chemical Biology, University of Michigan, Ann Arbor, MI, United States
[*]Corresponding author: e-mail address: jebridwe@umich.edu

Contents

Abstract

The cobalamin (Cbl)-dependent radical *S*-adenosylmethionine (SAM) enzymes use a [4Fe-4S] cluster, SAM, and Cbl to carry out remarkable catalytic feats in a large number of biosynthetic pathways. However, despite the abundance of annotated Cbl-dependent radical SAM enzymes, relatively few molecular details exist regarding how these enzymes function. Traditionally, challenges associated with purifying and

ISSN 0076-6879
https://doi.org/10.1016/bs.mie.2021.12.015

reconstituting Cbl-dependent radical SAM enzymes have hindered biochemical studies aimed at elucidating the structures and mechanisms of these enzymes. Herein, we describe a bottom-up approach that was used to crystallize OxsB, learn about the overall architecture of a Cbl-dependent radical SAM enzyme, and facilitate mechanistic studies. We report lessons learned from the crystallization of different states of OxsB, including the apo-, selenomethionine (SeMet)-labeled, and fully reconstituted form of OxsB that has a [4Fe-4S] cluster, SAM, and Cbl bound. Further, we suggest that, when appropriate, this bottom-up method can be used to facilitate studies on enzymes in this class for which there are challenges associated with purifying and reconstituting the active enzyme.

1. Introduction

The radical *S*-adenosylmethionine (SAM) enzymes are widespread in Nature and implicated in a plethora of reactions that give rise to medically and environmentally relevant natural products (Broderick, Duffus, Duschene, & Shepard, 2014). These enzymes are characterized by the presence of a Cys-X$_3$-Cys-X$_2$-Cys motif that is used to bind a [4Fe-4S] cluster (Sofia, Chen, Hetzler, Reyes-Spindola, & Miller, 2001). Most often, the one-electron reduced state of the [4Fe-4S] cluster is used to reductively cleave SAM and form methionine and a $5'$-deoxyadenosyl radical ($5'$-dAdo$^{\bullet}$) that can be used to initiate a reaction (Frey, 2010; Magnusson, Reed, & Frey, 2001). In many cases, radical SAM enzymes employ additional cofactors to extend their catalytic range (Berkovitch, Nicolet, Wan, Jarrett, & Drennan, 2004; Haft, 2011; Haft & Basu, 2011; McLaughlin et al., 2016; Shisler et al., 2017; Sofia et al., 2001; Wang, 2018). For example, based on sequence, a large number of radical SAM enzyme superfamily members are predicted to couple the canonical SAM binding [4Fe-4S] cluster with a vitamin B$_{12}$ (or cobalamin (Cbl)) cofactor (Sofia et al., 2001). The sheer number of radical SAM enzymes that use this cofactor combination to accomplish their chemical feats can be visualized in the sequence similarity network (SSN) shown in Fig. 1A (radicalsam.org). As observed in the SSN, there is a large disparity between the number of annotated Cbl-dependent radical SAM enzymes and the number that have been structurally characterized; several thousand sequences belong to this enzyme class but only three structures exist (Fig. 1A). The lack of structural information for these enzymes is likely attributable to the challenges of isolating soluble, homogenous, and reconstituted proteins using *E. coli* as a host for recombinant expression (Blaszczyk, Wang, & Booker, 2017; Lanz et al., 2018). In addition, unknown

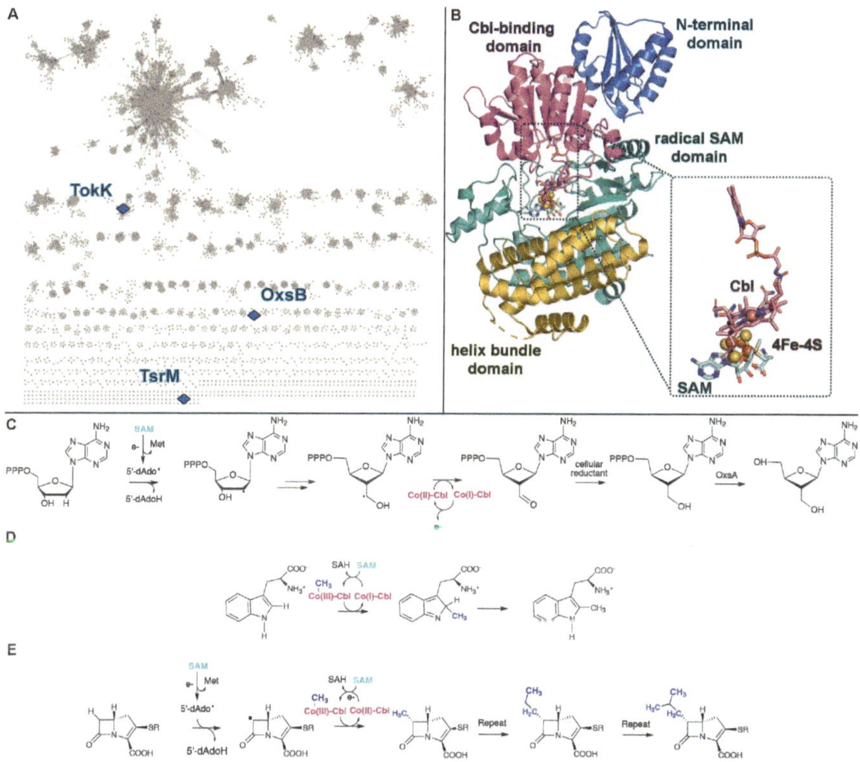

Fig. 1 Despite the large and diverse number of Cbl-dependent radical SAM enzymes that are found in Nature, there are only three available structures of these enzymes. (A) A sequence similarity network (SSN) for the Cbl-dependent radical SAM enzymes is shown with an alignment score of 70. The sequence IDs are from radicalSAM.org (Gerlt et al., 2015; Zallot, Oberg, & Gerlt, 2018, 2019) and are visualized with cytoscape (Shannon et al., 2003). Blue nodes represent the structurally characterized enzymes OxsB, TsrM, and TokK (Bridwell-Rabb, Zhong, Sun, Drennan, & Liu, 2017; Knox et al., 2021; Knox, Sinner, Townsend, Boal, & Booker, 2021). Note that these enzymes do not represent the largest cluster of nodes in the SSN that contains enzymes such as Fom3, BcpD/PhpK, and BchE, which along with OxsB, are founding members of the Cbl-dependent radical SAM enzyme class (Sofia et al., 2001). (B) OxsB was the first structurally characterized Cbl-dependent radical SAM enzyme. OxsB contain four domains, two of which, the Cbl-binding domain and radical SAM domain, look like prototypical Cbl-binding and radical SAM enzymes (PDB: 5UL4). (C) OxsB, with the help of a partner protein OxsA and a cellular reductant catalyzes a radical rearrangement to form the four-membered ring of the nucleoside analog oxetanocin A (Bridwell-Rabb et al., 2017; Bridwell-Rabb, Kang, Zhong, Liu, & Drennan, 2016). (D) TsrM converts L-tryptophan into 2-methyl-tryptophan (Blaszczyk, Wang, Silakov, Ho, & Booker, 2017; Knox, Chen, et al., 2021). (E) TokK catalyzes a series of three radical-dependent methylation reactions in asparenomycin A biosynthesis to form an isopropyl moiety (Knox, Sinner, et al., 2021).

physiological substrates, and a lack of knowledge regarding the necessary cofactors for activity have impeded these studies. Importantly, however, several recent advances including optimized purification protocols (Blaszczyk, Wang, & Booker, 2017; Lanz et al., 2018) and three published structures of Cbl-dependent radical SAM enzymes (Bridwell-Rabb et al., 2017; Knox, Chen, et al., 2021; Knox, Sinner, et al., 2021) now provide strategies for isolating these enzymes and a framework for thinking about how they catalyze their different reactions.

The first structurally characterized Cbl-dependent radical SAM enzyme OxsB catalyzes a radical-based deoxyribose ring contraction to produce the oxetane ring of oxetanocin A (Fig. 1B and C) (Bridwell-Rabb et al., 2017). This structure showed that OxsB contains four domains, a small N-terminal domain, a Cbl-binding domain that is reminiscent of that observed in the Cbl binding fragment of methionine synthase (Drennan, Huang, Drummond, Matthews, & Ludwig, 1994), a radical SAM domain that binds a [4Fe-4S] cluster and SAM, and a C-terminal helix bundle domain (Fig. 1B) (Bridwell-Rabb et al., 2017). This structure also revealed how SAM, the [4Fe-4S] cluster, and hydroxo (HO)Cbl are oriented in a single active site to facilitate catalysis. Finally, the structure of OxsB described a new, base-off, His-off mode of Cbl binding (Bridwell-Rabb et al., 2017). In this binding mode, the dimethylbenzimidazole (DMB) tail does not serve as a ligand to the Co-ion of Cbl nor is the Co-ion coordinated by a His residue from the protein (Bridwell-Rabb et al., 2017; Bridwell-Rabb, Grell, & Drennan, 2018; Dowling, Croft, & Drennan, 2012). Similar to OxsB, TsrM and TokK both use a similar arrangement of the radical SAM and Cbl-binding domains to house their metallocofactors. In addition, both TsrM and TokK use a base-off, His-off Cbl binding mode (Knox, Chen, et al., 2021; Knox, Sinner, et al., 2021). However, small modifications to the protein scaffold and the environment of the [4Fe-4S] cluster, SAM and Cbl appear to support the different reactivity of TsrM and TokK. For example, TsrM does not use 5′-dAdo$^{\bullet}$ (Fig. 1D) (Blaszczyk, Wang, Silakov, et al., 2017; Knox, Chen, et al., 2021). Rather, a fourth ligand to the [4Fe-4S] cluster occupies the traditional SAM binding location and supports the ability of TsrM to catalyze a non-radical methylation of L-tryptophan in the biosynthetic pathway for thiostrepton A (Blaszczyk, Wang, Silakov, et al., 2017; Knox, Chen, et al., 2021). TokK, on the other hand, catalyzes three sequential radical-based methylation reactions on a carbapenem scaffold to form the isopropyl sidechain of asparenomycin A

(Fig. 1E) (Knox, Sinner, et al., 2021). These two proteins, unlike OxsB employ methyl (Me)-Cbl to facilitate chemistry (Blaszczyk et al., 2016; Knox, Chen, et al., 2021; Knox, Sinner, et al., 2021).

Despite these advances, there is no structural information for several of the originally identified members of the Cbl-dependent radical SAM enzyme class, including Fom3, Fms7, and BcpD/PhpK, which catalyze single radical-based methylation reactions in the biosynthetic pathways of fosfomycin, fortimicin, and L-phosphinothricin, respectively (Allen & Wang, 2014a; Lanz et al., 2018; Sato, Kudo, Kim, Kuzuyama, & Eguchi, 2017; Sofia et al., 2001; Werner et al., 2011). Likewise, there are essentially no mechanistic details available for BchE, which is instead implicated in oxygen-atom insertion and radical-mediated cyclization (Gough, Petersen, & Duus, 2000; Ouchane, Steunou, Picaud, & Astier, 2004). Thus, even with the three available structures, based on the expansive uncharted Cbl-dependent radical SAM enzyme sequence space, it is clear there is still a critical gap in knowledge regarding the structure-function relationships in this class of enzymes (Fig. 1A).

Toward the goal of increasing the number of structurally characterized Cbl-dependent radical SAM enzymes, in this chapter, we describe strategies that can be used to over-produce, purify, and crystallize OxsB. Using OxsB as a model, we highlight some of the challenges to Cbl-dependent radical SAM enzyme crystallography and emphasize important considerations that allowed for the crystallization and structural determination of the apo-, selenomethionine (SeMet)-labeled, and reconstituted structures of OxsB. Furthermore, we suggest that in some cases, using a bottom-up approach, or a method that starts with purification of the cofactor-free form of the enzyme, is powerful in that it can inform correct protein cofactor stoichiometry and guide future structural and mechanistic studies.

2. Expression and purification of apo-OxsB

In contrast to several other Cbl-dependent radical SAM enzymes (Allen & Wang, 2014a, 2014b; Blaszczyk, Wang, & Booker, 2017; Kim et al., 2013; Werner et al., 2011), OxsB does not suffer from insolubility when expressed recombinantly in *E. coli*. Instead, OxsB is soluble and can be purified to high levels when recombinantly expressed as a His-maltose binding protein (MBP) gene fusion (Bridwell-Rabb et al., 2017). Similarly, previous work has shown that SeMet-labeled OxsB can be produced by expressing this same His-MBP construct in the methionine auxotroph *E. coli* B834 (DE3) using M9 minimal media that is supplemented with SeMet (Bridwell-Rabb et al., 2017). Here, we demonstrate that OxsB can also be expressed using a codon optimized construct of OxsB in a pET28a vector that contains only an N-terminal His-tag. The pET28a vector contains a kanamycin (Kan) resistance marker that allows for *in vivo* selection and an encoded thrombin tag-cleavage site for removal of the His-tag. In addition, we show that expression of the *oxsB* gene in this construct does not require the use of an additional construct to incorporate Cbl (Lanz et al., 2018), nor is it imperative to co-express with a construct such as pDB1282 that encodes proteins from the *Azotobacter vinelandii* Fe-S cluster assembly operon (Zheng, Cash, Flint, & Dean, 1998). Instead, OxsB can be purified in an aerobic environment and subsequently reconstituted to its active state.

2.1 Protocol for expression of the His-tagged *oxsB* gene

1. Transform 1 μg of the codon optimized pET28a-oxsB plasmid into *E. coli* BL21 (DE3) cells using standard procedures. Select transformants on Luria-Bertani (LB) Miller agar plates containing Kan at 60 μg/mL
2. Use a single colony to inoculate six 5 mL starters of LB Miller media containing Kan at 60 μg/mL. Allow these starter cultures to grow with shaking (180 rpm) for 16 h at 37 °C. Additionally, sterilize six 2 L flasks that each contain 1 L of LB Miller media.
3. To each of the six 2 L flasks, add 60 μg/mL Kan, 250 μL of a 25 mM aqua (Aq)-Cbl solution, and 1 mL of ethanolamine. This latter chemical has been previously suggested to facilitate the ability of *E. coli* to uptake Cbl from the media into the cell (Bandarian & Matthews, 2004).
4. Inoculate each of the 2 L flasks with 4 mL from a prepared starter culture (see step 2). Place the now-inoculated 2 L flasks into a constant-temperature incubator and allow cultures to incubate, with shaking (180 rpm) at 37 °C for approximately 5 h
5. At an OD_{600} of 0.6, cool the flasks in an ice bath for approximately 30 min and decrease the incubator temperature from 37 to 17 °C. To each 2 L flask, add 200 mg of ammonium ferric chloride ($FeCl_3$), 300 mg of ferrous ammonium sulfate hexahydrate ($Fe(NH_4)_2(SO_4)_2{\cdot}6H_2O$), 20 mg of L-cysteine, and isopropyl β-D-1-thiogalactopyranoside (IPTG) to a final concentration of 0.1 mM.

6. Place the cooled 2 L flasks back into the constant-temperature incubator and allow the cultures to incubate and shake (180 rpm) for 16–18 h at 17 °C
7. Harvest the bacterial cultures by centrifugation. Freeze the cell paste in liquid N_2 and store in −80 °C freezer. Typically, this process will yield approximately 2.3 g/L of cell paste.

2.2 Producing a pure, homogeneous, and tag-free sample of apo-OxsB

To purify OxsB, a method that capitalizes on the presence of an N-terminal His-tag and Ni^{2+}-affinity chromatography can be used (Crowe et al., 1994). The benefit of using a His-tag includes its small size, which often does not have significant effects on the structure or the function of the purified protein (Crowe et al., 1994). Thus, it is often not required for it to be removed following protein purification. In crystallography, it is known that the His-tag can help crystallization by facilitating packing (Carson, Johnson, McDonald, Brouillette, & Delucas, 2007; Derewenda, 2004). Alternatively, His-tags can prohibit crystallization due to their dynamic nature (Deller, Kong, & Rupp, 2016). Thus, crystal conditions are often screened using both tagged and un-tagged protein samples (Derewenda, 2004). For OxsB, the presence of a His-tag on the N-terminus greatly facilitates the ease of purification, but it is imperative to remove the His-tag prior to crystallization. When OxsB is expressed using the pET28a-construct, this task can be accomplished using thrombin protease.

1. Prepare 500 mL each of the following buffer solutions to be used for cell lysis, affinity purification, tag-cleavage, and size exclusion chromatography purification steps
   a. Buffer for lysis and Ni^{2+}-column washing: 50 mM 4-(2-hydroxyethyl)-1-piperazineethanesulfonic acid (HEPES, pH 8.0), 150 mM NaCl, 10% glycerol, and 20 mM imidazole
   b. Elution buffer for the Ni^{2+}-column: 50 mM HEPES (pH 8.0), 150 mM NaCl, 10% glycerol, and 200 mM imidazole
   c. Size exclusion column and His-tag cleavage buffer: 50 mM HEPES (pH 8.0), 100 mM NaCl, and 5% glycerol
2. Perform each step of the purification aerobically. In a 250 mL metal sonication cup, resuspend approximately 15 g of frozen cell paste in 100 mL of lysis buffer

3. Once the cell paste is resuspended in lysis buffer, place the metal sonication cup into an ice bucket to reduce heating of the sample. Sonicate using a Fisher brand Model 120 Sonic Dismembrator in pulse mode using a program that cycles between being on for 10s (25% amplitude) and off for 15s. The total sonication time of this program should be set at 30min, which equates to 12min of total on time and 18min of off time
4. Following completion of the 30min sonication program, pour the cell lysate into centrifuge tubes. Centrifuge the cell lysate at 12,000rpm for 1h at 4 °C. During centrifugation, equilibrate a column of Ni-NTA resin with Ni^{2+}-column washing buffer at 4 °C. This step can be performed using a gravity column or a pre-packed column and an FPLC
5. After centrifugation, carefully decant the supernatant into a clean 250mL flask. Remove a 100 μL sample of the supernatant and a sample of the pellet for analysis using sodium dodecyl sulfate–polyacrylamide (SDS-PAGE) gel electrophoresis. The latter sample should be resuspended in 1mL lysis buffer (Fig. 2A).
6. Load the supernatant from centrifugation onto the Ni^{2+}-column. Proceed by washing the column with six column volumes (CV) of Ni^{2+}-column washing buffer. Take a 100 μL sample of the

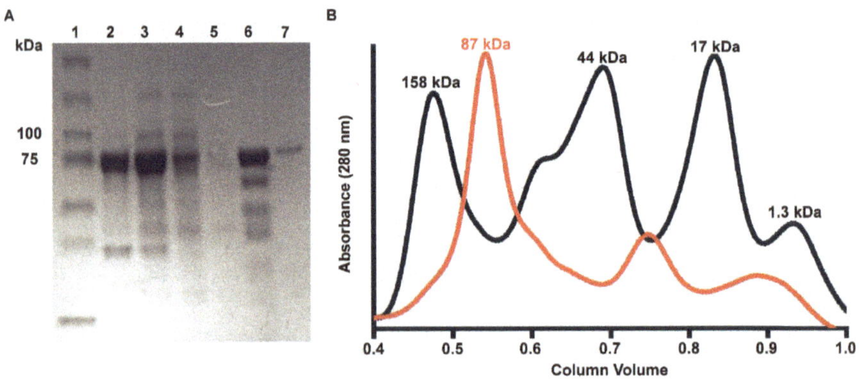

Fig. 2 His-tagged apo-OxsB (expressed in the pET28a vector) is purified using Ni^{2+}-affinity and size exclusion chromatography. (A) SDS-PAGE gel of the different OxsB purification steps. Lanes labeled from left to right; (1) molecular weight marker, (2–3) insoluble and soluble fractions following sonication and centrifugation, (4–6) flow-through, wash, and elution fractions from the Ni^{2+}-affinity column, and (7) fractions that correspond to size exclusion purified OxsB, which has a molecular weight of 87 kDa. (B) A chromatogram of the size exclusion column run for OxsB (pink) overlaid with that of a BioRad protein standard. Monomeric OxsB is the peak labeled 87kDa and the remaining peaks are impurities in the sample.

flow-through from both the loaded lysate and from the wash steps for SDS-PAGE analysis (Fig. 2A).

7. Elute the protein from the column by washing with 2–3 CV of the Ni^{2+}-column elution buffer. At this stage, even though the protein has been aerobically purified, the [4Fe-4S] cluster will be partially bound, leading to a partial brown color of the elution fractions, which should also be analyzed using SDS-PAGE (Fig. 2A).
8. Collect the elution fractions and add them to Amicon centrifugal filter devices (Millipore-Sigma) that have a 30 kDa molecular weight cutoff (MWCO). Centrifuge these tubes at a speed of 4200 × g and 4 °C to a final volume of approximately 2 mL
9. Measure the concentration of the now-concentrated, partially purified sample of His-tagged OxsB protein using the Bradford Protein Assay (Bradford, 1976).
10. Add 1 Unit of Thrombin per 1 mg of tagged OxsB. Place this solution in Thermo-Scientific SnakeSkin dialysis tubing that has a MWCO of 3.5 kDa. Suspend the dialysis tubing in 500 mL of His-tag cleavage buffer and stir slowly at 4 °C for 16 h
11. Equilibrate an S200 Cytiva HiLoad 16/600 Superdex 200 pg size exclusion column with 1.5 CV of tag cleavage/size exclusion column buffer.
12. Following equilibration of the size exclusion column, inject approximately 2 mL of the concentrated tag-cleaved sample of OxsB onto the column. Collect the fractions that elute at a CV consistent with the molecular weight of OxsB (744 amino acids, approximately 87 kDa). At this step, it is important to only collect the homogeneous monomeric fractions and dispose of the higher oligomeric state, aggregated protein samples or remaining impurities (Fig. 2B). Importantly, the homogeneous, monomeric OxsB samples will appear colorless, due to loss of the Fe-S cluster, unlike the Ni^{2+}-column purification fractions. Collect 20 μL of the size exclusion purified apo-OxsB and assess purity using SDS-PAGE analysis (Fig. 2A).
13. Concentrate the fractions containing apo-OxsB and place in an Amicon centrifugal filter device (30 kDa MWCO). Centrifuge these tubes at a speed of 4200 × *g* and 4 °C. Measure the protein concentration using Bradford Protein Assay (Bradford, 1976).

Continue concentrating the protein until it reaches a final concentration of 35–60 mg/mL. Aliquot and flash-freeze the pure and homogeneous apo-OxsB protein sample in liquid nitrogen before storing in −80 °C freezer.

3. Crystallization of apo- and selenomethionine (SeMet)-labeled OxsB

As with all protein crystallization experiments, the guiding principles of using a pure and homogeneous sample plays a key role in crystallizing the different (i.e., apo-, SeMet-labeled, and reconstituted) states of OxsB (McPherson & Gavira, 2014). Initial attempts at crystallizing a reconstituted sample of OxsB were complicated by the incorrect assumption that OxsB would bind more than one Fe-S cluster per monomer. This hypothesis was made based on the large number of Cys residues found in the primary structure of OxsB. With this assumption, the incorrect reconstitution of the apo-protein resulted in creation of a sample that was not amenable to crystallography. Nevertheless, the ability to purify apo-OxsB was powerful, as it allowed for generation of initial OxsB crystals and a method for incorporating SeMet into OxsB. This latter point allowed for growth of SeMet-labeled crystals, the ability to calculate phases and subsequently determine a structure that clearly showed that OxsB would bind one molecule of Cbl and one [4Fe-4S] cluster to form the active enzyme (Bridwell-Rabb et al., 2017). Having the structure of apo/SeMet-labeled OxsB thus informed efforts (and the protocol described below) to correctly reconstitute OxsB and determine a structure of OxsB that contained a [4Fe-4S] cluster, SAM, and Cbl (Bridwell-Rabb et al., 2017).

3.1 Protocol for growing and harvesting crystals of apo- and SeMet-labeled OxsB

1. As described above, following affinity purification and tag cleavage of OxsB, a final size exclusion step should be in place to remove aggregated protein and create a homogenous, monomeric sample of OxsB for crystallography
2. The peak that corresponds to the non-aggregated, monomeric OxsB should be pooled and concentrated to 35 mg/mL
3. Add 1 μL of apo-OxsB (35 mg/mL) to 1 μL of a precipitant solution that contains 0.1 M Bicine (pH 7.6), 10% *w*/*v* PEG 1500 (apo-OxsB) or PEG 3350 (SeMet-OxsB), and between 20 and 24% ethylene glycol
4. Store the crystal tray aerobically at a constant temperature of 21 °C overnight
5. Crystals of apo- and SeMet-labeled OxsB should appear within 24 h (Fig. 3A and B).

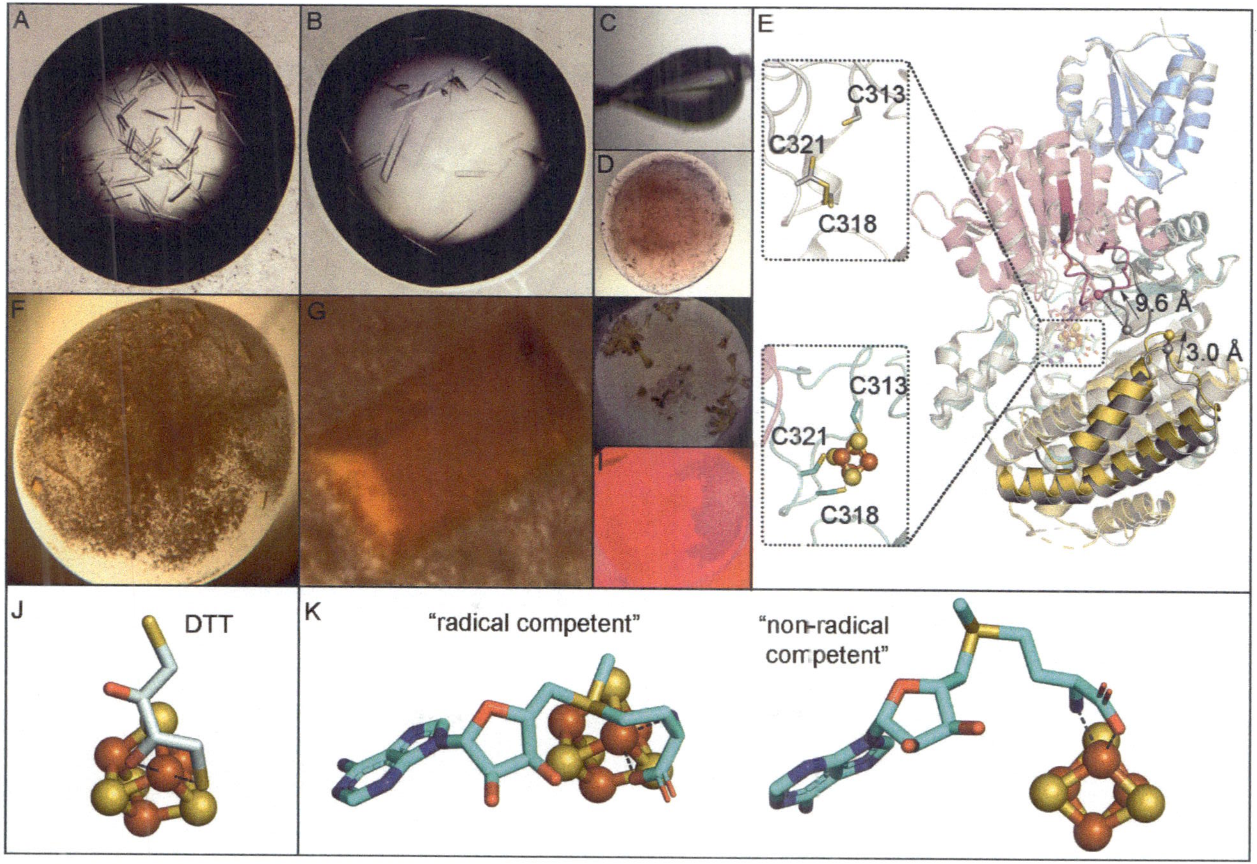

Fig. 3 See figure legend on next page.

6. Following crystal formation, harvest the crystals within 24h to avoid formation of a "skin" in the drop. The so-called skin, or layer of denatured protein (Luft, Wolfley, & Snell, 2011), will markedly affect the quality of data that can be collected from the crystal. Flash freeze and store the crystals in a liquid nitrogen dewar. At this step, there is no need for the addition of cryoprotectant as high levels of ethylene glycol have been included in the crystallization condition (Fig. 3C).

4. Reconstitution of OxsB and crystallization of different cofactor-bound states

Using the crystal conditions determined for apo-OxsB and SeMet-labeled OxsB, crystals that contain HOCbl can also be generated. However, these crystals are generally of lower quality than those grown for the former samples (Fig. 3D). To obtain a snapshot of the active form of OxsB, the protein sample used for crystallization must additionally

Fig. 3 Crystals of OxsB can be produced in different apo- and cofactor-reconstituted states. (A) Optimized crystals of apo-OxsB are colorless and grow within 12h (Bridwell-Rabb et al., 2017). (B) A very similar condition to that used to crystallize apo-OxsB can be used to grow crystals of SeMet-labeled OxsB. (C) Due to the high amounts of ethylene glycol in the crystallization condition, the addition of extra cryoprotectant is not needed. Crystals freeze well when harvested directly from the drop. (D) HOCbl-bound OxsB crystals can be grown, but they are of much lower quality than that of the apo-OxsB and SeMet-labeled OxsB samples. (E) An overlay of SeMet-labeled OxsB (gray) and reconstituted OxsB (colors) show changes mostly in the helix bundle domain and a loop, both of which move toward Cbl (PDB: 5UL2 and 5UL4). In the structure of SeMet-labeled OxsB, there is a partial occupancy disulfide bond formed between two of the Cys residues that constitute the [4Fe-4S] cluster binding site (top inset). In the reconstituted structure of OxsB, there is a [4Fe-4S] cluster bound in the equivalent site (bottom panel). (F) Crystals of reconstituted OxsB show more precipitation than observed for apo-OxsB and SeMet-labeled OxsB. (G) Slight optimization of reconstituted OxsB crystals by manipulation of the ethylene glycol concentration allows for growth of larger crystals. (H) Crystals of [4Fe-4S] cluster reconstituted OxsB can be produced, but these crystals are of much lower quality than those from the fully reconstituted protein. (I) Crystals of reconstituted OxsB with MeCbl bound can also be grown but must be imaged under red light. (J) The high levels of DTT used during reconstitution can compete with SAM for ligating the [4Fe-4S] cluster (PDB: 5UL3). (K) A correct SAM:DTT ratio permits crystallization of OxsB with SAM bound to the [4Fe-4S] cluster rather than DTT (PDB: 5UL4). In OxsB, SAM binds to the [4Fe-4S] cluster in two orientations, one of these orientations is the canonical "radical competent" mode and the other is not (Bridwell-Rabb et al., 2017).

contain a [4Fe-4S] cluster and SAM. Detailed methods for reconstituting OxsB with these cofactors are described below.

4.1 Reconstitution of OxsB with a [4Fe-4S] cluster and Cbl

Once apo-OxsB has been purified (see Section 2), it can be reconstituted with a [4Fe-4S] cluster, SAM, and Cbl using the following steps (Bridwell-Rabb et al., 2017).

1. Make concentrated stock solutions of 500 mM $FeCl_3$ or $Fe(NH_4)_2(SO_4)_2$ and 500 mM Na_2S in the anaerobic chamber. The stock of $FeCl_3$ should be prepared in 1 M HCl and $Fe(NH_4)_2(SO_4)_2$ should be prepared in 50 mM HCl. These concentrated stocks should be used to subsequently make 100 mM stocks of either $FeCl_3$ or $Fe(NH_4)_2(SO_4)_2$ and 100 mM Na_2S, by dilution with anaerobic water
2. Take 100 mL of an oxygen-free solution of 50 mM HEPES (pH 8.0), 100 mM NaCl and 500 μL of frozen 1 M dithiothreitol (DTT, stock made with water) into the anaerobic chamber
3. Take a 100 μL sample of a frozen size exclusion purified sample of OxsB (35 mg/mL) into the anaerobic chamber to thaw
4. Add DTT to a final concentration of 10 mM into the OxsB sample and let it incubate for 30–60 min. This and subsequent steps should be performed at low temperature, ideally 4 °C. Note that high concentrations of DTT are imperative for correct [4Fe-4S] cluster reconstitution and likely play a role in mediating cluster assembly and/or keeping the Cys residues from the cluster binding motif in the reduced state for Fe-S cluster assembly and/or insertion (Vranish et al., 2015). For OxsB, comparison of the apo- and reconstituted crystal structures reveals that a disulfide bond is found between two of the three Cys loop residues in the absence of the cluster, again necessitating the need for a reductant at this stage (Fig. 3E).
5. Following incubation with DTT, rapidly add 1.5 molar equivalents of the 100 mM stocks of $FeCl_3$ and Na_2S to the OxsB sample
6. Incubate the sample until a brown color appears, which could take up to 4 h. If at any stage the color starts to have a black hue, immediately begin the desalting step (see step 7).
7. Remove precipitation from the protein sample by centrifugation at 10,000 rpm for 5 min and desalt the reconstituted protein at least two times to remove the unincorporated iron and sulfide. This desalting step can be accomplished using either the Bio-Rad micro Bio-Spin or the Cytiva PD Minitrap G-25 chromatography columns

8. Use a Bradford assay and a bovine serum albumin (BSA) standard to determine the concentration of reconstituted OxsB (Bradford, 1976). Verify correct reconstitution of the [4Fe-4S] cluster in OxsB using UV–vis analysis and the ferrozine assay (Fig. 4A and B) (Fish, 1988; Stookey, 1970). The ferrozine assay relies on formation of a complex between ferrozine and Fe^{2+} that absorbs light at 562 nm (Fig. 4B). Using an extinction coefficient of $27{,}900\,M^{-1}\,cm^{-1}$, the concentration of Fe^{2+} in the protein sample can be calculated (Fish, 1988; Stookey, 1970).
9. Concentrate OxsB to approximately 35–50 mg/mL for crystallization and reduce the DTT concentration in the sample to 2 mM using a desalting column
10. Add 10–12 molar equivalents of (*S,S*)-SAM to the protein sample. This biologically active enantiomer of SAM can be produced by combination of purified SAM synthetase with its substrates, ATP and methionine (Fig. 4C) (Kamarthapu et al., 2008). Typical HPLC profiles obtained using these published methods to produce (*S,S*)-SAM can be visualized in Fig. 4D. Once purified, it is important to know that (*S,S*)-SAM can spontaneously convert into (*R,S*)-SAM in solution (Zhang & Klinman, 2015). Thus, SAM should be carefully stored in an acidic solution and at low temperature to maintain the correct enantiomeric state for crystallography (Kamarthapu et al., 2008; Matos & Wong, 1987; Zhang & Klinman, 2015). In addition, such storage conditions will prevent degradation of (*S,S*)-SAM into *S*-adenosylhomocysteine, adenine, or methylthioadenosine (Parks & Schlenk, 1958).
11. Add 1 molar equivalent of Cbl to the sample. Both MeCbl and HOCbl derivatives can be used at this stage. However, if MeCbl is used, due to its inherent light-sensitivity, it requires different handling than HOCbl and must be worked with in the dark or under red light

4.2 Additional methods for assessing reconstitution of Cbl-dependent radical SAM enzymes

In the above-described protocol for reconstituting OxsB, the Bradford assay is coupled with the ferrozine assay to determine the number of Fe^{2+}-ions per monomer. A supplementary method to facilitate the coupling of these methods includes using amino acid analysis to provide a correction factor for the Bradford assay and a more accurate measurement of protein concentration (Lanz et al., 2012). To complement these methods, sulfide

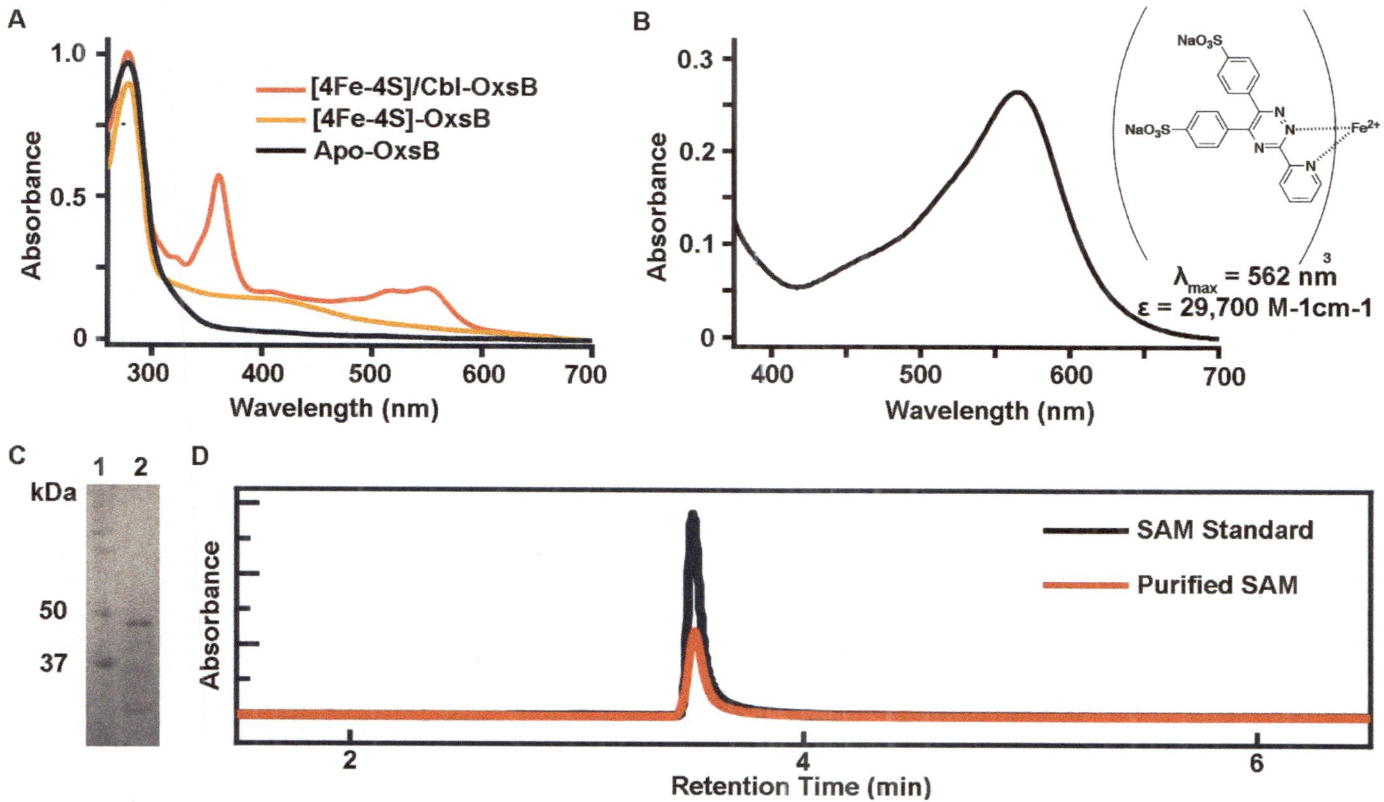

Fig. 4 See figure legend on next page.

incorporation can also be measured using previously described methods (Beinert, 1983). In place of the ferrozine assay, other colorimetric assays that instead rely on bathophenanthroline or phenanthroline can also be performed to measure iron incorporation (Fuss, Tsai, Ishida, & Tainer, 2015). Alternatively, a more sensitive method is inductively coupled plasma mass spectrometry (ICP-MS), which can detect iron in the picomolar range (Fuss et al., 2015). The intact [4Fe-4S] clusters can also be studied using a spectroscopic method that has been previously described (Lanz et al., 2012). Additionally, aside from confirming that iron is incorporated correctly, if needed, published protocols for quantifying Cbl content are available (Blaszczyk, Wang, & Booker, 2017). These methods require conversion of the incorporated Cbl species into a dicyano-Cbl adduct, which shows a characteristic absorbance peak at 367 nm and has an extinction coefficient of 30,800 M^{-1} cm^{-1} (Ljungdahl, LeGall, & Lee, 1973). Regardless of the methods chosen to quantify correct reconstitution of OxsB or another Cbl-dependent radical SAM enzyme, care must be taken to ensure that there is not extra iron, sulfide, or Cbl in the sample and that the reconstituted sample (like the apo-OxsB and SeMet-labeled OxsB samples) is pure and homogenous for crystallographic experiments.

4.3 Protocol for growing and harvesting crystals of reconstituted OxsB

As described for the crystals of apo- and SeMet-labeled OxsB, time is also an important variable for growing crystals of reconstituted OxsB (Bridwell-Rabb et al., 2017). Once OxsB has been correctly reconstituted with a

Fig. 4 Apo-OxsB can be reconstituted with a [4Fe-4S] cluster, SAM and Cbl. (A) Apo-OxsB was purified aerobically and as indicated by UV–vis analysis lacks both the [4Fe-4S] cluster and Cbl cofactor (black). The apo-OxsB protein sample can be chemically reconstituted to contain a [4Fe-4S] cluster as indicated by the characteristic absorption spectra (orange). Subsequent addition of HOCbl to this reconstituted sample shows the presence of both a bound [4Fe-4S] cluster and Cbl (pink). (B) Iron analysis of reconstituted OxsB using the ferrozine assay shows the characteristic peak at 562 nm (Fish, 1988; Stookey, 1970). This peak is indicative of complex formation between Fe^{2+} and ferrozine (right inset) and can be used to calculate iron concentration. (C) Lane 1 is a molecular weight marker and lane 2 is purified SAM synthetase, which has a molecular weight of 43 kDa. (D) Enantiomerically pure SAM can be produced by combination of ATP and methionine with a purified SAM synthetase enzyme. Isolation of pure SAM can be accomplished using published protocols and HPLC (Kamarthapu, Rao, Srinivas, Reddy, & Reddy, 2008). Shown here is an enantiomerically pure SAM standard (black) overlaid with SAM purified from the SAM synthetase reaction (red).

[4Fe-4S] cluster, SAM, and either HOCbl or MeCbl, crystal trays should be immediately set up using the following protocol (Bridwell-Rabb et al., 2017).

1. In an anaerobic chamber, combine 1 μL of reconstituted OxsB (35 mg/mL) with 1 μL of a precipitant solution that contains 0.1 M 2-(*N*-morpholino)-ethanesulfonic acid (MES) (pH 6.5), 6% *w/v* PEG 3350, and 20–24% ethylene glycol
2. Store the crystal tray in the anaerobic chamber at a constant temperature of 21 °C overnight
3. Crystals should appear within 24 h of storing the crystal tray (Fig. 3F and G). If crystals do not appear in this time window, there is likely a problem with the integrity of the sample that was used to set up the crystal tray. Careful assessment of purity, homogeneity, and correct [4Fe-4S] cluster reconstitution should be performed
4. As described for apo- and SeMet-labeled OxsB, once produced, the crystals of reconstituted OxsB must be harvested and cryocooled within 24 h to avoid formation of a "skin" in the drop. These steps should be performed in the anaerobic chamber. Again, as described for the apo- and SeMet-labeled OxsB crystals, due to the high levels of ethylene glycerol in the precipitant solution, there is no need to add additional cryoprotectant to the crystals prior to freezing
5. Crystals of OxsB with just a [4Fe-4S] cluster bound form haystacks and are of much lower quality than crystals of fully reconstituted OxsB (Fig. 3H). If crystals of OxsB contain a [4Fe-4S] cluster, SAM, and MeCbl, all imaging should be performed under red light (Fig. 3I). This latter point is easily accomplished using red filter paper to cover the microscope light (Fig. 3I).

5. Lessons applicable to crystallography of other Cbl-dependent radical SAM enzymes

5.1 The importance of modulating the DTT:SAM ratio in radical SAM enzymes

Many of the initial attempts to obtain a structure of OxsB that contained a [4Fe-4S] cluster, SAM, and Cbl resulted in determination of structures that lacked SAM. Instead, in these initial trials, DTT was found in place of SAM, ligating the [4Fe-4S] cluster using one hydroxyl- and one thiol-moiety (Fig. 3J). To obtain a SAM bound structure, the DTT concentration needs to be decreased following Fe-S cluster reconstitution such that the SAM:

DTT ratio is at least 2:1 (see Section 4.1, steps 9–10). For OxsB, SAM binds to the [4Fe-4S] cluster in two different orientations (Bridwell-Rabb et al., 2017). One of these orientations is the canonical orientation that is poised for formation of 5′-dAdo•. The second, alternative orientation of SAM ligates the [4Fe-4S] cluster similarly but is not poised to facilitate radical chemistry (Bridwell-Rabb et al., 2017) (Fig. 3K). The trapping of DTT at the [4Fe-4S] cluster revealed an important insight for obtaining a SAM-bound crystal structure. However, it may also provide relevant information for performing assays on radical SAM enzymes as the activity may be compromised by the ability of DTT to compete with SAM for binding the [4Fe-4S] cluster. This result is reminiscent of that described for the Cbl-dependent radical SAM enzyme GenK, which is not activated via reduction with the common-reducing agent sodium dithionite (Kim et al., 2013). Instead, it is proposed that GenK catalysis is inhibited by sodium dithionite due to the noted ability of SO_2^- to interact with and reduce the Co-ion of Cbl (Kim et al., 2013; Salnikov, Silaghi-Dumitrescu, Makarov, van Eldik, & Boss, 2011).

5.2 The importance and limitations of crystal packing is demonstrated by OxsB

In the process of crystallization, interactions between protein molecules must form to create a periodic crystal lattice (Rupp, 2010). These interactions are generally non-specific and can cause structural artifacts, including the locking down of flexible and dynamic protein regions (Rupp, 2010). Remarkably, the SeMet-, apo-, and reconstituted structures of OxsB were all crystallized using similar conditions in the same $P2_12_12_1$ space group with similar unit cell dimensions (Bridwell-Rabb et al., 2017). Inspection of the crystal packing reveals a similar arrangement of symmetry-related molecules and explains some of the curiosities experienced when working with OxsB. First, it was determined that cleavage of the tag from OxsB was critical to identify crystallization conditions, as even a small His-tag prevented crystallization of OxsB in the conditions described above (Fig. 3). From the crystal packing, it is observed that the N-terminal domain packs tightly among three symmetry-related molecules and makes several interactions with two of these molecules (Fig. 5A). It appears that any addition to the N-terminus of OxsB would not fit in the current crystal lattice and serves to emphasize why, as described in Section 2.2, crystal conditions should be screened in the presence and absence of the tag (Fig. 5A).

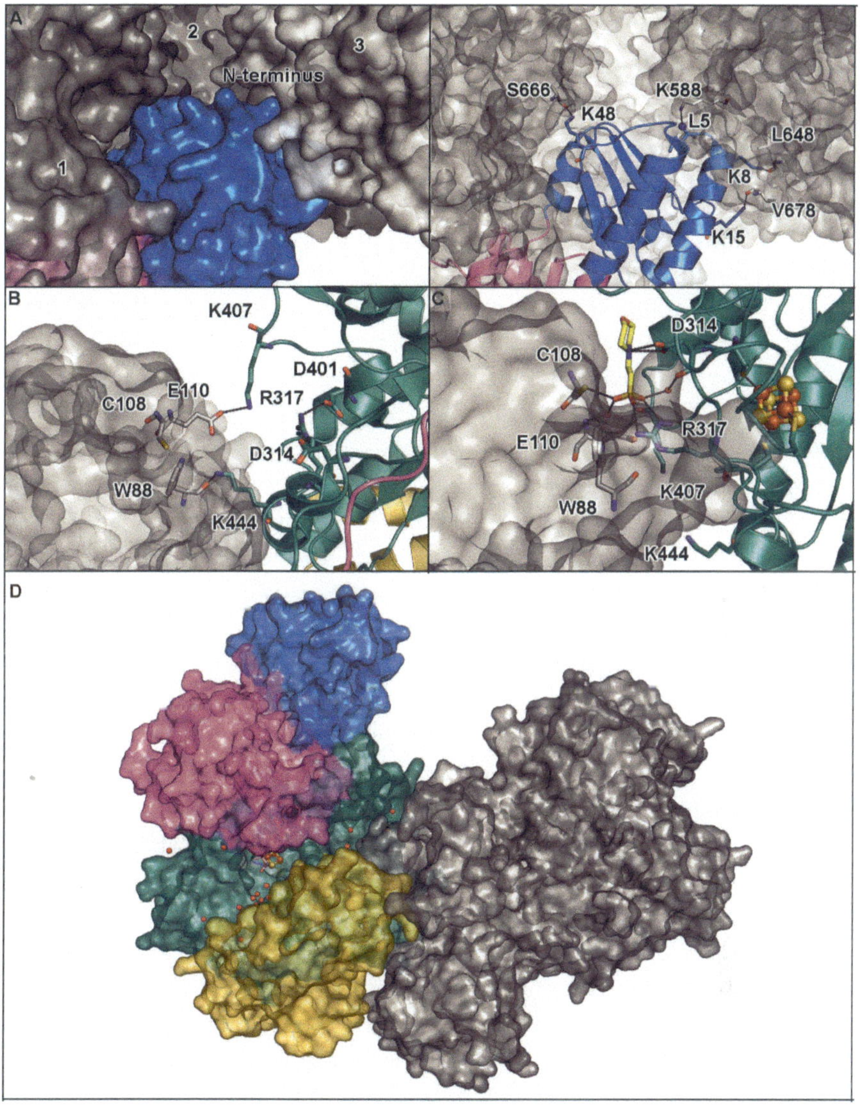

Fig. 5 The observed packing of symmetry related molecules in the OxsB structure explains some of the crystallographic challenges for OxsB. (A) The N-terminal domain of OxsB (blue) is tightly packed among three symmetry-related molecules (gray, left panel). The residues in the N-terminal domain make several interactions with the surrounding monomeric units. This tight packing near the N-terminal domain is presumably inhibited by the addition of a tag and may explain the reason why His-tagged OxsB will not crystallize in this condition (PDB: 5UL4). (B) The packing between OxsB and a symmetry related molecule near the [4Fe-4S] cluster is mediated by Lys residues in the apo and SeMet-labeled structures (PDB: 5UL2). (C) In the reconstituted structure of OxsB, interactions in the same region of the protein shown in (B) is mediated by a molecule of MES from the crystallization buffer (PDB: 5UL4). (D) The observed open conformation of OxsB shows a solvent-filled active site, which would likely prevent catalysis from occurring. The openness is also proposed to account for the inability to obtain substrate-bound crystal structure. Further movement of the helix bundle domain (yellow) toward Cbl is prohibited by a symmetry related molecule (gray) (Bridwell-Rabb et al., 2018, 2017).

Second, inspection of the crystallization conditions for the apo-, SeMet-labeled, and reconstituted OxsB protein samples reveals that the largest difference between the crystallization conditions is the buffer that is needed to grow the crystals. Bicine buffer is used to crystallize apo- and SeMet-labeled OxsB, whereas MES buffer is substituted in the reconstituted OxsB condition (Bridwell-Rabb et al., 2017). Inspection of these structures, as noted above and shown in Fig. 3E, reveals many similarities with respect to overall architecture. Small deviations, however, are observed in the regions that bind the [4Fe-4S] cluster and Cbl, as well as in the C-terminal helix bundle domain (Fig. 3E). With regards to the [4Fe-4S] cluster binding site, subtle changes in the position of the Cys motif and the cluster binding loop propagate to changes in the interface between OxsB and a symmetry-related molecule. In the apo/SeMet-labeled structure, two Lys residues interact with residues from the N-terminal domain of the adjacent symmetry-related monomer (Fig. 5B). In the reconstituted protein structure, the packing is relatively unchanged, but the equivalent interface houses a molecule of MES between the monomers and the Lys-mediated interactions are absent (Fig. 5C). The presence of MES at this interface and the movement of the Lys sidechains presumably explains why, despite extensive screening of conditions and substitution of the different buffer and precipitant solutions, only the condition described above for reconstituted OxsB yielded crystals (Fig. 3). It also emphasizes again that the architectures of the apo- and reconstituted proteins are globally similar.

Finally, despite the three available crystal structures of OxsB, there is a not a structure of OxsB with a substrate (dAMP, dADP, or dATP) bound. The lack of a substrate-bound structure is hypothesized to be due to the inability of the C-terminal helix bundle domain to completely close off the active site from solvent, which would be a pre-requisite for chemistry to occur (Bridwell-Rabb et al., 2018, 2017). The openness of the active site is due to a symmetry-related molecule which sits between the Cbl-binding and the C-terminal domain (Fig. 5D). In addition, as observed in the later-determined structures of TsrM and TokK, residues from the C-terminal domain in each case interact with the bound substrate (Knox, Chen, et al., 2021; Knox, Sinner, et al., 2021) (Fig. 5D). Such interactions are not possible in the open active site of OxsB (Bridwell-Rabb et al., 2017). These observations suggest that conditions should be rescreened using alternative constructs of OxsB to find a different crystal form.

5.3 The importance of reconstituting the active form of OxsB for crystallization

As described above, starting with the apo-form of OxsB was effective in that it allowed for insight into the necessary cofactors for activity and development of proper reconstitution methods. However, despite the utility of this approach for OxsB, there are also examples for which this bottom-up approach hasn't worked. Extensive attempts to express native and His-tagged Fom3 in *E. coli* resulted in protein that was exclusively found in inclusion bodies (Allen & Wang, 2014a). Similar insoluble expression of the Cbl-dependent radical SAM enzymes PhpK (Werner et al., 2011), SD1168 (Allen & Wang, 2014b) and GenK (Kim et al., 2013) necessitated methods for anaerobic protein refolding and reconstitution. Despite the ability of these methods to produce the proteins of interest, new methods to facilitate the ease of Cbl-dependent radical SAM enzyme purification have emerged. For example, co-expression of a pBAD42-*BtuCEDFB* plasmid with a Cbl-dependent radical SAM enzyme has been shown to enhance production of soluble PhpK and Fom3 (Lanz et al., 2018; McLaughlin & van der Donk, 2018). This pBAD42-*BtuCEDFB* plasmid encodes proteins involved in Cbl uptake and trafficking (Lanz et al., 2018). Similar to the pDB1282 plasmid, which encodes the [Fe-S] cluster biosynthetic genes from *Azotobacter vinelandii* and the Δ*iscR* Bl21-(DE3) cells, which change transcription of the *E. coli* [Fe-S] cluster assembly genes, the pBAD42-*BtuCEDFB* plasmid should facilitate reconstitution of the needed cofactors and ideally support folding, soluble expression, and purification of a homogeneous protein sample for crystallography (Akhtar & Jones, 2008; Lanz et al., 2018; Zheng et al., 1998).

6. Conclusions

The large and chemically diverse group of Cbl-dependent radical SAM enzymes is at this point represented by just three protein structures (Fig. 1A) (Bridwell-Rabb et al., 2017; Knox, Chen, et al., 2021; Knox, Sinner, et al., 2021). These structures are of enzymes that catalyze chemically distinct reactions and have already taught us a great deal about how these enzymes function (Bridwell-Rabb et al., 2018, 2017; Knox, Chen, et al., 2021; Knox, Sinner, et al., 2021). Historically, the challenges inherent to working with Cbl-dependent radical SAM enzymes have stymied biochemical studies aimed at elucidating the structure-function relationships in this enzyme class. Here, we emphasize that the bottom-up approach used for OxsB provided has provided valuable insights into its mechanism of actions, even in the absence of the metallocofactors (Bridwell-Rabb et al., 2017). We posit that due to the relatively small overall differences observed between the apo-, SeMet-labeled, and reconstituted structures of OxsB, that this bottom-up approach may be advantageous for learning about the structure, function, and mechanism of other Cbl-dependent radical SAM enzymes as well. Most importantly, this bottom-up approach should be capitalized on when limited information regarding the necessary cofactors for enzyme activity exists or when challenges associated with producing a fully reconstituted homogeneous sample cannot be overcome. In combination with newly developed methods for recombinant cofactor incorporation and methods to increase protein solubility (Lanz et al., 2018), we are hopeful that the future holds a more comprehensive understanding of how one enzyme class accomplishes such a long list of important chemical transformations.

Acknowledgments

This material is based upon work supported by the U.S. Department of Energy, Office of Science, Office of Basic Energy Sciences, under Award Number DE-SC0021240 (J.B.R) and the Searle Scholars Program (J.B.R).

References

Akhtar, M. K., & Jones, P. R. (2008). Deletion of iscR stimulates recombinant clostridial Fe-Fe hydrogenase activity and H2-accumulation in Escherichia coli BL21(DE3). *Applied Microbiology and Biotechnology*, *78*(5), 853–862. https://doi.org/10.1007/s00253-008-1377-6.

Allen, K. D., & Wang, S. C. (2014a). Initial characterization of Fom3 from Streptomyces wedmorensis: The methyltransferase in fosfomycin biosynthesis. *Archives of Biochemistry and Biophysics*, *543*, 67–73. https://doi.org/10.1016/j.abb.2013.12.004.

Allen, K. D., & Wang, S. C. (2014b). Spectroscopic characterization and mechanistic investigation of P-methyl transfer by a radical SAM enzyme from the marine bacterium Shewanella denitrificans OS217. *Biochimica et Biophysica Acta*, *1844*(12), 2135–2144. https://doi.org/10.1016/j.bbapap.2014.09.009.

Bandarian, V., & Matthews, R. G. (2004). Measurement of energetics of conformational change in cobalamin-dependent methionine synthase. *Methods in Enzymology*, *380*, 152–169. https://doi.org/10.1016/S0076-6879(04)80007-7.

Beinert, H. (1983). Semi-micro methods for analysis of labile sulfide and of labile sulfide plus sulfane sulfur in unusually stable iron-sulfur proteins. *Analytical Biochemistry*, *131*(2), 373–378. https://doi.org/10.1016/0003-2697(83)90186-0.

Berkovitch, F., Nicolet, Y., Wan, J. T., Jarrett, J. T., & Drennan, C. L. (2004). Crystal structure of biotin synthase, an S-adenosylmethionine-dependent radical enzyme. *Science*, *303*(5654), 76–79. https://doi.org/10.1126/science.1088493.

Blaszczyk, A. J., Silakov, A., Zhang, B., Maiocco, S. J., Lanz, N. D., Kelly, W. L., et al. (2016). Spectroscopic and electrochemical characterization of the Iron-sulfur and cobalamin cofactors of TsrM, an unusual radical S-Adenosylmethionine methylase. *Journal of the American Chemical Society*, *138*(10), 3416–3426. https://doi.org/10.1021/jacs.5b12592.

Blaszczyk, A. J., Wang, R. X., & Booker, S. J. (2017). TsrM as a model for purifying and characterizing cobalamin-dependent radical S-adenosylmethionine methylases. *Methods in Enzymology*, *595*, 303–329. https://doi.org/10.1016/bs.mie.2017.07.007.

Blaszczyk, A. J., Wang, B., Silakov, A., Ho, J. V., & Booker, S. J. (2017). Efficient methylation of C2 in l-tryptophan by the cobalamin-dependent radical S-adenosylmethionine methylase TsrM requires an unmodified N1 amine. *The Journal of Biological Chemistry*, *292*(37), 15456–15467. https://doi.org/10.1074/jbc.M117.778548.

Bradford, M. M. (1976). A rapid and sensitive method for the quantitation of microgram quantities of protein utilizing the principle of protein-dye binding. *Analytical Biochemistry*, *72*, 248–254. https://doi.org/10.1006/abio.1976.9999.

Bridwell-Rabb, J., Grell, T. A. J., & Drennan, C. L. (2018). A rich man, poor man story of S-adenosylmethionine and cobalamin revisited. *Annual Review of Biochemistry*, *87*, 555–584. https://doi.org/10.1146/annurev-biochem-062917-012500.

Bridwell-Rabb, J., Kang, G., Zhong, A., Liu, H. W., & Drennan, C. L. (2016). An HD domain phosphohydrolase active site tailored for oxetanocin-A biosynthesis. *Proceedings of the National Academy of Sciences of the United States of America*, *113*(48), 13750–13755.

Bridwell-Rabb, J., Zhong, A., Sun, H. G., Drennan, C. L., & Liu, H. W. (2017). A B12-dependent radical SAM enzyme involved in oxetanocin A biosynthesis. *Nature*, *544*(7650), 322–326. https://doi.org/10.1038/nature21689.

Broderick, J. B., Duffus, B. R., Duschene, K. S., & Shepard, E. M. (2014). Radical S-adenosylmethionine enzymes. *Chemical Reviews*, *114*(8), 4229–4317. https://doi.org/10.1021/cr4004709.

Carson, M., Johnson, D. H., McDonald, H., Brouillette, C., & Delucas, L. J. (2007). His-tag impact on structure. *Acta Crystallographica. Section D, Biological Crystallography*, *63*(Pt. 3), 295–301. https://doi.org/10.1107/S0907444906052024.

Crowe, J., Dobeli, H., Gentz, R., Hochuli, E., Stuber, D., & Henco, K. (1994). 6xHis-Ni-NTA chromatography as a superior technique in recombinant protein expression/purification. *Methods in Molecular Biology*, *31*, 371–387. https://doi.org/10.1385/0-89603-258-2:371.

Deller, M. C., Kong, L., & Rupp, B. (2016). Protein stability: A crystallographer's perspective. *Acta Crystallographica Section F: Structural Biology Communications*, *72*(Pt. 2), 72–95. https://doi.org/10.1107/S2053230X15024619.

Derewenda, Z. S. (2004). The use of recombinant methods and molecular engineering in protein crystallization. *Methods*, *34*(3), 354–363. https://doi.org/10.1016/j.ymeth.2004.03.024.

Dowling, D. P., Croft, A. K., & Drennan, C. L. (2012). Radical use of Rossmann and TIM barrel architectures for controlling coenzyme B12 chemistry. *Annual Review of Biophysics*, *41*, 403–427. https://doi.org/10.1146/annurev-biophys-050511-102225.

Drennan, C. L., Huang, S., Drummond, J. T., Matthews, R. G., & Ludwig, M. L. (1994). How a protein binds B12: A 3.0 A X-ray structure of B12-binding domains of methionine synthase. *Science*, *266*(5191), 1669–1674.

Fish, W. W. (1988). Rapid colorimetric micromethod for the quantitation of complexed iron in biological samples. *Methods in Enzymology*, *158*, 357–364. https://doi.org/10.1016/0076-6879(88)58067-9.

Frey, P. A. (2010). Comprehensive natural products II Chemistry and biology Vol. 7. In L. L. (eds Mander, H.-W.) 501-546 (Ed.): Elsevier.

Fuss, J. O., Tsai, C. L., Ishida, J. P., & Tainer, J. A. (2015). Emerging critical roles of Fe-S clusters in DNA replication and repair. *Biochimica et Biophysica Acta, 1853*(6), 1253–1271. https://doi.org/10.1016/j.bbamcr.2015.01.018.

Gerlt, J. A., Bouvier, J. T., Davidson, D. B., Imker, H. J., Sadkhin, B., Slater, D. R., et al. (2015). Enzyme function initiative-enzyme similarity tool (EFI-EST): A web tool for generating protein sequence similarity networks. *Biochimica et Biophysica Acta, 1854*(8), 1019–1037. https://doi.org/10.1016/j.bbapap.2015.04.015.

Gough, S. P., Petersen, B. O., & Duus, J. O. (2000). Anaerobic chlorophyll isocyclic ring formation in Rhodobacter capsulatus requires a cobalamin cofactor. *Proceedings of the National Academy of Sciences of the United States of America, 97*(12), 6908–6913.

Haft, D. H. (2011). Bioinformatic evidence for a widely distributed, ribosomally produced electron carrier precursor, its maturation proteins, and its nicotinoprotein redox partners. *BMC Genomics, 12*, 21. https://doi.org/10.1186/1471-2164-12-21.

Haft, D. H., & Basu, M. K. (2011). Biological systems discovery in silico: Radical S-adenosylmethionine protein families and their target peptides for posttranslational modification. *Journal of Bacteriology, 193*(11), 2745–2755. https://doi.org/10.1128/JB.00040-11.

Kamarthapu, V., Rao, K. V., Srinivas, P. N., Reddy, G. B., & Reddy, V. D. (2008). Structural and kinetic properties of Bacillus subtilis S-adenosylmethionine synthetase expressed in Escherichia coli. *Biochimica et Biophysica Acta, 1784*(12), 1949–1958. https://doi.org/10.1016/j.bbapap.2008.06.006.

Kim, H. J., McCarty, R. M., Ogasawara, Y., Liu, Y. N., Mansoorabadi, S. O., LeVieux, J., et al. (2013). GenK-catalyzed C-6′ methylation in the biosynthesis of gentamicin: Isolation and characterization of a cobalamin-dependent radical SAM enzyme. *Journal of the American Chemical Society, 135*(22), 8093–8096. https://doi.org/10.1021/ja312641f.

Knox, H. L., Chen, P. Y., Blaszczyk, A. J., Mukherjee, A., Grove, T. L., Schwalm, E. L., et al. (2021). Structural basis for non-radical catalysis by TsrM, a radical SAM methylase. *Nature Chemical Biology, 17*(4), 485–491. https://doi.org/10.1038/s41589-020-00717-y.

Knox, H. L., Sinner, E. K., Townsend, C. A., Boal, A. K., & Booker, S. J. (2021). Structural basis for carbapenam C-alkylation by TokK, a 746 B12-dependent radical SAM enzyme. *bioRxiv*. https://doi.org/10.1101/2021.07.10.451909. 2021.2007.2010.451909.

Lanz, N. D., Blaszczyk, A. J., McCarthy, E. L., Wang, B., Wang, R. X., Jones, B. S., et al. (2018). Enhanced solubilization of class B radical S-Adenosylmethionine methylases by improved cobalamin uptake in Escherichia coli. *Biochemistry, 57*(9), 1475–1490. https://doi.org/10.1021/acs.biochem.7b01205.

Lanz, N. D., Grove, T. L., Gogonea, C. B., Lee, K. H., Krebs, C., & Booker, S. J. (2012). RlmN and AtsB as models for the overproduction and characterization of radical SAM proteins. *Methods in Enzymology, 516*, 125–152. https://doi.org/10.1016/B978-0-12-394291-3.00030-7.

Ljungdahl, L. G., LeGall, J., & Lee, J. P. (1973). Isolation of a protein containing tightly bound 5- methoxybenzimidazolylcobamide (factor 3m) from Clostridium thermoaceticum. *Biochemistry, 12*(9), 1802–1808. https://doi.org/10.1021/bi00733a022.

Luft, J. R., Wolfley, J. R., & Snell, E. H. (2011). What's in a drop? Correlating observations and outcomes to guide macromolecular crystallization experiments. *Crystal Growth & Design, 11*(3), 651–663. https://doi.org/10.1021/cg1013945.

Magnusson, O. T., Reed, G. H., & Frey, P. A. (2001). Characterization of an allylic analogue of the 5′-deoxyadenosyl radical: An intermediate in the reaction of lysine 2,3-aminomutase. *Biochemistry, 40*(26), 7773–7782.

Matos, J. R., & Wong, C. H. (1987). S-Adenosylmethionine—Stability and stabilization. *Bioorganic Chemistry, 15*(1), 71–80. https://doi.org/10.1016/0045-2068(87)90008-3.

McLaughlin, M. I., Lanz, N. D., Goldman, P. J., Lee, K. H., Booker, S. J., & Drennan, C. L. (2016). Crystallographic snapshots of sulfur insertion by lipoyl synthase. *Proceedings of the National Academy of Sciences of the United States of America, 113*(34), 9446–9450. https://doi.org/10.1073/pnas.1602486113.

McLaughlin, M. I., & van der Donk, W. A. (2018). Stereospecific radical-mediated B12- dependent methyl transfer by the Fosfomycin biosynthesis enzyme Fom3. *Biochemistry, 57*(33), 4967–4971. https://doi.org/10.1021/acs.biochem.8b00616.

McPherson, A., & Gavira, J. A. (2014). Introduction to protein crystallization. *Acta Crystallographica Section F: Structural Biology Communications, 70*(Pt. 1), 2–20. https://doi.org/10.1107/S2053230X13033141.

Ouchane, S., Steunou, A. S., Picaud, M., & Astier, C. (2004). Aerobic and anaerobic Mg- protoporphyrin monomethyl ester cyclases in purple bacteria: A strategy adopted to bypass the repressive oxygen control system. *The Journal of Biological Chemistry, 279*(8), 6385–6394. https://doi.org/10.1074/jbc.M309851200.

Parks, L. W., & Schlenk, F. (1958). Stability and hydrolysis of S-Adenosylmethionine—Isolation of S-Ribosylmethionine. *Journal of Biological Chemistry, 230*(1), 295–305. Retrieved from <Go to ISI>://WOS:A1958WA43400030.

Rupp, B. (2010). Biomolecular crystallography: Principles, practice, and application to structural biology. *Biomolecular Crystallography: Principles, Practice, and Application to Structural Biology, 1*, 1–809. Retrieved from < Go to ISI >://WOS:000271460000014.

Salnikov, D. S., Silaghi-Dumitrescu, R., Makarov, S. V., van Eldik, R., & Boss, G. R. (2011). Cobalamin reduction by dithionite. Evidence for the formation of a six-coordinate cobalamin(II) complex. *Dalton Trans, 40*(38), 9831–9834. https://doi.org/10.1039/c1dt10219b.

Sato, S., Kudo, F., Kim, S.-Y., Kuzuyama, T., & Eguchi, T. (2017). Methylcobalamin-dependent radical SAM C-methyltransferase Fom3 recognizes cytidylyl-2-hydroxyethylphosphonate and catalyzes the nonstereoselective C-methylation in fosfomycin biosynthesis. *Biochemistry, 56*(28), 3519–3522.

Shannon, P., Markiel, A., Ozier, O., Baliga, N. S., Wang, J. T., Ramage, D., et al. (2003). Cytoscape: A software environment for integrated models of biomolecular interaction networks. *Genome Research, 13*(11), 2498–2504. https://doi.org/10.1101/gr.1239303.

Shisler, K. A., Hutcheson, R. U., Horitani, M., Duschene, K. S., Crain, A. V., Byer, A. S., et al. (2017). Monovalent cation activation of the radical SAM enzyme pyruvate formate-lyase activating enzyme. *Journal of the American Chemical Society, 139*(34), 11803–11813. https://doi.org/10.1021/jacs.7b04883.

Sofia, H. J., Chen, G., Hetzler, B. G., Reyes-Spindola, J. F., & Miller, N. E. (2001). Radical SAM, a novel protein superfamily linking unresolved steps in familiar biosynthetic pathways with radical mechanisms: Functional characterization using new analysis and information visualization methods. *Nucleic Acids Research, 29*(5), 1097–1106.

Stookey, L. L. (1970). Ferrozine—A new spectrophotometric reagent for iron. *Analytical Chemistry, 42*(7), 779. https://doi.org/10.1021/ac60289a016.

Vranish, J. N., Russell, W. K., Yu, L. E., Cox, R. M., Russell, D. H., & Barondeau, D. P. (2015). Fluorescent probes for tracking the transfer of iron-sulfur cluster and other metal cofactors in biosynthetic reaction pathways. *Journal of the American Chemical Society, 137*(1), 390–398. https://doi.org/10.1021/ja510998s.

Wang, S. C. (2018). Cobalamin-dependent radical S-adenosyl-l-methionine enzymes in natural product biosynthesis. *Natural Product Reports, 35*(8), 707–720. https://doi.org/10.1039/c7np00059f.

Werner, W. J., Allen, K. D., Hu, K., Helms, G. L., Chen, B. S., & Wang, S. C. (2011). In vitro phosphinate methylation by PhpK from Kitasatospora phosalacinea. *Biochemistry, 50*(42), 8986–8988. https://doi.org/10.1021/bi201220r.

Zallot, R., Oberg, N. O., & Gerlt, J. A. (2018). 'Democratized' genomic enzymology web tools for functional assignment. *Current Opinion in Chemical Biology*, *47*, 77–85. https://doi.org/10.1016/j.cbpa.2018.09.009.

Zallot, R., Oberg, N., & Gerlt, J. A. (2019). The EFI web resource for genomic enzymology tools: Leveraging protein, genome, and metagenome databases to discover novel enzymes and metabolic pathways. *Biochemistry*, *58*(41), 4169–4182. https://doi.org/10.1021/acs.biochem.9b00735.

Zhang, J., & Klinman, J. P. (2015). High-performance liquid chromatography separation of the (S,S)- and (R,S)-forms of S-adenosyl-L-methionine. *Analytical Biochemistry*, *476*, 81–83. https://doi.org/10.1016/j.ab.2015.02.004.

Zheng, L., Cash, V. L., Flint, D. H., & Dean, D. R. (1998). Assembly of iron-sulfur clusters. Identification of an iscSUA-hscBA-fdx gene cluster from *Azotobacter vinelandii*. *Journal of Biological Chemistry*, *273*(21), 13264–13272. https://doi.org/10.1074/jbc.273.21.13264.

SECTION IV

Methods for studying the mechanisms of B_{12} enzymes

CHAPTER SIX

Computational investigations of B_{12}-dependent enzymatic reactions

Megan J. Toda, Arghya P. Ghosh, Saurav Parmar, and Pawel M. Kozlowski*
Department of Chemistry, University of Louisville, Louisville, KY, United States
*Corresponding author: e-mail address: pawel.kozlowski@louisville.edu

Contents

Abstract

Nature employs two biologically active forms of vitamin B_{12}, adenosylcobalamin (or coenzyme B_{12}) and methylcobalamin, as cofactors in molecular transformations both in bacteria and mammals. Computational chemistry, guided by experimental data, has been used to explore fundamental characteristics of these enzymatic reactions. In particular, the quantum mechanics/molecular mechanics (QM/MM) method has proven to be a powerful tool in elucidating important characteristics of B_{12}-dependent enzymatic reactions. Herein, we will present a brief tutorial in conducting QM/MM calculations for B_{12} enzymatic reactions. We will summarize recent contributions that target the use of QM/MM calculations in both photochemical and enzymatic reactions including AdoCbl-dependent ethanolamine ammonia lyase, glutamate mutase, and photoreceptor CarH.

Methods in Enzymology, Volume 669
ISSN 0076-6879
https://doi.org/10.1016/bs.mie.2022.01.002

1. Introduction

B_{12} derivatives, commonly referred to as cobalamins (Cbls), catalyze various biological reactions, which are classified into three main categories. First, methylcobalamin (MeCbl)-dependent methyltransferases (MeTrs) catalyze CH_3 transfer reactions; second, adenosylcobalamin (AdoCbl)-dependent isomerases catalyze 1,2-rearrangement reactions; and third, reductive dehalogenases, catalyze dehalogenation reactions (Banerjee & Ragsdale, 2003). Among the most well-studied Cbl-dependent enzymes are those which employ MeCbl and AdoCbl as their cofactors. MeCbl is the cofactor in methionine synthase (MetH), a MeTrs that catalyzes the transfer of the Me group from methyltetrahydrofolate (CH_3-THF) to the thiolate of homocysteine. This occurs via demethylation of the tetrahydrofolate (THF) to generate methionine (Met) and a Co(I)Cbl intermediate. The resulting Co(I)Cbl is further remethylated by the CH_3-THF substrate to generate the MeCbl and THF. AdoCbl is the cofactor for a variety of enzymes which are delineated into two groups: eliminases and mutases. The eliminase class includes AdoCbl cofactors which are found in the base-on conformation; ethanolamine ammonia lyase (EAL) and diol dehydratase (DDH) are two representative examples. The mutase class includes enzymes such as glutamate mutase (GLM) (Gruber, Reitzer, & Kratky, 2001) and methylmalonyl-CoA mutase (MCM) where the AdoCbl cofactor is found in the base-off/His-on conformation.

The key step in B_{12}-dependent enzymatic catalysis, involving either MeCbl or AdoCbl as the cofactor, is the cleavage of the Co—C bond (Banerjee, 1999; Banerjee & Ragsdale, 2003; Dolphin, 1982). In MeCbl-dependent MeTrs, the cleavage is heterolytic leading to the formation of Co(I)Cbl. AdoCbl is known for its radical chemistry and cleavage of the Co—$C_{5'}$ bond in the enzymatic environment is homolytic. Unlike the MeCbl-dependent MetH, the general catalytic process of the AdoCbl-dependent enzymes begins with the homolytic cleavage of the Co—$C_{5'}$ bond forming the Co(II)/5′-deoxy-5′-adenosyl$^{\bullet}$ (Ado$^{\bullet}$) radical pair (RP) (Fig. 1). The primary Ado$^{\bullet}$ then abstracts an H-atom from the substrate generating a substrate radical and 5′-deoxyadenosine. The radical substrate undergoes a 1,2-rearrangement and re-abstracts an H-atom from 5′-deoxyadenosine to produce Ado$^{\bullet}$, which recombines with Co(II)Cbl. The reformation of the Co—$C_{5'}$ bond closes the catalytic cycle.

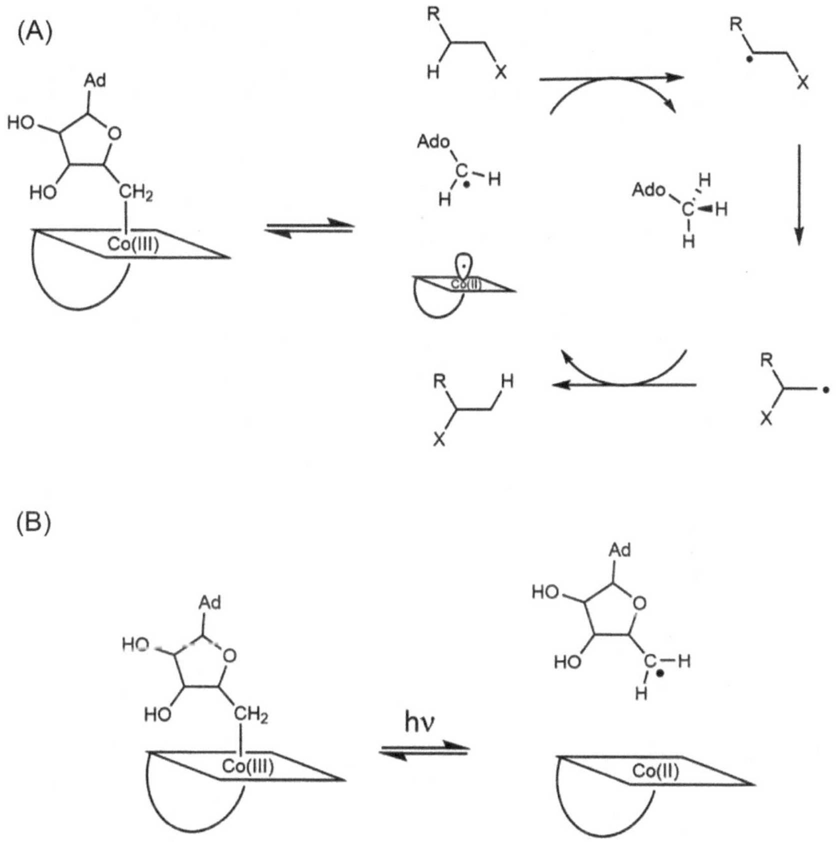

Fig. 1 (A) Scheme depicting the general mechanism of 1,2-rearrangements in AdoCbl-dependent mutases and eliminases and (B) light-induced radical pair formation in AdoCbl (Ad = adenine).

The cleavage of the Co—$C_{5'}$ bond in B_{12}-dependent mutases occurs on the millisecond timescale whereas the homolysis of the Co—$C_{5'}$ bond of isolated AdoCbl in solution is very slow (Hay & Finke, 1986, 1987). Based on kinetic studies, it has been found that the homolysis of the Co—$C_{5'}$ bond in several coenzyme B_{12}-dependent mutases attains a high catalytic rate (k_{cat}) of 2–300 s^{-1} (Hay & Finke, 1986). Thus, in AdoCbl-dependent enzymatic reactions, the homolysis of the Co—$C_{5'}$ bond experiences a trillion-fold rate enhancement compared to the thermal homolysis in solution (Finke & Martin, 1990; Hay & Finke, 1987; Martin & Finke, 1990). Understanding the source of this remarkable rate enhancement is the most pressing issue in B_{12} chemistry.

While the main step in the enzymatic reactions involving coenzyme B_{12} as their cofactor is the cleavage of the Co—$C_{5'}$ bond, this bond can also be broken via light (Fig. 1B). The cleavage of the Co—$C_{5'}$ bond in the enzymatic environment is homolytic and results in the formation of the Co(II)/Ado$^{\bullet}$ RP. The activation of the Co—$C_{5'}$ bond in such systems is related to substrate binding. Similarly, the Co—$C_{5'}$ bond of AdoCbl can be cleaved with light which also results in the formation of Co(II)/Ado$^{\bullet}$ RPs (Walker, Shiang, Anderson, Pullen, & Sension, 1998; Yoder, Cole, Walker, & Sension, 2001). Accordingly, the hypothesis that the photochemical cleavage of the Co—$C_{5'}$ bond can be used to elucidate aspects of the corresponding cleavage in the enzymatic environment, has been a target for recent computational analyses (Mamun, Toda, & Kozlowski, 2019). This study will be discussed in detail later in the chapter.

The enzymatic reactions of B_{12}-dependent systems have been extensively discussed in the appended references and we encourage the curious reader to consult these key texts (Banerjee, 1997, 1999; Banerjee & Ragsdale, 2003; Brown, 2005; Dolphin, 1982; Matthews, 2009). In addition, several recent works have discussed the major strides in the area of B_{12} photochemistry (Jones, 2017; Kozlowski, Garabato, Lodowski, & Jaworska, 2016; Toda, Kozlowski, & Andruniów, 2019; Toda, Lodowski, Mamun, Jaworska, & Kozlowski, 2019). Computational chemistry has played an important role in both of these major research areas. The main target of the present chapter is to describe the use of QM/MM methodology in elucidating the details of the reaction mechanisms of B_{12}-dependent enzymes. This chapter will provide a basic tutorial for how QM/MM calculations can be used to study enzymatic reactions of B_{12}-dependent systems, with a focus on the enzymes which employ the biologically active cofactors, MeCbl and AdoCbl.

2. Computational approach

2.1 Introduction to QM/MM

The application of computational methods has become an invaluable instrument that is frequently utilized in the disciplines of physics, chemistry, and biology to validate experimental results. In computational chemistry, there are various methodologies based on various levels of theory to study complex biological and bioinorganic systems. Methods used to study enzymes include molecular dynamics (MD), molecular mechanics (MM), and QM/MM. MD simulation is commonly used to study the motions of molecules in complicated biological systems using classical mechanical force fields. MM is geared toward determinations of molecular equilibrium structures. Molecular modelers make use of a force field to predict the geometry of a novel molecule, usually containing atoms more than ~10,000 atoms, by generating data related to bond parameters from other related molecules. The force field equations are empirical and derived on the assumption that atoms in a molecule behave as balls joined by springs. A large molecule is treated as a system consisting of the same features, which are present in small molecules, but put together in different ways. MM methods are very versatile primarily because the derived force field parameters are transferable from molecule to molecule. However, there are limitations in the use of mechanical force fields and changes in the electronic structure including bond breaking and formation, electron transfer, and electronic excitations cannot be characterized by MM and MD methods.

QM/MM is a method which combines QM (electronic structure methods) and MM (empirical force field) to study large systems including enzymes (Lin & Truhlar, 2006; Senn & Thiel, 2009; Sousa, Fernandes, & Ramos, 2012; van der Kamp & Mulholland, 2013). QM methods are used to determine the electronic structure of molecules whereas MM methods are used to determine the potential energy as a function of nuclear coordinates. In general, a QM/MM set-up can be described as follows, the chemically important part of the system such as the active site is treated with a QM method and the remainder of the system is treated with an MM method. The major advantage in the QM/MM method on large systems is that it can deliver the relatively low computational cost compared to a hitherto impossible calculation with QM methods alone and with accuracy which cannot be matched by typical MM methods.

The era of QM/MM methods for enzymatic reactions commenced with the seminal paper by Warshel and Levitt in which the method was applied to model carbocation intermediate stabilization in the active site of lysozyme (Warshel & Levitt, 1976). The development of the QM/MM method, which was presented by Warshel and Levitt in 1976, can be traced to the contributions from Honig and Karplus (Honig & Karplus, 1971) as well as Warshel and Karplus (Warshel & Karplus, 1972). Honig and Karplus presented an approach to combine QM calculations with MM calculations to determine the details of the ground state geometry of retinal and to assign the features of the experimental spectrum with the corresponding excited states (Honig & Karplus, 1971). The early contribution of Warshel and Karplus was the development of a formulation for the consistent calculation of the potential energy surface (PES) of the ground and excited states of conjugated molecules (Warshel & Karplus, 1972). The method, technically, was based on separation of molecular electrons into σ and π-electrons to be approximated by empirical potential function and semi-empirical Parrisar-Parr-Pople (PPP) level of theory, respectively (Warshel & Karplus, 1972). Warshel, Levitt and Karplus were awarded the 2013 Nobel Prize in chemistry for their achievements in multiscale modeling. Among its major achievements, the QM/MM method when complemented with MD simulations, can help resolve crucial questions regarding transition state (TS) structures which are of prime importance in any enzymatic reaction but cannot be studied directly by experimental methods (Carloni, Rothlisberger, & Parrinello, 2002; Hollingsworth & Dror, 2018; van der Kamp & Mulholland, 2013). Some of the successes of the method include investigations into the catalytic cycle of a heme protein cytochrome P450, understanding bacterial drug-resistance due to lactamase protein and determining the enzymatic mechanism of the anti-AIDS target HIV-1 protease. Our research group, among others, have been involved in investigations of Cbl-dependent catalysis and photochemistry using various levels of theoretical calculations including QM/MM (Brunk, Kellett, Richards, & Rothlisberger, 2014; Kwiecien et al., 2006; Schopf, Mills, & Warshel, 2015; Sharma, Chu, Olsson, & Warshel, 2007; Vreven et al., 2006). While there are potentially other approaches that may be taken to implement QM/MM calculations, the general steps undertaken by our research group using the ONIOM method are described below.

2.2 From the crystal structure to the QM/MM input

The crystal structure of B_{12}-dependent enzymes can be obtained from the protein data bank (PDB). Care should be taken when selecting the proper file for the desired QM/MM study. Typically, the crystal structures of B_{12}-dependent enzymes contain a structure with the substrate and a separate structure without the substrate. If the research target is related to the enzymatic reaction, i.e., activation of the Co—$C_{5'}$ bond via a substrate binding, the PDB file which contains the substrate should be used. Alternatively, if the research target is related to the photochemistry of a system where the substrate is not needed for Co—$C_{5'}$ activation, if available, one should download the crystal structure that does not contain the substrate.

Once the proper file has been selected, the downloaded file (.pdb) should be initially inspected with protein visualization tools such as Chimera

(Pettersen et al., 2004), Maestro (Schrodinger Release 2021-3: Maestro Schrodinger, LLC, 2021), etc. The structure may contain more than one chain (e.g., homodimers in MCM enzymes) with labels A, B, etc.; typically, only one of the chains is retained for further model development and the others can be removed if molecular changes in the active site are the desired target instead of, for example, subunit interactions in the tetramer assembly. While all the amino acids are aggregated together as standard residues, the cofactor (typically delineated separately in the file as the B_{12} portion and the Ado portion), the substrate, and crystal waters therein are clumped together as non-standard residues. The non-standard residues should be inspected closely for a variety of characteristics related mostly to geometry.

First, one should check whether the crystal structure is representative of intact AdoCbl or representative of the moment after Co—$C_{5'}$ bond cleavage (the general molecular structure of Cbls is shown in Fig. 2). It is also necessary to ascertain whether the Cbl is Class I or Class II, i.e., whether the DBI base connected to the nucleotide loop or one of the standard His residues is present as lower axial ligand to Co ion (Banerjee, 1999; Banerjee & Ragsdale, 2003). Determining whether the Cbl is in the base-on or base-off/His-on configuration is essential for proper partitioning and verifying that the protonation step properly determines the protonation state of the DBI or His which is bound on the alpha face to the Co. The Co—$N_{\text{DBI or His}}$ bond length should also be checked as the X-ray method notoriously leads a Co—$N_{\text{DBI or His}}$ bond that is too long. The Ado ligand contains a ribose ring and an adenine portion connected via a *N*-glycosidic bond (Fig. 3). For DDH, rotation around the *N*-glycosidic bond is a key player in the hydrogen abstraction step. So one should be aware of the specific orientation of the adenine portion of the Ado group relative to the ribose ring. Additionally, the various conformations of the ribose portion of the Ado ligand can be captured (Fig. 4). For instance, in the case of coenzyme B_{12}-dependent GLM, the crystal structures contain two orientations of the ribose portion of the Ado ligand, namely, 2′-endo and 3′-endo. Dependent on what the research target is, the unwanted Ado ligand conformation should be removed from the .pdb file. Additionally, contaminants from the crystallization process such as unnecessary metal and organic ions can also be removed for the QM/MM study via the visualization software or manually. At this step, one can also decide whether to remove the crystal waters in the file (the crystal waters can also be removed automatically in the next step involving protonation via the PDB2PQR web service)

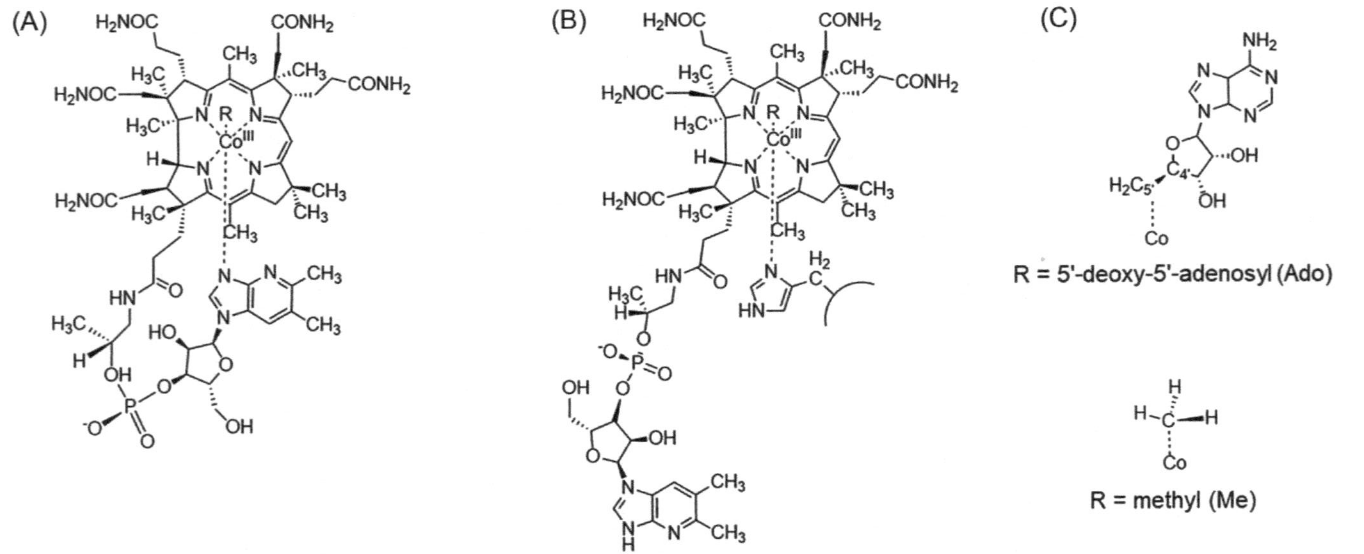

Fig. 2 General molecular structure of (A) base-on Cbls, (B) base-off/His-on Cbls, and (C) upper ligands of the biologically active Cbls described in text.

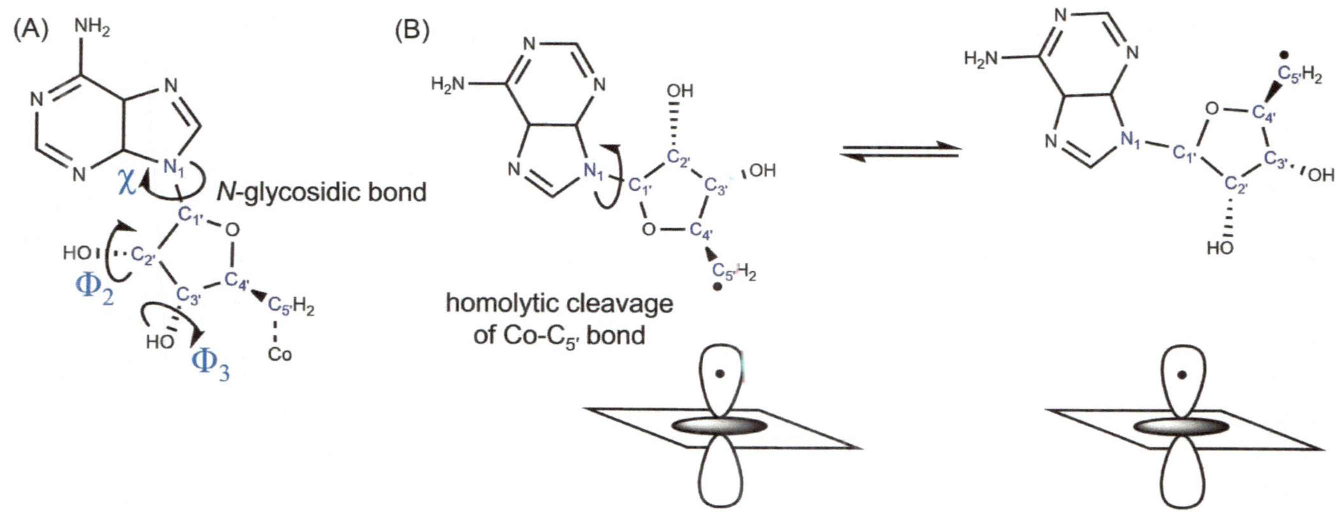

Fig. 3 (A) Atom notation used in determining ribose conformation (highlighted in blue) and key rotation angles designated. (B) RP formation after Co—$C_{5'}$ bond cleavage and subsequent rotation around the *N*-glycosidic bond in B_{12}-dependent isomerases.

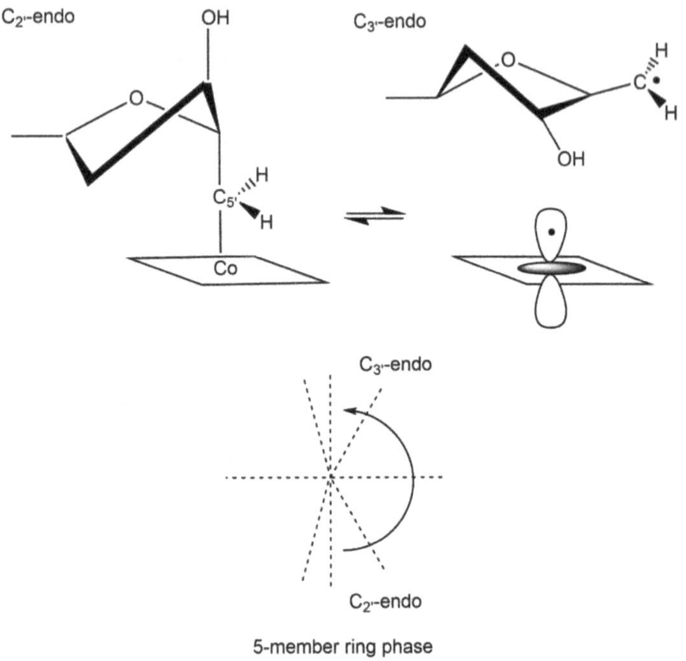

Fig. 4 2′-endo to 3′-endo conformational switch after Co—$C_{5'}$ bond cleavage in AdoCbl-dependent mutases.

(Jurrus et al., 2018; Olsson, Søndergaard, Rostkowski, & Jensen, 2011). For reference, key geometric parameters for several B_{12}-dependent enzymes are listed in Tables 1 and 2.

Next, the protonation states of titratable residues should be determined. There are a variety of ways that this can be achieved. One straightforward way is to use the PDB2PQR webserver to determine protonation at a desired pH using PROPKA (Jurrus et al., 2018; Olsson et al., 2011). The webserver provides several options for forcefields and, additionally, the option to use a user-defined force field. The software allows the user to select a variety of additional options including (1) Ensure that new atoms are not rebuilt too close to existing atoms; (2) Optimize the hydrogen bonding network; (3) Assign charges to the ligand specified in a MOL2 file; (4) Create an APBS input file; (5) Add/keep chain IDs in the PQR file; (6) Insert whitespaces between atom name and residue name, between x and y, and between y and z; (7) Make the protein's N-terminus neutral (requires PARSE forcefield); (8) Make the protein's C-terminus neutral (requires PARSE forcefield); (9) Remove the waters from the output file.

Table 1 Geometric parameters for QM/MM optimized AdoCbl structure bound to CarH, EAL, GLM, and MCM.

	Optimized structure				Crystal structure			
					CarH	EAL	GLM	MCM
	CarH	EAL	GLM	MCM	5C8E	3ABS	1I9C	6OXC
Bond Length (Å)								
Co-$C_{5'}$	2.00	2.05	2.08	2.05	2.03	2.01	3.17	2.53
Co-N_{Im}	2.28	2.37	2.27	2.22	2.25	2.61	2.22	2.27
Co-N_{21}	1.92	1.93	1.92	1.89	1.88	1.82	1.88	1.89
Co-N_{22}	1.93	1.96	1.94	1.94	1.92	1.85	2.02	1.92
Co-N_{23}	1.94	1.94	1.94	1.93	1.92	1.83	1.93	1.89
Co-N_{24}	1.89	1.91	1.90	1.91	1.89	1.83	1.90	1.90
Bond Angle (°)								
$C_{5'}$-Co-N_{Im}	174.8	171.5	163.8	177.1	166.7	177.1	170.5	177.9
N_{21}-Co-$C_{5'}$	82.6	92.3	101.9	91.0	94.5	89.4	90.8	95.8
N_{22}-Co-$C_{5'}$	87.8	94.7	88.3	88.9	88.9	90.4	80.4	98.5
N_{23}-Co-$C_{5'}$	97.7	93.1	81.8	92.6	86.6	95.1	82.5	86.8
N_{24}-Co-$C_{5'}$	89.9	84.4	92.0	87.5	83.7	84.7	92.4	87.9
Torsional Angle (°)								
N_{21}-N_{22}-N_{23}-Co	−0.1	−4.4	−2.5	−2.4	−0.7	−3.2	5.3	−2.5
N_{21}-N_{22}-N_{23}-N_{24}	−1.1	−3.3	−3.1	−4.9	−6.1	−6.2	−0.1	1.8
Co-N_{22}-C_9-C_{10}	−11.9	−3.9	−10.2	−3.2	−6.4	−11.5	−9.4	6.0
Co-N_{22}-C_6-C_5	−5.9	−1.0	3.3	−2.2	−12.6	−2.1	8.7	−23.7
Co-N_{23}-C_{14}-C_{15}	−2.4	−9.8	−3.4	−15.2	−8.8	−19.0	−28.3	−35.4

Corresponding values from the crystal structures included.

These should be de-selected/selected dependent on the program that will be used to implement the QM/MM job and the desired project. Once the protonation is complete, a PQR file will be available for download.

One thing to look out for is, since the force field does not contain charges for non-standard residues such as B_{12}, these residues may get automatically deleted from the obtained PQR output file. Should this occur, the non-standard residues can be replaced into the PQR file from the original PDB file through a text editor. The PQR file from the webserver can finally

Table 2 Exocyclic and endocyclic of the Ado ligand for QM/MM optimized models of CarH, EAL, GLM, and MCM with the corresponding values from crystal structures.

		Optimized structure				Crystal structure			
						CarH	EAL	GLM	MCM
		CarH	EAL	GLM	MCM	5C8E	3ABS	1I9C	6OXC
Exocyclic Angle (°)									
Φ_2	$C_{1'}$-$C_{2'}$-$O_{2'}$-$O_{2'}H$	88.11	−30.25	167.33	158.08	175.17	N/A	170.80	59.76
Φ_3	$C_{2'}$-$C_{3'}$-$O_{3'}$-$O_{3'}H$	1.00	9.11	−6.68	51.55	63.00	N/A	173.50	51.25
χ_{CN}	$O_{1'}$-$C_{1'}$-$N_{1'}$-C_8	87.44	−72.46	−11.57	−87.09	75.85	N/A	17.04	−75.55
Endocyclic Angle (°)									
θ_0	$C_{1'}$-$C_{2'}$-$C_{3'}$-C_4	30.49	9.87	1.64	−36.51	15.79	N/A	−45.21	−37.33
θ_1	$C_{2'}$-$C_{3'}$-$C_{4'}$-O_1	−35.99	−35.86	−27.05	−21.93	−33.93	N/A	29.28	38.10
θ_2	$C_{3'}$-$C_{4'}$-$O_{1'}$-C_1	26.53	49.72	44.01	2.88	40.10	N/A	0.73	−20.81
θ_3	$C_{4'}$-$O_{1'}$-$C_{1'}$-C_2	−6.53	−43.30	−42.76	−26.99	−29.89	N/A	−29.83	−0.45
θ_4	$O_{1'}$-$C_{1'}$-$C_{2'}$-C_3	−15.98	19.54	23.88	39.58	7.61	N/A	46.65	25.33
P	Pseudorotation Phase	29.49	78.48	87.91	157.07	66.45	N/A	152.70	196.09
θ_m	Pseudorotation Amplitude	35.03	49.42	52.90	39.64	39.53	N/A	51.37	38.85
	Ribose Conformation	3′-endo	3′-endo	3′-endo	2′-endo	3′-endo	N/A	2′-endo/3′-endo	2′-endo

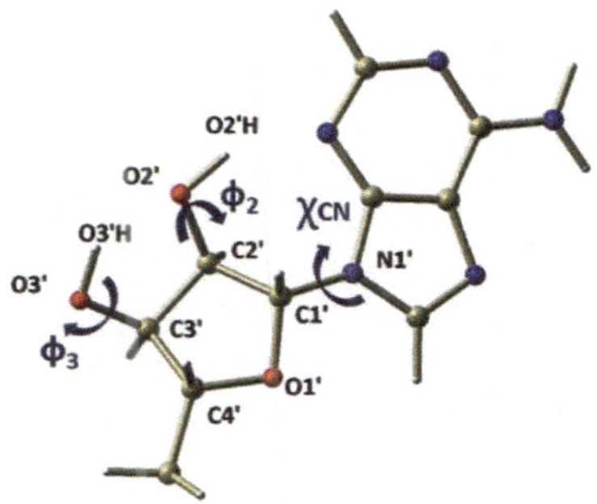

$$P = arctan\left(\frac{\theta_2 + \theta_4 - \theta_1 - \theta_3}{2\theta_0(\sin 36° + \sin 72°)}\right) + X$$

$$\theta_m = \frac{\theta_0}{\cos P}$$

$$\theta_0 = C1' - C2' - C3' - C4'$$
$$\theta_1 = C2' - C3' - C4' - O1'$$
$$\theta_2 = C3' - C4' - O1' - C1'$$
$$\theta_3 = C4' - O1' - C1' - C2'$$
$$\theta_4 = O1' - C1' - C2' - C3'$$

Note: For EAL (PDB ID: 3ABS) the cofactor is adeninylpentylcobalamin (AdePeCbl) instead of AdoCbl. AdePeCbl does not contain a ribose ring. Pseudorotation phase (P) and amplitude (Θm) as well as the conformation of ribose included. Inset: Structure of Ado ligand with glycosyl rotation angle $\chi CN = O_{1'}\text{-}C_{1'}\text{-}N_{1'}\text{-}C_{8'}$, exocyclic ribose angle $\Phi 2 = C_{1'}\text{-}C_{2'}\text{-}O_{2'}\text{-}O_{2'}H$, and exocyclic ribose angle $\Phi 3 = C_{2'}\text{-}C_{3'}\text{-}O_{3'}\text{-}O_{3'}H$. If $\Theta_0 > 0$ and $\Theta_2 + \Theta_4 - \Theta_1 - \Theta_3 > 0$, $X = 0°$. If $\Theta_0 > 0$ and $\Theta_2 + \Theta_4 - \Theta_1 - \Theta_3 < 0$, then $X = 360°$. If $\Theta_0 < 0$ then $X = 180°$. P values between $-90°$ and $90°$ correspond to the $3'$-endo conformation. P values between $90°$ and $270°$ correspond to the $2'$-endo conformation.

be viewed in Chimera (Pettersen et al., 2004) and saved as a .pdb file. For convenience of operation, since the different strands of non-standard residues aren't automatically given a designation such as Chain A, B, etc., in the saved PDB file the user can define for themselves chain labels for components such as B_{12}, Ado, or the substrate to distinguish these from the standard residues. This can again be accomplished in the text editor and will certainly aid in visualization as well as in figure preparation. Finally, if the non-standard residues were not protonated in the original PDB file, they can be protonated using the Chimera interface via the minimization process. In fact, one can choose to protonate the entire system using this function instead of the PropKa webserver. As a point of caution, protons that might be inserted at unreasonable sites in the molecule, such as those attached to the equatorial nitrogen atoms of corrin-ring, should be removed.

2.3 Minimization

Regardless of how the user chooses to protonate the model, the protonated structure should be subjected to minimization prior to beginning QM/MM calculations. Additionally, one can choose to perform MD simulations at this step to further optimize the protein environment, if necessary. Nevertheless, the protein minimization step can also be implemented within Chimera. In the studies of the B_{12}-dependent enzymes which will be described below, the MM level of theory using the Amber ff99sb force field was used for the minimization step (Lindorff-Larsen et al., 2010). In Chimera, the Co atom should be temporarily removed while the minimization is running. The missing Co atom can be replaced into the minimized structure after the job has completed. The resulting structure should be inspected with a visualization software such as Chimera (Pettersen et al., 2004) or GaussView (Roy Dennington & Millam, 2019) to ensure all atoms are present and to determine if the cofactor has been correctly protonated. The user should also inspect residues such as arginine (Arg), histidine (His), lysine (Lys), aspartic acid (Asp), glutamic acid (Glu), tyrosine (Tyr), etc. to ensure that the expected protonation state was achieved.

2.4 Classical MD simulation

The crystal structure used for QM/MM model preparation may suffer from artifacts because of crystal lattice packing of certain atoms/organic groups or an absence of a part of the cofactor or substrate. Such unwanted artifacts can be corrected by running an appropriate MD simulation with necessary corrections to the protein structure which allows for relaxation to a more favorable conformation (Hollingsworth & Dror, 2018). For instance, the crystal structure of PDB ID: 4REQ was chosen to investigate Co—$C_{5'}$ reductive bond cleavage of AdoCbl-dependent MCM enzyme (Kumar, Bucher, & Kozlowski, 2019), and the structure consists of a mixture of both the substrate and product and the Ado moiety is already separated from the Co ion of B_{12} macrocycle. While preparing structural model from the 4REQ crystal structure, the ribose conformation in the dissociated Ado moiety was considered problematic since the stability of ribose conformation changes at longer Co—C bond length. Thus, as a good initial guess regarding 4REQ structural model, the Ado moiety was assumed to be bonded to the Co ion, but the conformation of ribose was extracted from 3REQ crystal structure where the Ado group remains bonded to Co ion. The active site was prepared by superimposing the ribose part of the 3REQ structure with 4REQ and retaining the substrate from crystal structure. Finally, the prepared active site was employed to perform standard MD calculation. Based on root mean square deviation (RMSD) analysis using Visual Molecular Dynamics (VMD) (Humphrey, Dalke, & Schulten, 1996) program, certain snapshots of more stable structures were chosen for the QM/MM calculation setup.

2.5 Partitioning of cofactor and QM/MM setup

Although large biomolecular systems like B_{12} enzymes consist of several thousand atoms, electronic changes involved in bond-breakage and formation are typically limited to a small part of the enzyme, namely, the active site. Thus, the molecular structure can be partitioned into different "layers" in a divide and conquer approach where each layer is treated with a different level of theory. The electronically prominent region, which typically includes the key components of the active site such as the cofactor and substrate, is treated with a QM method and is referred to as the high layer (HL). The low layer (LL) is typically significantly larger than the HL and contains the protein residues and potentially some portions of a large substrate of cofactor. The LL is typically treated with a classical MM method. In addition, a middle layer might be defined depending on the system. The ML layer might be treated with a semi-empirical method such as PM6. For example, in the QM/MM study of EAL which will be discussed later, three layers were used.

In B_{12} systems, the HL typically consists of the corrin macrocycle along with the central Co ion, the upper ligand (Ado or Me depending on the enzyme) and the Im portion of lower axial ligand (Fig. 5) (Ghosh, Mamun, Lodowski, Jaworska, & Kozlowski, 2018; Ghosh, Toda, & Kozlowski, 2021; Mamun et al., 2019; Toda, Lodowski, Mamun, & Kozlowski Pawel, 2021; Toda, Mamun, Lodowski, & Kozlowski, 2020). If desired, the HL can also contain a substrate or inhibitor molecule in an enzyme which is substrate-bound (Fig. 6). The Im may be from a His residue from the enzymatic environment in the base-off/His-on conformation. For example, a His residue is the lower ligand in MCM, GLM, and CarH. Alternatively, the Im can be a portion of the DBI ligand, which is the lower axial ligand in base-on systems such as EAL. As a side note, the user should ensure that all His residues are properly protonated and one should be familiar with the AMBER notation for His residues. There are three possibilities for the protonation states of His and are defined as HID, HIE, and HIP where the delta, epsilon, or both nitrogens are protonated, respectively.

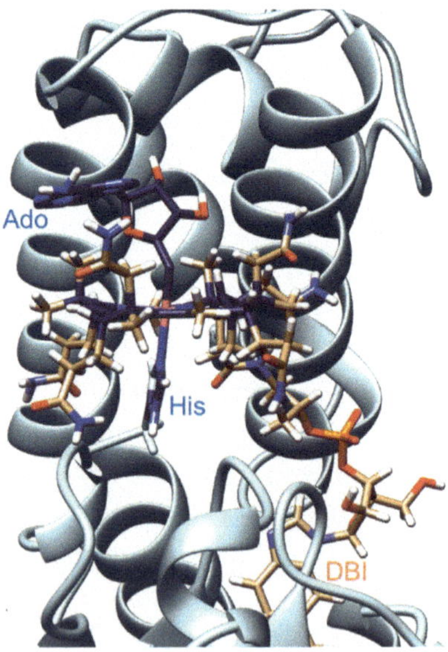

Fig. 5 QM/MM partitioning in base-off/His-on AdoCbl-dependent systems. The HL contains the Ado ligand, Co, corrin ring, and the Im portion of the lower axial histidine (highlighted in navy). The side chains attached to the macrocycle and the nucleotide loop ending in the DBI ligand are highlighted in tan. These can be placed in either the ML or LL. The protein environment is depicted here in ribbon form and these residues can be placed in the LL.

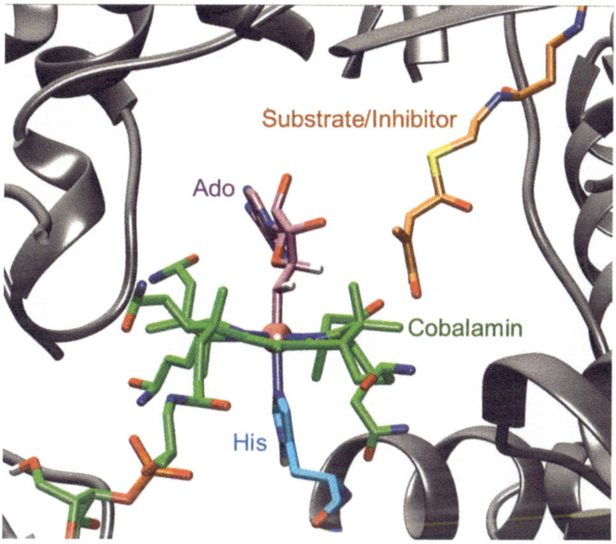

Fig. 6 QM/MM partitioning in base-off/His-on AdoCbl-dependent system with substrate (PDB ID: 6OXD used for representation). The HL should contain, at a minimum, the Ado ligand (purple), Co, corrin ring (green), the Im portion of the lower axial histidine (highlighted in blue), and the portion on the substrate or inhibitor that is being investigated. The side chains attached to the macrocycle can be placed in the HL, ML, or LL depending on the research question. The protein environment is depicted here in ribbon form (gray) and the majority of these residues should be placed in the LL while residues that may be involved in catalysis might be placed in the HL.

Typically in .pdb files, His residues are labeled as HIS, however, if the Cbl is in the base-off/His-on form, the His which is the lower ligand should be HID. The user may also choose to include standard residues in the HL that are suspected to be involved in bond breaking/forming reactions in the active site.

The LL is made of that part of the molecular system where no formation and breaking of bonds takes place and thus typically consists of all the standard residues (except of course the Im of lower axial HID), the side chains of the corrin ring including the nucleotide loop (and the terminal DBI attached to the loop) as well as the part of the substrate far from active site. However, if DBI moiety acts as lower ligand to Co ion as in EAL, then only the five-membered imidazole section of the moiety forms part of the QM layer. The remainder of the DBI ligand can be imparted into the ML or LL layer. Similarly, for a His acting as lower ligand, the boundary between the QM and MM layers is between C_β and C_γ atoms. The scheme of truncation of the cofactor is accomplished with the help of the "Atom groups/Atom selection" options inside the GaussView GUI. In ONIOM, the QM-MM boundary created by cutting the covalent bond during truncation of the cofactor is treated by the link-atom method. In this methodology, the free valence at QM atom is capped by an additional atom, usually H, covalently bonded to the QM atom and is treated at the QM level of theory.

Once the partitioning has been determined a few additional things should be checked. In the ONIOM approach atoms can be selected to be frozen or unfrozen. In other words, the user can select certain regions to remain fixed throughout the optimization. In most of the QM/MM studies that will be described here, all residues and components of the cofactor that are within 20 Å from Co ion are allowed to move whereas the rest of the protein part remains frozen throughout the course of the optimization (Fig. 7).

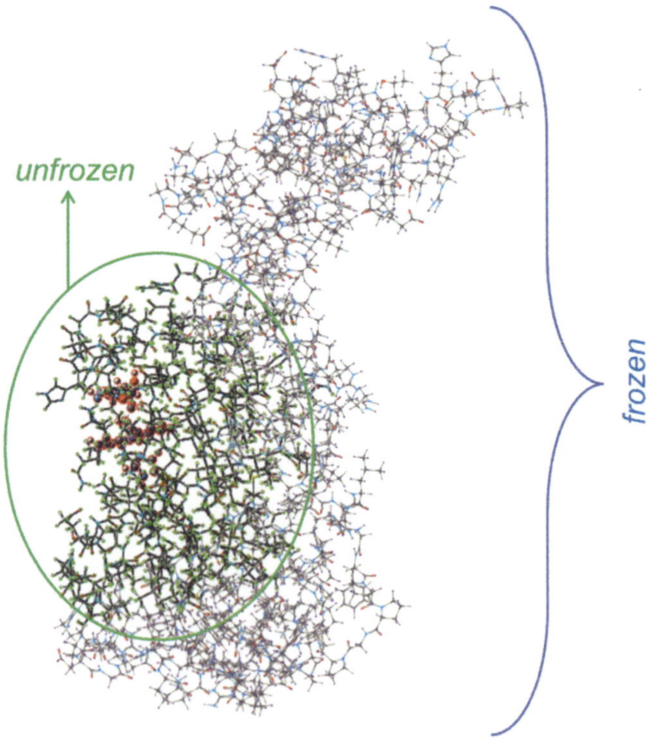

Fig. 7 Molecular structure of CarH QM/MM input with the unfrozen and frozen portions of the enzyme depicted. Atoms within 20 Å of the Co are kept unfrozen while atoms >20 Å away from the Co are kept frozen.

The final step in the preparation of the ONIOM input is to ensure that the charges of each layer are properly determined. The user should also count all charged amino acid residues to ensure that the charges in the LL are properly accounted for.

2.6 Computational methodology

Density functional theory (DFT) and its counterpart, time-dependent density functional theory (TD-DFT) used to study electronic excitations, are commonly employed QM methods for their ease in implementation as well as their reasonable balance between computational cost and accuracy. Computationally, it has been reported that DFT is best suited to study QM layer of B_{12} systems in terms of both speed and accuracy (Toda, Kozlowski, & Andruniów, 2019; Toda, Lodowski, et al., 2019). The DFT functional employed for investigating various aspects of B_{12} catalysis and photochemistry must uphold utmost accuracy in determining the Co—C bond dissociation energy (BDE). Since hybrid DFT functionals, including B3LYP, underestimate the experimentally observed Co—C BDE they are not recommended for study of B_{12} catalysis (Kozlowski et al., 2012; Toda, Kozlowski, & Andruniów, 2019; Toda, Lodowski, et al., 2019). Rather, pure GGA's functionals are recommended for B_{12}-dependent systems and, to date, BP86 is the most widely used and accepted functional for B_{12} studies. The use of BP86 is discussed extensively elsewhere and the curious reader can consult these citations for a more thorough discussion (Toda, Kozlowski, & Andruniów, 2019; Toda, Lodowski, et al., 2019). A TZVPP basis set for Co, C, N, and O atoms and a TZVP basis set for H are reliable for B_{12} systems. The LL is treated with MM and many B_{12} studies have employed the ff99sb force field (Lindorff-Larsen et al., 2010). It is expected that future studies of Cbl systems might explore the use of newer force fields. As a side note, AMBER parameters should be properly added to the input depending on the implementation of the QM/MM calculation. We recommend the TAO package as an excellent resource for adding AMBER parameters or missing AMBER parameters to the QM/MM input (Tao et al., 2009). AMBER parameters for the AdoCbl cofactor can be obtained from Marques et al. (Marques, Ngoma, Egan, & Brown, 2001).

For a typical B_{12} system partitioned into two layers, subtractive QM/MM as implemented in ONIOM involves three calculations and the final energy of the entire system is given by following equation,

$$E_{total} = E_{MM,real} + E_{QM,model} - E_{MM,model} \quad (1)$$

First, energy of the overall real system is calculated at MM level of theory. Then the energy of model system is calculated both at QM and MM levels of theory. Finally, to avoid double counting, energy of model system calculated at MM level of theory is subtracted. The subtractive QM/MM scheme is very straightforward in terms of implementation. Specifically, no explicit QM-MM coupling terms are needed since, in this scheme, automatic mechanical embedding (ME) is employed. In ME, all the bonded (stretching, bending and tortional) and non-bonded (electrostatic and van der Waals) interactions between the two layers are treated at the MM level of theory. Successful implementation of ME requires a highly accurate set of MM parameters in terms of atom-centered point charges for all the atoms including the active site QM layer. As the reaction progresses, charge distribution in QM layer undergoes a change and thus calculated PES may accrue serious error if the same set of MM parameters are used throughout. The problem is compounded by the fact that ME forbids the perturbation of QM layer charge distribution by QM-MM boundary interactions. The drawback can be dealt effectively in two ways: one may, either, calculate the atom centered point charges for the QM layer atoms as the system evolves along the reaction path or perform a single point electrostatic embedding (EE) along the PES (Chung et al., 2015). The derivation of electrostatic potential (esp) charges and fitting it over QM layer atoms all along PES increases the computational cost significantly and remains impractical at best. Whereas in EE scheme QM calculation on model system is carried out in presence of rest of the system by incorporating certain terms in the Hamiltonian as one-electron operators that accounts for electrostatic interaction between the two layers such that electrostatistics in the LL polarize the wavefunction in the QM layer. Herein, all the bonded and non-bonded interactions, except of course electrostatic, between the two layers are treated at MM level of theory again. EE is implemented with ONIOM = EmbedCharge in single point or optimization after successful mechanical embedding calculation.

One of the advantages of the ONIOM approach is that calculations can be carried out using multiple layers. In case of a three-layer calculation, further simplification of the active site is rendered utilizing an additional middle layer which may contain all the side chains of corrin macrocycle. The medium layer, formed of middle and small layers, can be treated with MM or an inexpensive semiempirical QM level of theory such as PM6. For a three-layer ONIOM calculation, the energy of the entire system is given by the following equation,

$$E_{total} = E_{MM,real} + E_{SQM,medium} + E_{QM,model} - E_{MM,medium} - E_{SQM,model} \quad (2)$$

where real denotes the whole biomolecular system (i.e., full 3-layer system), medium denotes the two innermost layers and model denotes only the innermost layer.

3. Model systems

3.1 EAL

While the cleavage of the Co—C bond in either MeCbl- or AdoCbl-dependent enzymes involves substrate binding, the Co—C bond can also be cleaved with light. Specifically, for AdoCbl-dependent enzymes, the cleavage of the Co—$C_{5'}$ bond of AdoCbl results in RP formation regardless of whether the activation is via substrate binding or light. This leads to an important question, is there a connection that can be drawn between photolytic cleavage of the Co—$C_{5'}$ bond and cleavage involving substrate binding? or can photochemical investigations of B_{12}-dependent systems be used to probe key unanswered questions in native B_{12}-catalysis? The coenzyme B_{12} dependent enzyme, EAL provides an excellent test case with which to examine these questions (Mamun et al., 2019).

The mechanism of RP formation in EAL via photolytic cleavage was proposed by Kozlowski and coworkers (Mamun, Toda, Lodowski, Jaworska, & Kozlowski, 2018), and a connection between photolysis and native enzymatic catalysis was proposed based on QM/MM calculations (Mamun et al., 2019). The S_1 PES was used to determine the mechanism of photodissociation in EAL and the process for constructing such PESs is described in detail in these references. Two possible routes for photodissociation (i.e., RP formation) were identified and called Path A and Path B. Path A involves the initial elongation of the Co—$C_{5'}$ bond followed by elongation of the Co—N_{Im} bond which results in the population of the LF region of the S_1 PES. Path B involves the initial elongation of the Co—N_{Im} bond followed by the Co—$C_{5'}$ bond. Both pathways result in the LF region of the S_1 PES, where RP formation occurs. The active pathway is determined based on the relative energetics between the MLCT minimum and the LF region and in EAL, Path B favored over Path A.

The TD-DFT results led to a different interpretation of the model proposed by Robertson and Warncke (Robertson, Wang, & Warncke, 2011). The geometric coordinates involved in photolysis of both AdoCbl in solution and in the enzymatic environment are the axial bond lengths. This cooperativity between the axial bond distances and photodissociation was also observed in XANES measurements and with several other theoretical studies. On the basis of this evidence, it was proposed that the second coordinate should be the Co—N_{DBI} bond distance, instead of protein configuration, for photochemical formation of the Co(II)/Ado$^{\bullet}$ RP. Nevertheless, no direct connection between RP formation via photolysis and RP formation via the native catalytic cycle was immediately apparent.

In a follow-up computational analysis, determining whether photodissociation could be a model or mimic for the native catalytic cycle was the main target (Mamun et al., 2019). A combined QM/MM approach was used to conclude that there is no direct comparison that can be made between photolysis and enzymatic cleavage because different electronic states are associated with each mechanism. This led to exploring the hypothesis that the connection between photolysis and the native catalytic cycle involves the one-electron reduced form of the AdoCbl cofactor. Accordingly, the S_1 and one-electron reduced (D_1) PESs were compared (Fig. 8). The S_0 and D_1 PES exhibited similar topology with two minima regions located in each. For the S_1 PES, the photolysis pathway involves the MLCT to LF regions. For the D_1 PES, the path towards Co—$C_{5'}$ bond cleavage involves traversing through two minima regions that are characterized as π^* and σ^*. So, the S_1 and D_1 PESs are not only similar in that both contain two minima regions but also in that the character of the electronic states of each region is similar.

3.2 Mutases

AdoCbl-dependent mutases have been the topic of many experimental and computational studies and we will summarize some key findings here. Many proposals have been put forth regarding the labilization of the Co—$C_{5'}$ bond, and these studies tend to focus on GLM and MCM, which are both Class I (base-off/His-on) isomerases (Dölker, Maseras, & Siegbahn, 2004; Jensen & Ryde, 2005; Kozlowski, Kamachi, Toraya, & Yoshizawa, 2007; Kumar et al., 2019; Kwiecien et al., 2006; Sharma et al., 2007). As discussed throughout, the major focus of these studies is to unearth the origin of the catalytic power of AdoCbl-dependent enzymes that results in the trillion-fold rate enhancement for cleavage of the Co—$C_{5'}$ bond in the enzyme compared to the isolated cofactor in solution. Warshel and coworkers suggest that the origin of the catalytic power in AdoCbl-dependent enzymes is associated with the electrostatic interaction between the ribose of the Ado group and the amino acid residues of the enzyme active site (Sharma et al., 2007). In addition to these, it has also been suggested that the catalytic power originates from binding the substrate molecule to the apoenzyme. It has further been postulated that the binding of the substrate triggers the radical

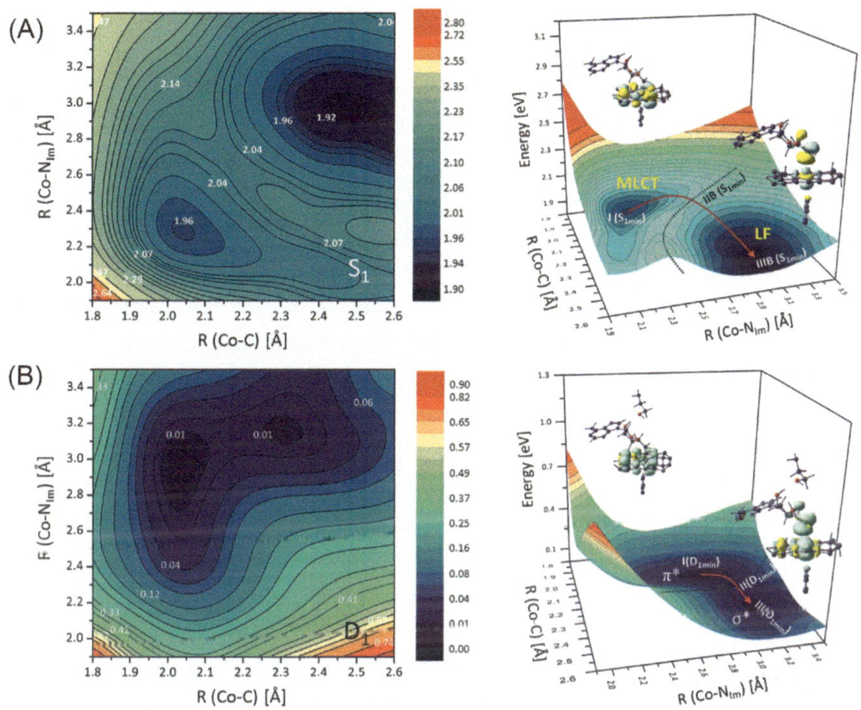

Fig. 8 The PESs of S_1 and D_1 state along with the contour representation. The relative energies were in eV and shown in the color scale. (A) The excited state (S_1) PES of the model system without the substrate molecule. TD-DFT/PM6/MM method were applied to construct the S_1 surface. Characteristic points MLCT and LF shown in the surface plot along with the spin densities. The spin density of these two states were shown right above the minima regions. (B) The doublet ground state (D_1) PES of the model system with the substrate molecule calculated using DFT/PM6/MM. One extra electron was added to reduce the cofactor of that model. The D_1 PES was also characterized as having two minima regions. *Reprinted with permission from Mamun, A. A., Toda, M. J., & Kozlowski, P. M. (2019). Can photolysis of the Co-C bond in coenzyme B_{12}-dependent enzymes be used to mimic the native reaction?* Journal of Photochemistry and Photobiology B: Biology, 191, *175–184. Elsevier.*

formation by inducing the destabilization of the Co—$C_{5'}$ bond (Mancia & Evans, 1998). However, no consensus was reached based on this mechanism. Due do the varying hypotheses and differing perspectives, it remains difficult to draw a succinct conclusion as to how substrate binding affects the catalytic activity and the unique labilization of the Co—$C_{5'}$ bond of AdoCbl-dependent enzymes.

The quest to determine the origin of the incredible catalytic power to AdoCbl-dependent enzymes received a boost in 2019 when new crystal structures for MCM (PDB ID: 6OXC and 6OXD) were deposited in the PDB (Ruetz et al., 2019). The 6OXC crystal structure does not contain a substrate. The 6OXD crystal structure contains an inhibitor (itaconyl-CoA, I-CoA), instead of the native substrate methlymalonyl-CoA. I-CoA is chemically similar to the native substrate. The inhibitor I-CoA in MCM works by a radical suicide inactivation mechanism and, like the native reaction, cleavage of the Co—$C_{5'}$ bond results in the formation of the Co(II)/Ado$^{\bullet}$ RP. The difference in the inhibitor-bound MCM and the native reaction is that in MCM-[I-CoA] is that the Ado$^{\bullet}$ forms a stable C—C bond with the I-CoA which results in a stable tertiary carbon radical.

A major challenge in studying AdoCbl radical reactions is decoupling the RP formation with the subsequent H-atom abstraction step. The primary Ado$^{\bullet}$ is highly reactive and very difficult to capture experimentally, nevertheless the major benefit of these MCM crystal structures is the capture of the air-stable diradical which is formed via cleavage of the Co—$C_{5'}$ bond. Put another way, the MCM-[I-CoA] crystal structure captures the stage in the mechanism just after bond cleavage and prior to the H-atom abstraction step. This has allowed for reinvestigation of the reaction catalyzed by MCM.

The 6OXC and 6OXD crystal structures were used in a QM/MM study where the activation of the Co—$C_{5'}$ bond was investigated by evaluation of the PECs along the reaction path (Ghosh et al., 2021). Comparison of the BDE of the Co—$C_{5'}$ of the substrate-free and I-CoA bound MCM revealed that the BDE is insensitivity to substrate binding. Specifically, the energy barrier for cleavage for substrate-free MCM and MCM-[I-CoA] is 22.2 and 22.7 kcal/mol, respectively. This result is not in-line with the previously reported hypotheses that substrate-induced conformational changes reduce the BDE in the labilization of the Co—$C_{5'}$ bond. When compared to isolated AdoCbl, the BDE for homolytic cleavage of the Co—$C_{5'}$ bond in both substrate-free MCM and MCM-[I-CoA] is lower by ~8 kcal/mol. While the results indicate that the protein environment is involved in the reduction of the Co—$C_{5'}$ BDE, the reduction is too small to fully explain the trillion-fold rate acceleration. During the homolytic cleavage of the Co—C bond in coenzyme B_{12}-dependent enzymes, the Ado group undergoes specific conformational changes, either through puckering of the ribose ring or rotation of the glycosidic bond (Banerjee & Ragsdale, 2003). In the case of MCM-[I-CoA], the ribose group undergoes a conformational change during the catalytic activation of the Co—$C_{5'}$ bond. It appears that when the Ado group is bound to the Cbl, it adopts a 2′-endo conformation, and it adopts a 3′-endo conformation after a partial cleavage of the Co—$C_{5'}$ bond

(Ghosh et al., 2021). This change in ribose conformation facilitates the migration of Ado radical during the catalytic reaction.

In the MCM-[I-CoA], the involvement of Tyr residue as a redox center in the activation of the Co—$C_{5'}$ bond has also been explored (Fig. 9). The Tyr residue was found to reduce the AdoCbl cofactor and generate the diradical state [AdoCbl]$^{\bullet}$-Tyr$^{\bullet}$ (Ghosh et al., 2021; Kozlowski, Kamachi, Kumar, Nakayama, & Yoshizawa, 2010; Kumar & Kozlowski, 2009; Kumar, Liu, & Kozlowski, 2012). Furthermore, because this diradical is the enzyme's lowest electronic state (Ghosh et al., 2021; Kozlowski et al., 2010), the Tyr105 residue may play a key role in activating the cleavage of the Co—$C_{5'}$ bond in the MCM-[I-CoA] by enabling an ET to the AdoCbl cofactor. Thus, in the case of AdoCbl-dependent enzymatic catalysis, Tyr serves as a redox center, which induces an ET to the AdoCbl, where the PT from the Tyr residue coupled with the ET process via a PCET resulting in the formation of AdoCbl/Tyr diradical. In other words, the formation of Tyr$^{\bullet}$, through the deprotonation of the phenoxyl group by the substrate/inhibitor moiety, will facilitate the transfer of an electron to the AdoCbl cofactor and form a corrin-ring-based radical anion ([AdoCbl]$^{\bullet -}$). It appears plausible that a Tyr105 residue can play a crucial role in triggering the cleavage of the Co—$C_{5'}$ bond in the MCM-[I-CoA] by facilitating an ET to the AdoCbl cofactor via a PCET mechanism.

3.3 CarH

In addition to the Cbl-dependent enzymes, an entirely new field of Cbl chemistry, namely, B_{12} photoreceptors, has emerged with the discovery

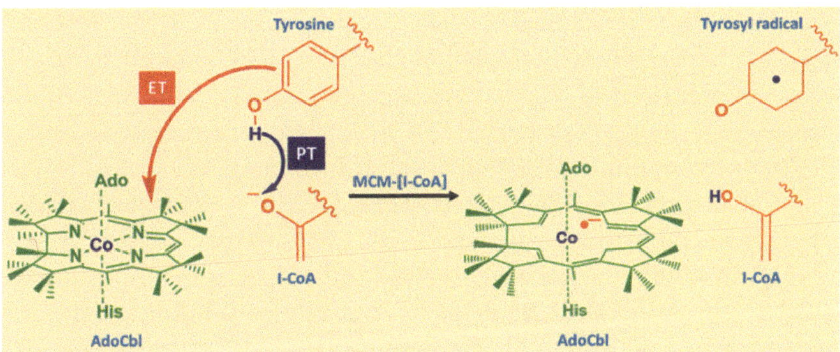

Fig. 9 PCET-Inspired mechanistic scheme for AdoCbl-dependent enzymatic catalysis with inhibitor bound MCM (MCM-[I-CoA]) as the sample system.

and X-ray structure determination of CarH (Jost et al., 2015). CarH is a coenzyme B_{12}-dependent transcription regulator which employs light to perform its catalytic function (Fig. 10) (Padmanabhan, Jost, Drennan, & Elías-Arnanz, 2017). This was an exciting finding for the B_{12} community because, although the light sensitivity of the Co—$C_{5'}$ bond in coenzyme B_{12} has been known for five decades, CarH was among the first examples of a physiological role for the light-sensing ability of AdoCbl. Computational modeling of CarH has provided deeper understanding of this complex system.

To our knowledge, the first example of a computational analysis of CarH involved MD simulations (Kutta et al., 2015). A homology model for the CarH ground state (CarH-GS) was developed which demonstrated that the active site contained base-off/His-on AdoCbl, in which the DBI ligand was displaced by His177 (Fig. 11). It was further shown that two His residues, namely, His132 and His142, were candidates for the residue which takes the place of the Ado ligand after photoproduct formation. Upon *bis*-HisCbl formation, CarH is described as CarH-light state (CarH-LS). Later it was shown, through experiment, that His132 was confirmed as the candidate for *bis*-HisCbl (Padmanabhan et al., 2017). The MD analyses also included comparison of conformational changes between a CarH-GS model system with and without the Ado ligand. It was shown that when the Ado ligand is absent, there is a large structural change in the conformation of the protein relative to its native state that causes His132 to be positioned directly above the Co ion. It was also proposed that a Glu129 may be involved in abstracting a proton from either the δ or ε nitrogen of His132 to allow for *bis*-HisCbl formation.

While QM/MM investigations of CarH and B_{12}-photoreceptors are in the early stages, one study compared CarH to other AdoCbl-dependent enzymes to determine why CarH operates with light-induced cleavage of the Co—$C_{5'}$ instead of the involvement of a substrate in the activation of the Co—$C_{5'}$ bond (Toda et al., 2020). Analyses of the crystal structures and QM/MM optimized models of the AdoCbl cofactor in EAL, GLM, and CarH revealed an important observation that various conformations may arise based on rotation around the $C_{5'}$—$C_{4'}$ single bond of AdoCbl (Fig. 12). In CarH, the hydrogen bound to the $C_{4'}$ (-$C_{4'}$—H) is directed toward the macrocycle of AdoCbl. With the -$C_{4'}$—H directed this way, it would be feasible to transfer the hydrogen to the Co(II) after Co—$C_{5'}$ bond cleavage. In EAL and GLM, the -$C_{4'}$—H is projected away from the macrocycle, which is likely to prevent against unwanted side reactions including transfer of the -$C_{4'}$—H to the Co(II).

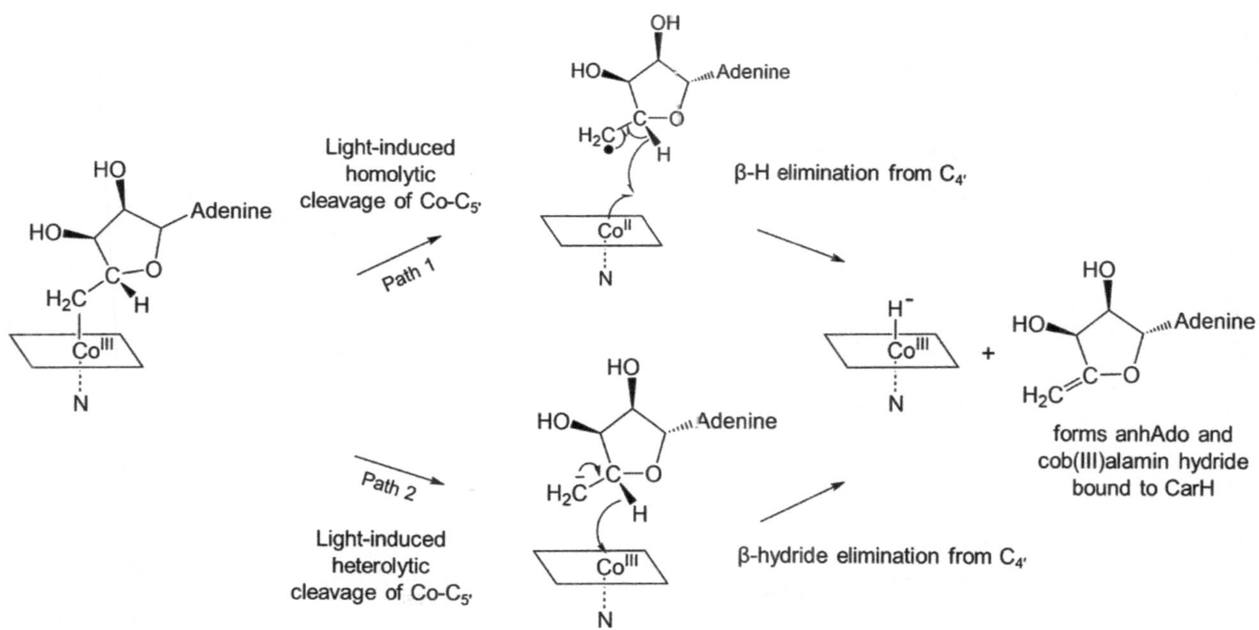

Fig. 10 Possible mechanisms for photocleavage of the Co—$C_{5'}$ bond of AdoCbl in CarH based on the proposal from Drennan and coworkers.

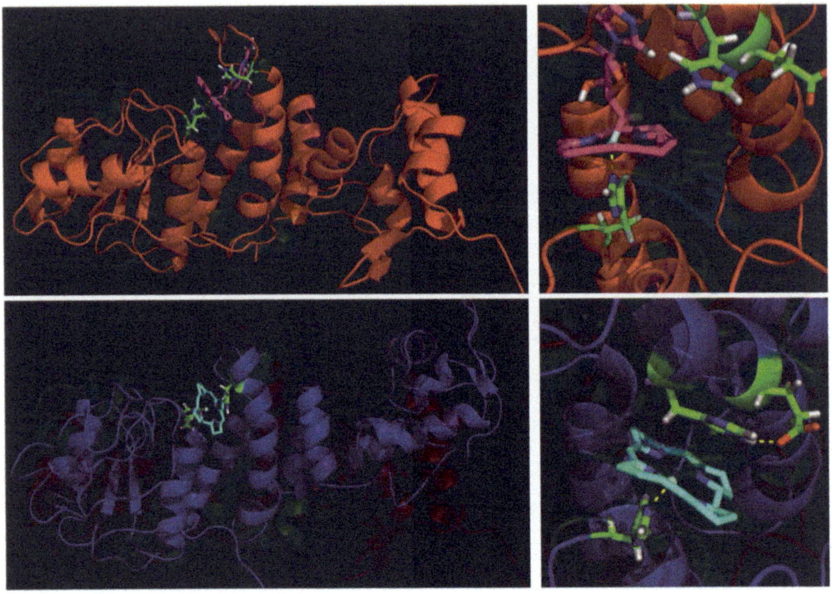

Fig. 11 MD simulations of CarH photoconversion based on a homology model. AdoCbl is represented in magenta and active site residues H177 (lower axial coordination), H132 and E129 in green. MD simulations were conducted with (red structure) and without (blue structure) the Ado ligand. *From Kutta, R. J., Hardman, S. J., Johannissen, L. O., Bellina, B., Messiha, H. L., Ortiz-Guerrero, J. M., et al. (2015). The photochemical mechanism of a B_{12}-dependent photoreceptor protein.* Nature Communications, 6, *7907. SpringerNature.*

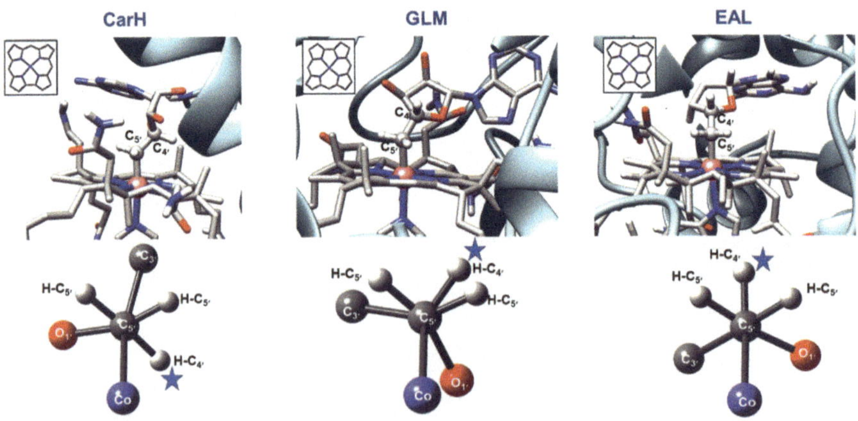

Fig. 12 Optimized ground state equilibrium structures of AdoCbl bound to CarH, GLM, and EAL (from left to right). $C_{5'}$ and $C_{4'}$ carbons and their corresponding hydrogens are represented in ball and stick form. The stereochemistry of β-lactam system along $C_{5'}$—$C_{4'}$ bond is presented in gray circle.

4. Conclusions

The application of QM/MM toward understanding enzymatic reactions has led to many developments since the seminal achievement of the 1970s. QM/MM is an excellent method for studying the complex enzymatic of B_{12}-dependent systems and many strides have been made in understanding these complex systems. While many hypotheses related to the trillion-fold rate acceleration have been tested, it seems that still there is much more to understand about the origin of this feat of nature. The elucidation of new high-resolution crystal structures such as the recent MCM structures have allowed for reinvestigation of AdoCbl-dependent enzymes, especially as the cofactor itself is well-resolved in the new structures (Ghosh et al., 2021; Ruetz et al., 2019). Furthermore, now that the elusive $Ado^{\bullet}$ has been detected in a SAM-dependent enzyme (Yang et al., 2019), future QM/MM studies of SAM-dependent enzymes will no doubt be forthcoming.

References

Banerjee, R. (1997). The Yin-Yang of cobalamin biochemistry. *Chemistry & Biology*, *4*, 175–186.

Banerjee, R. (1999). In R. Banerjee (Ed.), *Chemistry and biochemistry of B_{12}*. New York: Wiley.

Banerjee, R., & Ragsdale, S. W. (2003). The many faces of vitamin B_{12}: Catalysis by cobalamin-dependent enzymes. *Annual Review of Biochemistry*, *72*, 209–247.

Brown, K. L. (2005). Chemistry and enzymology of vitamin B_{12}. *Chemical Reviews*, *105*, 2075–2150.

Brunk, E., Kellett, W. F., Richards, N. G. J., & Rothlisberger, U. (2014). A mechanochemical switch to control radical intermediates. *Biochemistry*, *53*, 3830–3838.

Carloni, P., Rothlisberger, U., & Parrinello, M. (2002). The role and perspective of ab initio molecular dynamics in the study of biological systems. *Accounts of Chemical Research*, *35*, 455–464.

Chung, L. W., Sameera, W. M. C., Ramozzi, R., Page, A. J., Hatanaka, M., Petrova, G. P., et al. (2015). The ONIOM method and its applications. *Chemical Reviews*, *115*, 5678–5796.

Dölker, N., Maseras, F., & Siegbahn, P. E. M. (2004). Stabilization of the adenosyl radical in coenzyme B_{12} – A theoretical study. *Chemical Physics Letters*, *386*, 174–178.

Dolphin, D. (1982). *B_{12} volume 1: Chemistry (vol. 1)*. New York: John Wiley & Sons.

Finke, R. G., & Martin, B. D. (1990). Coenzyme $AdoB_{12}$ vs $AdoB_{12}$.-homolytic co-C cleavage following electron transfer: A rate enhancement greater than or equal to 10^{12}. *Journal of Inorganic Biochemistry*, *40*, 19–22.

Ghosh, A. P., Mamun, A. A., Lodowski, P., Jaworska, M., & Kozlowski, P. M. (2018). Mechanism of the photo-induced activation of co-C bond in methylcobalamin-dependent methionine synthase. *Journal of Photochemistry and Photobiology B: Biology*, *189*, 306–317.

Ghosh, A. P., Toda, M. J., & Kozlowski, P. M. (2021). What triggers the cleavage of the co–$C_{5'}$ bond in coenzyme B_{12}-dependent itaconyl-CoA methylmalonyl-CoA mutase? *ACS Catalysis*, *11*, 7943–7955.

Gruber, K., Reitzer, R., & Kratky, C. (2001). Radical shuttling in a protein: Ribose pseudorotation controls alkyl-radical transfer in the coenzyme B_{12} dependent enzyme glutamate mutase. *Angewandte Chemie International Edition, 40*, 3377–3380.

Hay, B. P., & Finke, R. G. (1986). Thermolysis of the cobalt-carbon bond of adenosylcobalamin. 2. Products, kinetics, and cobalt-carbon bond dissociation energy in aqueous solution. *Journal of the American Chemical Society, 108*, 4820–4829.

Hay, B. P., & Finke, R. G. (1987). Thermolysis of the cobalt-carbon bond in adenosylcorrins. 3. Quantification of the axial base effect in adenosylcobalamin by the synthesis and thermolysis of axial base-free adenosylcobinamide. Insights into the energetics of enzyme-assisted cobalt-carbon bond homolysis. *Journal of the American Chemical Society, 109*(26), 8012–8018.

Hollingsworth, S. A., & Dror, R. O. (2018). Molecular dynamics simulation for all. *Neuron, 99*, 1129–1143.

Honig, B., & Karplus, M. (1971). Implications of torsional potential of retinal isomers for visual excitation. *Nature, 229*, 558–560.

Humphrey, W., Dalke, A., & Schulten, K. (1996). VMD: Visual molecular dynamics. *Journal of Molecular Graphics, 14*, 33–38.

Jensen, K. P., & Ryde, U. (2005). How the co − C bond is cleaved in coenzyme B_{12} enzymes: A theoretical study. *Journal of the American Chemical Society, 127*, 9117–9128.

Jones, A. R. (2017). The photochemistry and photobiology of vitamin B_{12}. *Photochemical & Photobiological Sciences, 16*, 820–834.

Jost, M., Fernandez-Zapata, J., Polanco, M. C., Ortiz-Guerrero, J. M., Chen, P. Y., Kang, G., et al. (2015). Structural basis for gene regulation by a B_{12}-dependent photoreceptor. *Nature, 526*, 536–541.

Jurrus, E., Engel, D., Star, K., Monson, K., Brandi, J., Felberg, L. E., et al. (2018). Improvements to the APBS biomolecular solvation software suite. *Protein Science, 27*, 112–128.

Kozlowski, P. M., Garabato, B. D., Lodowski, P., & Jaworska, M. (2016). Photolytic properties of cobalamins: A theoretical perspective. *Dalton Transactions, 45*, 4457–4470.

Kozlowski, P. M., Kamachi, T., Kumar, M., Nakayama, T., & Yoshizawa, K. (2010). Theoretical analysis of the diradical nature of adenosylcobalamin cofactor − tyrosine complex in B12-dependent mutases: Inspiring PCET-driven enzymatic catalysis. *The Journal of Physical Chemistry B, 114*, 5928–5939.

Kozlowski, P. M., Kamachi, T., Toraya, T., & Yoshizawa, K. (2007). Does cob(II)alamin act as a conductor in coenzyme B_{12} dependent mutases? *Angewandte Chemie International Edition, 46*, 980–983.

Kozlowski, P. M., Kumar, M., Piecuch, P., Li, W., Bauman, N. P., Hansen, J. A., et al. (2012). The cobalt-methyl bond dissociation in methylcobalamin: New benchmark analysis based on density functional theory and completely renormalized coupled-cluster calculations. *Journal of Chemical Theory and Computations, 8*, 1870–1894.

Kumar, N., Bucher, D., & Kozlowski, P. M. (2019). Mechanistic implications of reductive co–C bond cleavage in B_{12}-dependent methylmalonyl CoA mutase. *The Journal of Physical Chemistry B, 123*, 2210–2216.

Kumar, M., & Kozlowski, P. M. (2009). Role of tyrosine residue in the activation of co − C bond in coenzyme B_{12}-dependent enzymes: Another case of proton-coupled electron transfer? *The Journal of Physical Chemistry B, 113*, 9050–9054.

Kumar, N., Liu, S., & Kozlowski, P. M. (2012). Charge separation propensity of the coenzyme B12–tyrosine complex in adenosylcobalamin-dependent methylmalonyl–CoA mutase enzyme. *The Journal of Physical Chemistry Letters, 3*, 1035–1038.

Kutta, R. J., Hardman, S. J., Johannissen, L. O., Bellina, B., Messiha, H. L., Ortiz-Guerrero, J. M., et al. (2015). The photochemical mechanism of a B_{12}-dependent photoreceptor protein. *Nature Communications, 6*, 7907.

Kwiecien, R. A., Khavrutskii, I. V., Musaev, D. G., Morokuma, K., Banerjee, R., & Paneth, P. (2006). Computational insights into the mechanism of radical generation in B_{12}-dependent methylmalonyl-CoA mutase. *Journal of the American Chemical Society*, *128*, 1287–1292.

Lin, H., & Truhlar, D. G. (2006). QM/MM: What have we learned, where are we, and where do we go from here? *Theoretical Chemistry Accounts*, *117*, 185.

Lindorff-Larsen, K., Piana, S., Palmo, K., Maragakis, P., Klepeis, J. L., Dror, R. O., et al. (2010). Improved side-chain torsion potentials for the Amber ff99SB protein force field. *Proteins: Structure, Function, and Bioinformatics*, *78*, 1950–1958.

Mamun, A. A., Toda, M. J., & Kozlowski, P. M. (2019). Can photolysis of the co-C bond in coenzyme B_{12}-dependent enzymes be used to mimic the native reaction? *Journal of Photochemistry and Photobiology B: Biology*, *191*, 175–184.

Mamun, A. A., Toda, M. J., Lodowski, P., Jaworska, M., & Kozlowski, P. M. (2018). Mechanism of light induced radical pair formation in coenzyme B_{12}-dependent ethanolamine ammonia-lyase. *ACS Catalysis*, *8*, 7164–7178.

Mancia, F., & Evans, P. R. (1998). Conformational changes on substrate binding to methylmalonyl CoA mutase and new insights into the free radical mechanism. *Structure*, *6*, 711–720.

Marques, H. M., Ngoma, B., Egan, T. J., & Brown, K. L. (2001). Parameters for the amber force field for the molecular mechanics modeling of the cobalt corrinoids. *Journal of Molecular Structure*, *561*, 71–91.

Martin, B. D., & Finke, R. G. (1990). Cobalt-carbon homolysis and bond dissociation energy studies of biological alkylcobalamins: Methylcobalamin, including $\geq 10^{15}$ co-CH_3 homolysis rate enhancement at 25° C following one-electron reduction. *Journal of the American Chemical Society*, *112*, 2419–2420.

Matthews, R. G. (2009). Cobalamin- and corrinoid-dependent enzymes. *Metal Ions in Life Sciences*, *6*, 53–114.

Olsson, M. H. M., Søndergaard, C. R., Rostkowski, M., & Jensen, J. H. (2011). PROPKA3: Consistent treatment of internal and surface residues in empirical pKa predictions. *Journal of Chemical Theory and Computation*, *7*, 525–537.

Padmanabhan, S., Jost, M., Drennan, C. L., & Elías-Arnanz, M. (2017). A new facet of vitamin B_{12}: Gene regulation by cobalamin-based photoreceptors. *Annual Review of Biochemistry*, *86*, 485–514.

Pettersen, E. F., Goddard, T. D., Huang, C. C., Couch, G. S., Greenblatt, D. M., Meng, E. C., et al. (2004). UCSF chimera—A visualization system for exploratory research and analysis. *Journal of Computational Chemistry*, *25*, 1605–1612.

Robertson, W. D., Wang, M., & Warncke, K. (2011). Characterization of protein contributions to cobalt-carbon bond cleavage catalysis in adenosylcobalamin-dependent ethanolamine ammonia-lyase by using photolysis in the ternary complex. *Journal of the American Chemical Society*, *133*, 6968–6977.

Roy Dennington, T. K., & Millam, J. (2019). *GaussView, Version 6.1.1*. Shawnee Mission, Kansas: Semichem Inc.

Ruetz, M., Campanello, G. C., Purchal, M., Shen, H., McDevitt, L., Gouda, H., et al. (2019). Itaconyl-CoA forms a stable biradical in methylmalonyl-CoA mutase and derails its activity and repair. *Science*, *366*, 589.

Schopf, P., Mills, M. J. L., & Warshel, A. (2015). The entropic contributions in vitamin B_{12} enzymes still reflect the electrostatic paradigm. *Proceedings of the National Academy of Sciences*, *112*, 4328–4333.

Schrodinger Release 2021-3. (2021). *Maestro Schrodinger*. New York, NY: LLC.

Senn, H. M., & Thiel, W. (2009). QM/MM methods for biomolecular systems. *Angewandte Chemie International Edition*, *48*, 1198–1229.

Sharma, P. K., Chu, Z. T., Olsson, M. H. M., & Warshel, A. (2007). A new paradigm for electrostatic catalysis of radical reactions in vitamin B_{12} enzymes. *Proceedings of the National Academy of Sciences*, *104*, 9661–9666.

Sousa, S. F., Fernandes, P. A., & Ramos, M. J. (2012). Computational enzymatic catalysis – Clarifying enzymatic mechanisms with the help of computers. *Physical Chemistry Chemical Physics*, *14*, 12431–12441.

Tao, P., Fisher, J. F., Shi, Q., Vreven, T., Mobashery, S., & Schlegel, H. B. (2009). Matrix metalloproteinase 2 inhibition: Combined quantum mechanics and molecular mechanics studies of the inhibition mechanism of (4-phenoxyphenylsulfonyl)methylthiirane and its oxirane analogue. *Biochemistry*, *48*, 9839–9847.

Toda, M. J., Kozlowski, P. M., & Andruniów, T. (2019). Assessing electronically excited states of cobalamins via absorption spectroscopy and time-dependent density functional theory. In E. Broclawik, T. Borowski, & M. Radoń (Eds.), *Transition metals in coordination environments: Computational chemistry and catalysis viewpoints* (pp. 219–258). Cham: Springer International Publishing.

Toda, M. J., Lodowski, P., Mamun, A. A., Jaworska, M., & Kozlowski, P. M. (2019). Photolytic properties of the biologically active forms of vitamin B_{12}. *Coordination Chemistry Reviews*, *385*, 20–43.

Toda, M. J., Lodowski, P., Mamun, A. A., & Kozlowski Pawel, M. (2021). Elelctronic and photolytic properties of hydridocobalamin. *Journal of Photochemistry and Photobiology, B: Biology*, *224*, 112295.

Toda, M. J., Mamun, A. A., Lodowski, P., & Kozlowski, P. M. (2020). Why is CarH photolytically active in comparison to other B_{12}-dependent enzymes? *Journal of Photochemistry and Photobiology B: Biology*, *209*, 111919.

van der Kamp, M. W., & Mulholland, A. J. (2013). Combined quantum mechanics/molecular mechanics (QM/MM) methods in computational enzymology. *Biochemistry*, *52*, 2708–2728.

Vreven, T., Byun, K. S., Komáromi, I., Dapprich, S., Montgomery, J. A., Morokuma, K., et al. (2006). Combining quantum mechanics methods with molecular mechanics methods in ONIOM. *Journal of Chemical Theory and Computation*, *2*, 815–826.

Walker, L. A., Shiang, J. J., Anderson, N. A., Pullen, S. H., & Sension, R. J. (1998). Time-resolved spectroscopic studies of B_{12} coenzymes: The photolysis and geminate recombination of adenosylcobalamin. *Journal of the American Chemical Society*, *120*, 7286–7292.

Warshel, A., & Karplus, M. (1972). Calculation of ground and excited state potential surfaces of conjugated molecules. I. Formulation and parametrization. *Journal of the American Chemical Society*, *94*, 5612–5625.

Warshel, A., & Levitt, M. (1976). Theoretical studies of enzymic reactions: Dielectric, electrostatic and steric stabilization of the carbonium ion in the reaction of lysozyme. *Journal of Molecular Biology*, *103*, 227–249.

Yang, H., McDaniel, E. C., Impano, S., Byer, A. S., Jodts, R. J., Yokoyama, K., et al. (2019). The elusive 5′-deoxyadenosyl radical: Captured and characterized by electron paramagnetic resonance and electron nuclear double resonance spectroscopies. *Journal of the American Chemical Society*, *141*, 12139–12146.

Yoder, L. M., Cole, A. G., Walker, L. A., & Sension, R. J. (2001). Time-resolved spectroscopic studies of B_{12} coenzymes: Influence of solvent on the photolysis of adenosylcobalamin. *The Journal of Physical Chemistry B*, *105*, 12180–12188.

CHAPTER SEVEN

Using kinetic isotope effects to probe the mechanism of adenosylcobalamin-dependent enzymes

E. Neil G. Marsh[a,b,*]

[a]Department of Chemistry, University of Michigan, Ann Arbor, MI, United States
[b]Department of Biological Chemistry, University of Michigan, Ann Arbor, MI, United States
*Corresponding author: e-mail address: nmarsh@umich.edu

Contents

Abstract

Adenosylcobalamin- (AdoCbl) dependent enzyme reactions involved the transfer of hydrogen atoms between the 5′-carbon of the coenzyme and the substrates and products of the reaction. Tritium and deuterium kinetic isotope effect measurements are, therefore, a valuable tool to probe the mechanisms of AdoCbl-dependent enzymes, as they can provide information about the reaction pathway and the rate-determining step. Furthermore, if the intrinsic kinetic isotope effect can be isolated, information on the nature of the transition state associated with hydrogen transfer can be obtained. In this chapter I present methods for the preparation of isotopically-labeled AdoCbl and their use in rapid chemical quench experiments that allow isotope effects on specific

Methods in Enzymology, Volume 669
ISSN 0076-6879
https://doi.org/10.1016/bs.mie.2021.12.018

steps in the reaction to be isolated. These techniques are illustrated with examples from my laboratory's studies on the AdoCbl dependent enzyme, glutamate mutase.

Abbreviations

5′-dAH 5′-deoxyadenosine
AdoCbl adenosylcobalamin
KIE kinetic isotope effect

1. Introduction

Isotopically-labeled substrates and coenzymes are an invaluable tool for investigating the reaction mechanisms of enzymes. As discussed in this chapter, isotope experiments have proved particularly useful to dissect the mechanisms of AdoCbl-dependent enzymes. Isotope experiments conceptually divide into two types. In *tracer* isotope experiments, a substrate is labeled at a specific position with an isotope and, after reaction with the enzyme, its position in the product (or sometimes cofactor) is determined. This type of experiment can provide valuable information on the chemical mechanism by which the enzyme catalyzes the reaction. On the other hand, *kinetic* isotope experiments measure the effect of incorporating a heavy isotope at a specific position in the substrate on the rate of reaction. This type of experiment can yield information on the rate-determining step of the reaction and, in certain cases, information on the nature of the transition state for the reaction if it involves breaking a bond to the isotopically labeled atom (Schramm, 2011, 2015). Both types of experiments have played important roles in elucidating the mechanisms of cobalamin-dependent enzymes (Marsh & Melendez, 2012). Historically, radioactive isotopes, e.g., ^{3}H or ^{14}C, were used in these types of experiments, however, advances in mass-spectrometry and the general availability of sensitive, high resolution LC-MS instruments has meant that, increasingly, substrates labeled with stable isotopes, e.g., ^{2}H, ^{13}C or ^{15}N, are the preferred way to design such experiments.

The mechanisms of all AdoCbl-dependent enzymes involve (at least) two isotopically sensitive steps (Fig. 1): in the first step a hydrogen atom from the substrate is transferred to the 5′-carbon of the adenosyl moiety, whereas in the last step a hydrogen atom is transferred from the 5′-carbon back to the product. Indeed, this mechanistic paradigm was primarily established in a classic series of experiments by Robert Abeles and coworkers on diol dehydratase, which accepts both ethylene glycol and propan-1,2-diol as substrates. Using a mixture of tritium-labeled propan-1,2-diol and unlabeled

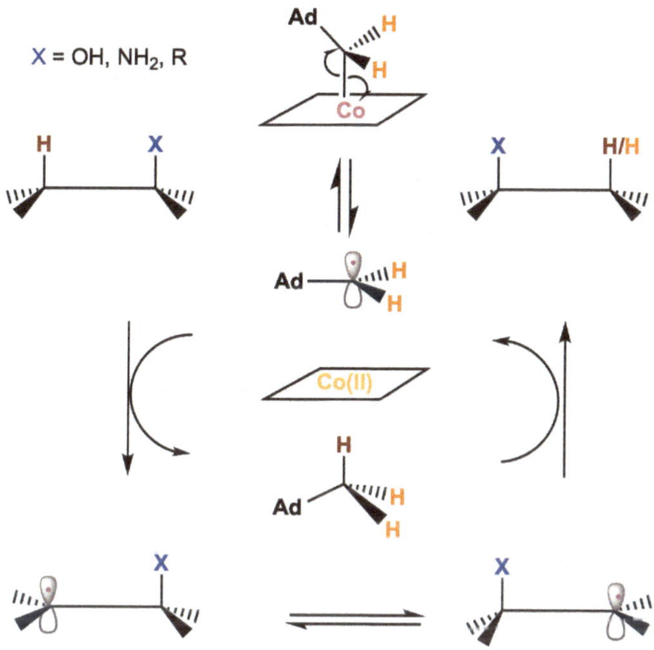

Fig. 1 The minimal mechanistic scheme for the 1,2-rearrangements catalyzed by AdoCbl-dependent isomerases. A defining feature is the transfer of a hydrogen atom between the substrate and the 5′ position of the cofactor.

ethylene glycol it was found that diol dehydratase catalyzes the transfer of tritium from propan-1,2-diol to ethylene glycol (Abeles & Zagalak, 1966). This observation established that the migration of hydrogen from one carbon to another, which is a central feature of all AdoCbl-dependent isomerizations, is an *inter*molecular reaction, rather than an *intra*molecular reaction as at first thought. Subsequently, it was demonstrated that tritium from propan-1,2-diol is transferred to the 5′-carbon of AdoCbl, and that when AdoCbl labeled with tritium was used in the reaction, tritium was transferred to the product aldehyde (Frey & Abeles, 1966; Frey, Essenber, & Abeles, 1967). Key evidence for the intermediacy of 5′-deoxyadenosine (5′-dAH), again using diol dehydratase, was obtained using deuterated and tritiated substrates, where it was shown that there was always a finite chance that hydrogen from the substrate could be transferred to the same product molecule (Essenberg, Frey, & Abeles, 1971).

Work in our laboratory has focused on elucidating the mechanism of glutamate mutase, which has the historical distinction of being the first AdoCbl-dependent enzyme to be discovered (Barker, Iodice, Rooze, & Suzuki, 1964). The enzyme catalyzes the conversion of L-glutamate to

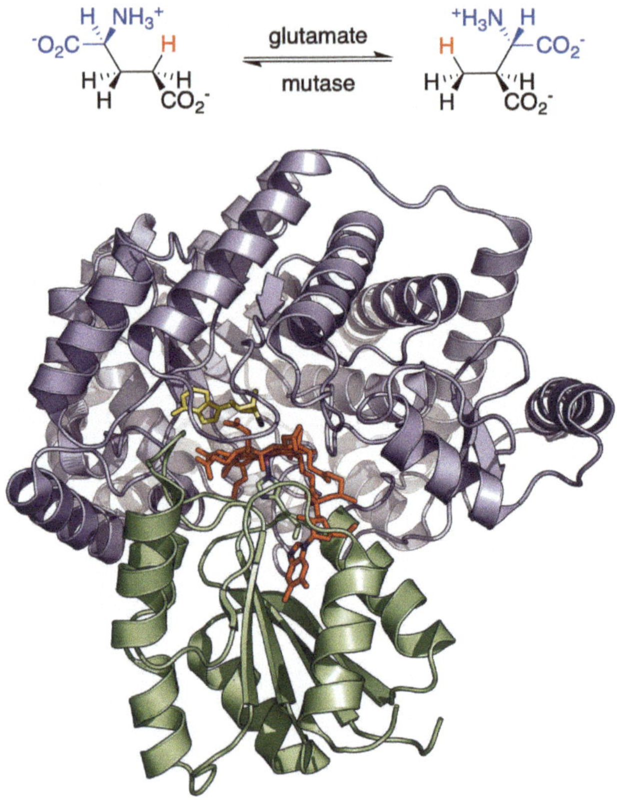

Fig. 2 *Top:* The carbon skeleton rearrangement catalyzed by glutamate mutase. *Bottom:* The structure of glutamate mutase (PDB 1I9C). The E subunit (gray) contains the active site while the S subunit (green) binds the lower face of the cofactor and coordinates cobalt through a histidine sidechain. The cobalamin portion of AdoCbl is shown in red and the adenosyl ligand is in yellow (the bond between the ligand and cobalt is broken in the structure.

L-*threo*-methylaspartate (Fig. 2) as the first step in the fermentation of glutamate by various clostridial bacteria (Buckel & Golding, 1996; Gruber & Kratky, 2002; Marsh, 2000). The enzyme has a number of attributes that make it an attractive system with which to study not only the mechanism of AdoCbl-catalyzed reactions, but also the broader question of how enzymes generate and control highly reactive free radicals. The simple, stable structure of the substrates, together with the commercial availability of various isotopically-labeled glutamates, make isotope experiments ideally suited to probing the mechanism of this enzyme. As discussed in more detail below, the enzyme also provides a very convenient method to prepare 5′-deuterated or 5′-tritiated AdoCbl.

Wildtype glutamate mutase is a hetero-tetramer in its active form, comprising a dimeric E-subunit that encompasses the active site and binds the "top" face of AdoCbl, and a freely dissociable, monomeric S-subunit that binds AdoCbl through the coenzyme's dimethylbenzamidazole "tail" (Gruber, Reitzer, & Kratky, 2001; Holloway & Marsh, 1993, 1994; Marsh & Holloway, 1992; Reitzer et al., 1999) (Fig. 2). This unusual subunit arrangement complicates analysis of the enzyme and therefore we engineered a single-subunit version, designated GlmES, in which the C-terminus of the E-subunit is genetically fused to the S-subunit through a flexible linker sequence (Chen & Marsh, 1997). GlmES has almost identical activity to wildtype glutamate mutase and is the preferred construct for mechanistic analyses. We typically over-express GlmES from a codon-optimized gene in *E. coli* and purify the protein straightforwardly as the apo-enzyme (*E. coli* does not synthesize AdoCbl) through an N-terminal 6-His tag (Roman-Melendez, von Glehn, Harvey, Mulholland, & Marsh, 2014).

In this chapter, I will first describe the preparation of $5'$-^{2}H- and $5'$-^{3}H-labeled AdoCbl using glutamate mutase, and the nonenzymatic synthesis of ^{14}C-labeled AdoCbl. Together with commercially available isotopically labeled substrates, these compounds allow a variety of mechanistically informative experiments to be undertaken. I will also describe some examples from our laboratory in which isotope effects have been used to elucidate the mechanism of glutamate mutase. It should evident that similar approaches can be applied to investigate the mechanisms of other AdoCbl-dependent enzymes—indeed some similar isotope effect experiments have been used to study the mechanism of the related AdoCbl enzyme, methylmalonyl-CoA mutase (Chowdhury & Banerjee, 2000; Meier, Thoma, & Leadlay, 1996; Padmakumar & Banerjee, 1997). These approaches may further be useful to study the mechanisms of the much larger family of radical SAM enzymes, (Broderick, Duffus, Duschene, & Shepard, 2014; Wang et al., 2014) which similarly generate adenosyl radical (in this case by the reductive cleavage of SAM) as a key intermediate in the reactions they catalyze.

2. Methods

2.1 Preparation of [$5'$-^{3}H]-AdoCbl and [$5'$-$^{2}H_2$]-AdoCbl

Glutamate mutase provides a very convenient way to synthesize $5'$-tritiated and $5'$-deuterated AdoCbl from commercially available L-[G-^{3}H]-glutamate (or L-[3,4-^{3}H]-glutamate) and [2,3,3,4,4-$^2H_5$]-glutamate respectively (Marsh, 1995). Although we have previously described several different glutamate mutase constructs, we currently produce the enzyme by over-expression of the GlmES fusion protein equipped with an N-terminal His-tag from a codon-optimized gene (Roman-Melendez et al., 2014). The gene is housed in the expression vector pET28b and over-expressed in the *E. coli* strain BL21 DE3 by standard methods. Using this expression system, large amounts of apo-enzyme can be easily purified by NTA-affinity chromatography, again by standard methods. The enzyme is stable for months when stored at −20 °C near neutral pH, e.g., 10 mM potassium phosphate, pH 7.0, or 10 mM Tris/Cl, pH 8.0, containing 20% glycerol.

A typical tritiation reaction is set up as follows: Reactions are performed in 10 mM Tris-HCl buffer (pH 8.0) containing 10% glycerol, 50 μM GlmES, 250 μM AdoCbl, 100 μM unlabeled L-glutamate and 75 μCi of L-[3,4-^{3}H]-glutamic acid (specific activity of 184 Ci/mol at the exchangeable position) in a volume of 1.2 mL. Assays are performed under dim light in an anaerobic glovebox (or under an argon blanket) to minimize exposure to oxygen and light-induced photolysis. Reactions are started by addition of 100 μL of substrate via a syringe and allowed to proceed at 37 °C, for ~1 h. Reactions are quenched on ice by addition of 25 μL of 500 mM HCl. Samples can be frozen in liquid nitrogen and stored at −20 °C prior to the recovery of AdoCbl by reverse-phase (HPLC). Deuteration reactions may be performed analogously by substituting 10 mM [2,3,3,4,4-$^2H_5$]-glutamate in the reaction. If labeled L-glutamate is not available, racemic compounds may be substituted as the D-isomer does not inhibit the enzyme (Fig. 3).

2.2 Preparation of [8-^{14}C]-AdoCbl

For some experiments it is desirable to have AdoCbl labeled at an unreactive position to serve as an internal standard; e.g., measurement of tritium isotope

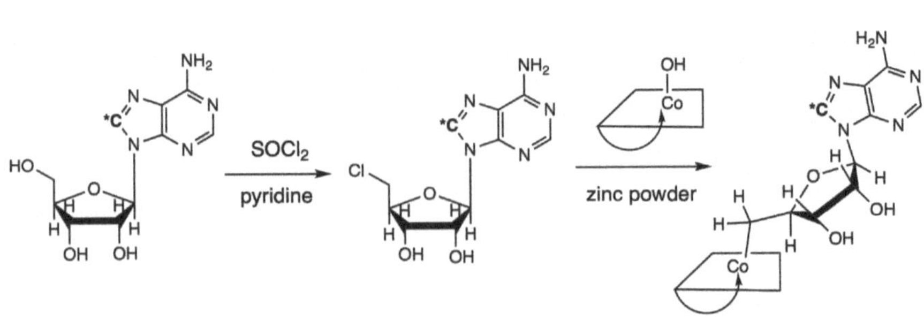

Fig. 3 Scheme for the synthesis of [8-^{14}C]-AdoCbl.

effects by $^{3}H/^{14}C$ ratio counting. [8-^{14}C]-AdoCbl may conveniently be synthesized from commercially available [8-^{14}C]-adenosine by first converting it to 5′-chloro-5′deoxyadenosine, which is used to alkylate Cob(I) alamin generated by in situ reduction of HOCbl with zinc metal. In a typical preparation, 50 mCi of [8-^{14}C]-adenosine is first mixed with 2.8 mg of unlabeled adenosine in 1 mL of water and the water removed by lyophilization. The dried mixture is resuspended in 750 μL of dry acetonitrile and 80 mL of a solution comprising 50 μL/mL of pyridine and 100 μL/mL $SOCl_2$ in dry acetonitrile is then added. The mixture is stirred overnight in the dark with argon purging. After removing the acetonitrile under a stream of argon gas, the residue is washed 3 times with 100 μL of tetrahydrofuran and finally dissolved in 500 μL of a solvent mixture comprising methanol:water:ammonium hydroxide, 5:2:1 (vol/vol). The 5′-chloro-5′deoxyadenosine is purified by HPLC using a reverse phase C-18 column.

The purified 5′-chloro-5'deoxyadenosine is dissolved in 500 μL of 10% NH_4Cl and introduced into a small septum-sealed vial or flask and deoxygenated by passing a stream of argon gas through it for ~1 h. In a separate septum-sealed vial or flask, 30 mg of fresh zinc powder and 5 mg HOCbl are dissolved/suspended in 500 μL of 10% NH_4Cl and similarly deoxygenated by passing a stream of argon gas through it for ~1 h. During this time HOCbl is reduced to cob(I)alamin and the solution becomes a very dark blue/green. The 5′-chloro-5'deoxyadenosine solution is then syringed into the cob(I)alamin solution (it should turn deep red) and the reaction incubated at 45 °C in the dark for 45 min. The resulting [8-^{14}C]-AdoCbl is then purified by reverse phase HPLC (Cheng & Marsh, 2004).

2.3 HPLC purification of AdoCbl and 5′-dAH

These compounds are fairly easily separated from each other and other components of the reactions in which they are formed, with 5′-dAH eluting shortly before AdoCbl. We commonly use a 250 mm × 4.6 mm C-18 reverse phase column, with a 5 μm particle size, pre-equilibrated in 0.1% trifluoroacetic acid (TFA). Samples are injected on to the column, the column washed with 5 mL of 0.1% TFA and compounds eluted with a linear gradient of acetonitrile containing 0.1% TFA, typically 0%–45% over 20 mL at a flow rate of 1 mL min^{-1}, and compounds detected by monitoring absorbance at 260 nm. 5′-dAH elutes at ~10%, and AdoCbl at ~45% acetonitrile. Under the acidic conditions used in chromatography, AdoCbl elutes as the yellow "base-off" form and should be neutralized by the addition of a buffer such as potassium phosphate, pH 7.0 to 0.1 M final concentration. Labeled AdoCbl may be stored at −80 °C for several months.

3. Using isotope effects to probe the mechanism of glutamate mutase

Over the years our laboratory has made extensive use of isotope effects to probe the mechanism of glutamate mutase (Cheng & Marsh, 2004, 2005, 2007; Chih & Marsh, 2001; Madhavapeddi, Ballou, & Marsh, 2002; Marsh, 1995; Marsh & Ballou, 1998; Yoon, Kalli, Lee, Hakansson, & Marsh, 2007; Yoon, Song, Hakansson, & Marsh, 2010). These measurements have relied on rapid quench techniques to follow the transfer of isotope at short time points, generally in the pre-steady state regime. This approach requires some specialized apparatus and, compared with steady state measurements, requires a large amount of enzyme. However, it allows detailed information to be obtained regarding the free energy profile of the reaction and has provided information on the nature of the transition states associated with hydrogen atom transfer that could not otherwise be obtained.

3.1 General considerations

We have previously published detailed protocols for conducting rapid chemical quench experiments with glutamate mutase and the subsequent analysis of the reaction products (Chih, Roymoulik, Huhta, Madhavapeddi, & Marsh, 2002). However, some general notes on experimental design may be helpful.

Constraints on mixing mean that the shortest aging times in rapid quench experiments are ~5 ms; there is essentially no upper limit on aging times—as mixing and quenching can be performed manually! As a guide to how much enzyme is needed, I note that a rapid quench instrument typically consumes 10–50 μL of enzyme solution per shot and may require 25–50 shots (or more), depending on the experiment, to complete a full time course. For our experiments with glutamate mutase, the enzyme solution is 100–200 μM and contains a 20% excess of AdoCbl to ensure that all active sites contain cofactor. The enzyme solution is diluted twofold by mixing with the substrate solution and, after aging, the reaction is quenched by acid denaturation of the enzyme. Typically, we use 5% TFA as the quenching agent as this strong acid ensures that the protein unfolds rapidly.

In our experiments we have found that it is vitally important to include an appropriate internal standard in the experiment. Otherwise, errors arising from small differences in mixing volumes introduced by the quench flow instrument, pipetting error associate with sample analysis, and sample loss during HPLC analysis can confound the interpretation of the results. A good internal standard does not interact with the enzyme, undergoes similar derivatization chemistry as the sample (if needed) and chromatographs similarly to the sample. We typically use L-leucine as an internal standard for experiments that require the amount of L-glutamate and/or L-*threo*-methylaspartate to be quantified. All 3 amino acids undergo reaction with dansyl chloride and can then be separated by HPLC (Chih & Marsh, 2001). The amount of L-leucine in the sample is determined by comparison with a standard curve constructed with authentic dansyl-leucine. Losses due to sample handling can thus be determined and corrected for. For experiments that require AdoCbl or 5′-dAH to be quantified, we have found that L-tryptophan is a convenient internal standard due to its strong u.v. absorbance and chromatographic properties.

3.2 Using tritium partitioning to examine the free energy profile of glutamate mutase

The partitioning of hydrogen isotopes from an enzyme-bound intermediate between products and substrates provides an elegant way of investigating the relative heights of energetic barriers in an enzyme-catalyzed reaction. We examined the partitioning of 5′-^{3}H-AdoCbl between glutamate and methylaspartate to determine the relative heights of the energy barriers associated with hydrogen transfer and substrate-radical rearrangement in glutamate mutase (Chih & Marsh, 2001). Similar experiments have also been performed by Leadlay and coworkers to investigate the mechanism of methylmalonyl-CoA mutase (Meier et al., 1996).

An important consideration in designing a partitioning experiment is to ensure that tritium transfer from the coenzyme to either the substrate or product is effectively irreversible. If this is not the case, and the tritium can make multiple passages through the enzyme, then the distribution of tritium will tend to reflect the relative concentrations of the reactants in the mixture. To avoid this problem, one needs to consider the timescale on which measurements are made and the concentrations of substrate and enzyme used in the reaction.

Thus, experiments on glutamate mutase employ relatively high concentrations of enzyme and coenzyme (the effective concentration of holoenzyme is ~50 μM after mixing, which is >20-fold higher than the K_d for AdoCbl = 2 μM) to avoid possible complications due to the dissociation of AdoCbl from the protein. To ensure that the reaction does not approach equilibrium during the experiment, high substrate concentrations and short reaction times are used. In this example, holo-glutamate mutase was reacted with either 10 mM L-glutamate (17 × K_M) or 2 mM L-*threo*-3-methylaspartate (14 × K_M) (final concentrations), so that the enzyme was effectively saturated with substrate. A rapid quenched flow apparatus was used to follow the reaction for times between ~35 and 250 ms. Glutamate and methylaspartate were then converted to their dansyl-derivatives to facilitate their separation by HPLC, as described in detail previously (Chih et al., 2002) and the tritium content determined by scintillation counting. On this time scale, less than 0.5% of the glutamate and 2.5% of the methylaspartate were consumed during the reaction, ensuring that the reaction remained far from equilibrium and that the reverse reaction was negligible. Also, on this timescale less than 20% of the tritium is transferred from 5′-dAH to substrate or product, meaning that the reaction did not deviate significantly from pseudo first-order conditions.

Analysis of the partitioning data (Fig. 4) shows that the partitioning coefficient is very similar regardless of the substrate, and is close to unity. (The partitioning coefficient is calculated as the ratio of the velocities (v) with which tritium is transferred to glutamate and to methylaspartate; it is therefore time-independent.) When glutamate is the substrate, the partition coefficient, vGlu/vMeAsp = 1.1 ± 0.15, whereas when methylaspartate is the substrate, vGlu/vMeAsp = 1.3 ± 0.15. Similarly, the rate at which tritium is lost from 5′-dAH is similar regardless of the substrate and approximates the rate at which tritium appears in glutamate and methylaspartate. Overall, these results indicate that the barriers for the transfer of tritium between 5′-dAH and either methylaspartyl radical or glutamyl radical are of very similar energy (Fig. 4C). Furthermore, since the partitioning ratios are independent of the direction in which the reaction is proceeding, the inter-conversion of methylaspartyl and glutamyl radicals [which occurs through the intermediacy of glycyl radical and acrylate (Chih & Marsh, 2000)] must be rapid compared with tritium transfer.

3.2.1 Determination of tritium kinetic isotope effects

The partitioning experiment also allows the KIEs associated with tritium transfer to glutamate and methylaspartate to be measured. The rate of product formation is determined by integrating the areas of the either the

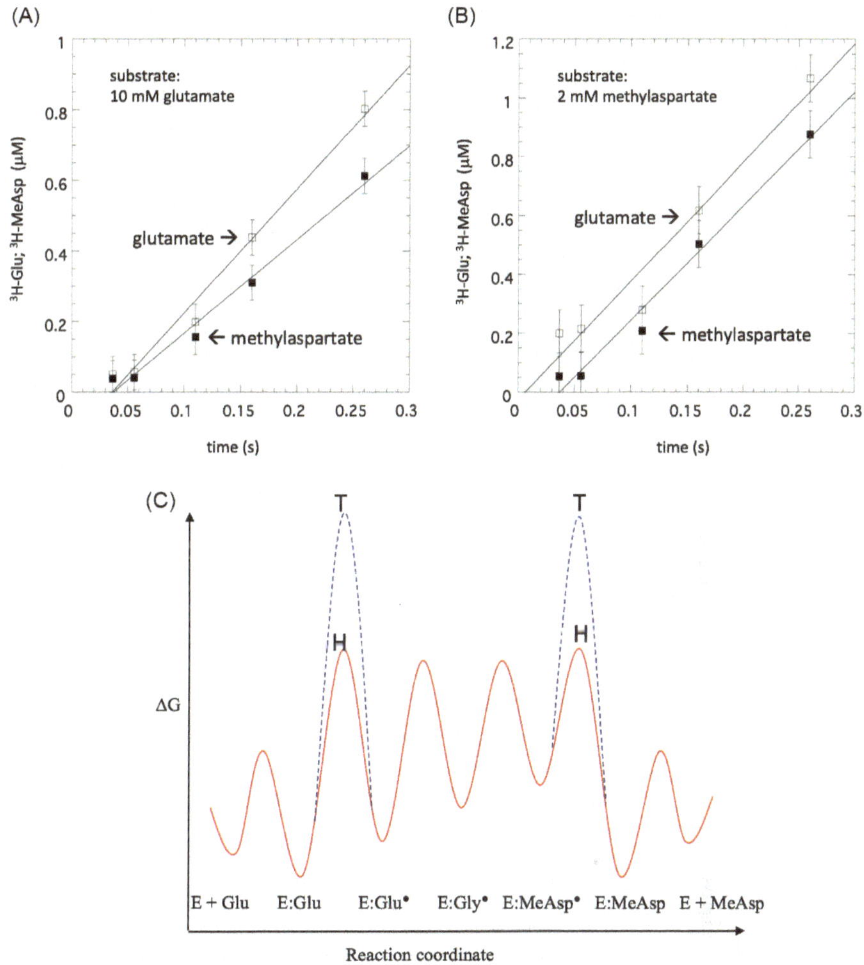

Fig. 4 The partitioning of ^3H from ^3H-labeled AdoCbl between substrate and product in the isomerization catalyzed by glutamate mutase was followed by rapid chemical quench methodology. (A) Partitioning when glutamate is the substrate. (B) Partitioning when methylaspartate is the substrate. In this case, the partitioning ratio is close to 1 and independent of the direction of reaction. This analysis leads to the construction of the free energy profile shown in (C).

dansyl-glutamate or dansyl-methylaspartate product peaks in the HPLC chromatogram and determining the amount of compound using a standard curve. This allows the specific activity of the tritiated products to be calculated. The tritium KIE is then simply the ratio of the specific activity of product divided by specific activity of 5′-dAH, which is assumed not to change significantly under the conditions of the experiment. This ratio must be corrected by a statistical factor to account for the fact that there are 3 exchangeable hydrogen atoms in the methyl group of 5′-dAH. Because tritium is present only in tracer amounts in 5′-dAH, and therefore the major flux of reactants through the enzyme will encounter completely unlabeled 5′-dAH, the statistical factor should be 3. [If the isotope were not in tracer amounts e.g. deuterium-labeled cofactor, then the statistical factor should be 2; see Chih and Marsh (2001) for a more detailed discussion of this issue.] In this case, the tritium KIEs we measured were 21 for transfer of tritium to glutamate and 19 for transfer of tritium to methylaspartate, indicating that hydrogen transfer is substantially rate determining in glutamate mutase.

3.3 Measuring intrinsic secondary tritium isotope effects associated with AdoCbl homolysis and the formation of 5′-deoxyadensosine

A secondary isotope effect refers to a change in either the rate constant or equilibrium constant for a reaction that arises from isotopic substitution at a bond *adjacent* to bond that is cleaved in the reaction; i.e., the bond to the isotope is not broken during the reaction. Secondary isotope effects arise from changes in the stiffness of the isotopically-substituted bond that occur upon reaction and, unlike primary isotope effects, may be normal, i.e., greater that one, or inverse, i.e., less than one. Secondary isotope effects are always much smaller than primary isotope effects: at 25 °C values of 0.8–1.2 and 0.65–1.35 are typical for ^{2}H–C and ^{3}H–C secondary equilibrium isotope effects respectively. However, they provide a powerful tool to probe the nature of transition states in both enzyme-catalyzed and non-enzymatic reactions (Schramm, 2011, 2015). In the absence of tunneling or coupled motion, the value of the secondary kinetic isotope effect should lie between unity and the value of the secondary equilibrium isotope effect. Secondary isotope effects are particularly sensitive to changes in hybridization at the carbon atom undergoing reaction. Generally, they are normal if the rate-determining step involves a change from sp^{3} to sp^{2} hybridization and, correspondingly, inverse in the opposite direction. It should be noted however that, especially in the context of AdoCbl-dependent enzymes, this generalization may not hold true because of the influence of the cobalt d-orbitals on the stretching frequencies of the hydrogen atoms attached to 5′-carbon.

The small size of secondary isotope effects makes them technically challenging to measure. To ensure the necessary level of precision, measurements are usually performed as competition experiments, with both the heavy and light isotope in the same reaction. Even then, non-isotopically sensitive steps, e.g., product release or substrate binding, can obscure the isotope effect if they are rate determining. To overcome these limitations we have used rapid chemical quench techniques to measure the secondary tritium kinetic and equilibrium isotope effects associated with homolysis of the Co–C bond and hydrogen atom abstraction to form Cob(*II*)alamin and 5′-dAH.

For these experiments, we use enzymatically synthesized [5′-^{3}H]-AdoCbl to place the isotope label at the secondary position and chemically synthesized [8-^{14}C]-AdoCbl as a remote label to follow the reaction of AdoCbl containing ^{1}H at the 5′-position (Cheng & Marsh, 2004). Using a rapid quench apparatus, glutamate mutase is reacted with 10 mM L-glutamate at 10 °C. Under these conditions, substrate binding is rapid and the chemical steps in the reaction are independent of substrate concentration. The reaction is quenched at various times between 7 and 100 ms, by which time the concentration of 5′-dAH has essentially reached internal equilibrium on the enzyme. For each time point, the 5′-dAH is recovered by HPLC and the ratio of ^{3}H- to ^{14}C-5'dA determined by scintillation counting. Note that to obtain the best accuracy in dual-label scintillation counting, the specific activities of the labeled AdoCbl molecules should be adjusted so that there are approximately equal counts in the ^{3}H and ^{14}C channels. Data from a typical experiment are shown in Fig. 5. The time course for formation of both ^{H}5'-dAH and ^{T}5'-dAH are described by single exponential functions, as would be expected for a 1st order reaction.

3.3.1 Calculation of secondary isotope effects

Most measurements of tritium isotope effects rely on *competition* between labeled and unlabeled molecules for the same enzyme active site. However, by making the measurements under single turnover conditions, the tritiated and unlabeled AdoCbl molecules react in *parallel* under identical conditions. The concentration of 5′-dAH (either labeled or unlabeled) as a function of time, $[5'\text{-dAH}]_t$, is given by Eq. (1), which describes the approach to equilibrium of a first-order, reversible reaction (Cheng & Marsh, 2004). Here k_f and k_r are the forward and reverse rate constants associated with the equilibrium between the Enzyme:AdoCbl:Substrate and Enzyme:Cob(II):5′-dAH: substrate-radical complexes, (labeled I and III in Fig. 5) and $[\text{AdoCbl}]_0$ is the initial concentration of enzyme-bound AdoCbl.

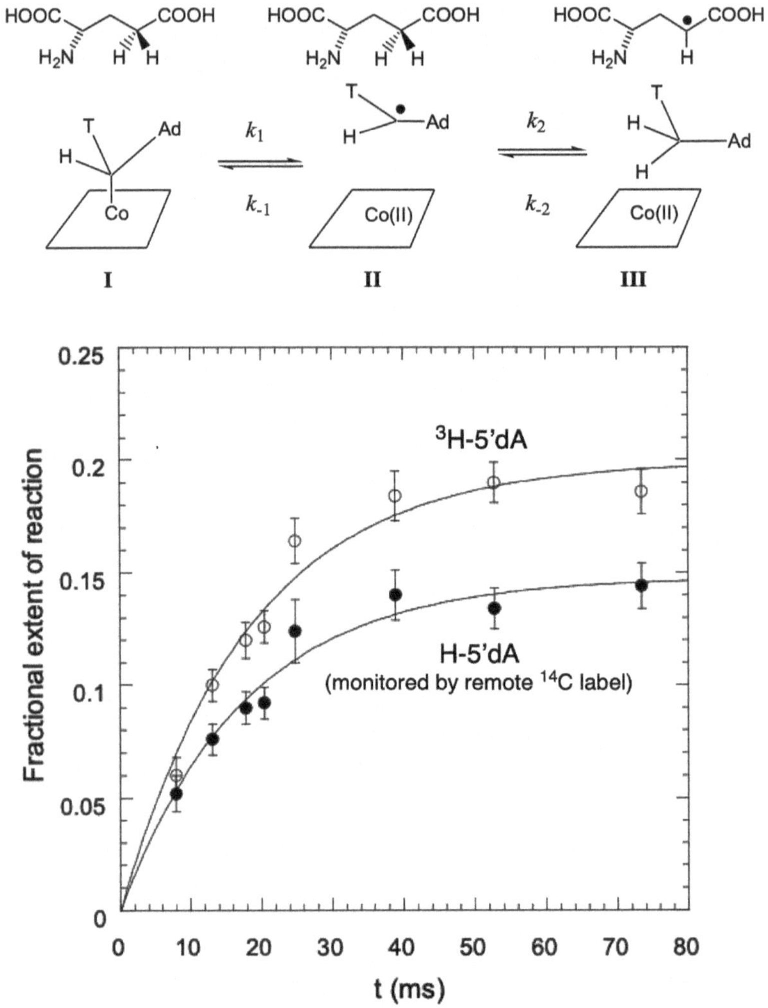

Fig. 5 *Top*: Mechanistic scheme showing bond-making and braking events that are subject to the secondary tritium KIE. *Bottom*: The formation of tritiated and non-tritiated 5′-dAH as a function of time. The data are fitted to a single exponential function.

$$[5' - dA]_t = \frac{[AdoCbl]_0 k_f}{k_f + k_r}\left(1 - e^{-(k_f + k_r)t}\right) \qquad (1)$$

From Eq. (1) it follows that the ratio ${}^{H}[5'\text{-dAH}]_t$:${}^{T}[5'\text{-dAH}]_t$ as a function of time given by Eq. (2), in which the superscripts denote whether a term refers to a tritium- or protium-bearing molecule. From Eq. (2) it can be seen that the secondary KIE, ${}^{T}k_f/{}^{H}k_f$, on 5'-dAH formation can be determined if the ratio of $^{H}[5'\text{-dAH}]$ to $^{T}[5'\text{-dAH}]$ can be measured, together with the corresponding rate constants ($k_f + k_r$) for their formation.

$$\frac{^{H}[5' - dA]_t}{^{T}[5' - dA]_t} = \frac{^{H}[AdoCbl]_0\,{}^{H}k_f\left({}^{T}k_f + {}^{T}k_r\right)\left(1 - e^{-\left({}^{H}k_f + {}^{H}k_r\right)t}\right)}{^{T}[AdoCbl]_0\,{}^{T}k_f\left({}^{H}k_f + {}^{H}k_r\right)\left(1 - e^{-\left({}^{T}k_f + {}^{T}k_r\right)t}\right)} \tag{2}$$

The secondary *equilibrium* isotope effect, $^{H}K/^{T}K$ is obtained from Eq. (2) by allowing t to approach infinity. This leads to Eq. (3), in which $^{H}[5'\text{-dAH}]_{eq}$ and $^{T}[5'\text{-dAH}]_{eq}$ are the equilibrium concentrations of unlabeled and tritiated and 5′-dAH respectively.

$$\frac{^{H}[5' - dA]_{eq}}{\left({}^{H}[AdoCbl]_0 - {}^{H}[5' - dA]_{eq}\right)}\;\frac{\left({}^{T}[AdoCbl]_0 - {}^{T}[5' - dA]_{eq}\right)}{^{T}[5' - dA]_{eq}} = \frac{^{H}k_f\,{}^{T}k_r}{^{T}k_f\,{}^{H}k_r} = \frac{^{H}K}{^{T}K} \tag{3}$$

The results of this experiment, performed on glutamate mutase, are shown in Fig. 5. Fitting the data to Eq. (1), the rate constant for the formation of ^{H}5'-dAH was $^{H}(k_f + k_r) = 53 \pm 5\,\text{s}^{-1}$. Fitting the tritium data to Eq. (1) gave $^{T}(k_f + k_r) = 53 \pm 5\,\text{s}^{-1}$, i.e., a result identical within error to that obtained for the formation of ^{H}5′-dAH. The fraction of enzyme active sites containing ^{H}5′-dAH is determined from the amplitude of the exponential and in this case is 15.2±0.6%. However, when tritium is at the secondary position the fraction of enzyme active sites containing ^{T}5′-dAH at equilibrium is 20±0.8%. This result indicates that there is a large *inverse* equilibrium secondary tritium isotope effect on the formation of 5′-dAH.

In this case, $^{T}(k_f + k_r) = {}^{H}(k_f + k_r)$ which simplifies the analysis because these terms cancel in Eq. (2), and therefore $^{H}[5'\text{-dAH}]_t/^{T}[5'\text{-dAH}]_t$ should be constant with time, as Fig. 5 shows, and is equal to $^{T}k_f/^{H}k_f$. The secondary kinetic isotope effect $^{H}k_f/^{T}k_f$ is then best calculated from the average of all the measurements. For glutamate mutase $^{H}k_f/^{T}k_f = 0.76 \pm 0.02$, representing a large *inverse* kinetic effect. The secondary equilibrium isotope effect, $^{H}K/^{T}K = 0.72 \pm 0.04$, calculated from Eq. (3), is also inverse and very close to the kinetic isotope effect, indicating a late transition state for the reaction.

A potential source of systematic error can arise if tritium is transferred to the substrate by reversal of the reaction after formation of 5′-dAH. This problem will be most significant at longer times. But because a large primary isotope effect operates against this process, only a small percentage (<5%) will be lost from the experiment before the concentration of 5'-dAH reaches equilibrium. This will lead to a slight under-estimate of the equilibrium isotope effect but is unlikely to be a major source of experimental error.

3.4 Measuring intrinsic deuterium kinetic isotope effects in glutamate mutase using intra-molecular competition

For small molecules reacting in solution, the KIE measured usually represents the intrinsic value, i.e., the isotope effect undiminished by other isotope-insensitive steps that may contribute to the overall rate of reaction. However, for enzymes it is notoriously hard to ensure that the intrinsic KIE is being measured. We have therefore devised a method to determine the intrinsic KIE for hydrogen atom transfer from the substrate, L-*threo*-3-methylaspartate, to 5′dA catalyzed by glutamate mutase (Lee, Yoon, & Marsh, 2007; Yoon et al., 2007). This experiment may be readily adapted to any enzyme in which hydrogen transfer from a methyl group occurs.

This experiment involves labeling the methyl carbon with one or two deuterium atoms and analyzing the isotopic composition of the reaction products. The measurement relies on the fact that for hydrogen transfer from a methyl group, the intrinsic isotope effect can be measured, even when the isotopically sensitive step is *not* rate determining, because it arises from *intra*-molecular competition between protium and deuterium atoms. The hydrogen isotopes remain chemically equivalent, even in the enzyme active site, due to the rotation of the methyl group, which is very rapid compared to the timescale of hydrogen transfer.

The principle of the experiment as implemented for glutamate mutase is shown in Fig. 6.

As with the experiments described above, we implemented this experiment using our standard rapid quench methodology. In this case, we reacted glutamate mutase with methylaspartate that is specifically mono-deuterated in the methyl group. [The synthesis of this molecule is detailed elsewhere (Lee et al., 2007)]. During the reaction, the 5′-deoxyadenosyl radical is confronted with the choice of abstracting either protium or deuterium from the methyl group of the *same* substrate molecule. Hydrogen or deuterium atom abstraction generates methylaspartyl radical that rapidly rearranges to the more stable glutamyl radical. Thus, at sufficiently short times (<100 ms) the reaction is effectively irreversible, and the intrinsic isotope effect is manifested (Yoon et al., 2007).

In a typical experiment, a solution containing 275 μM glutamate mutase and 275 μM AdoCbl in 50 mM phosphate buffer, pH 7.0 is reacted with

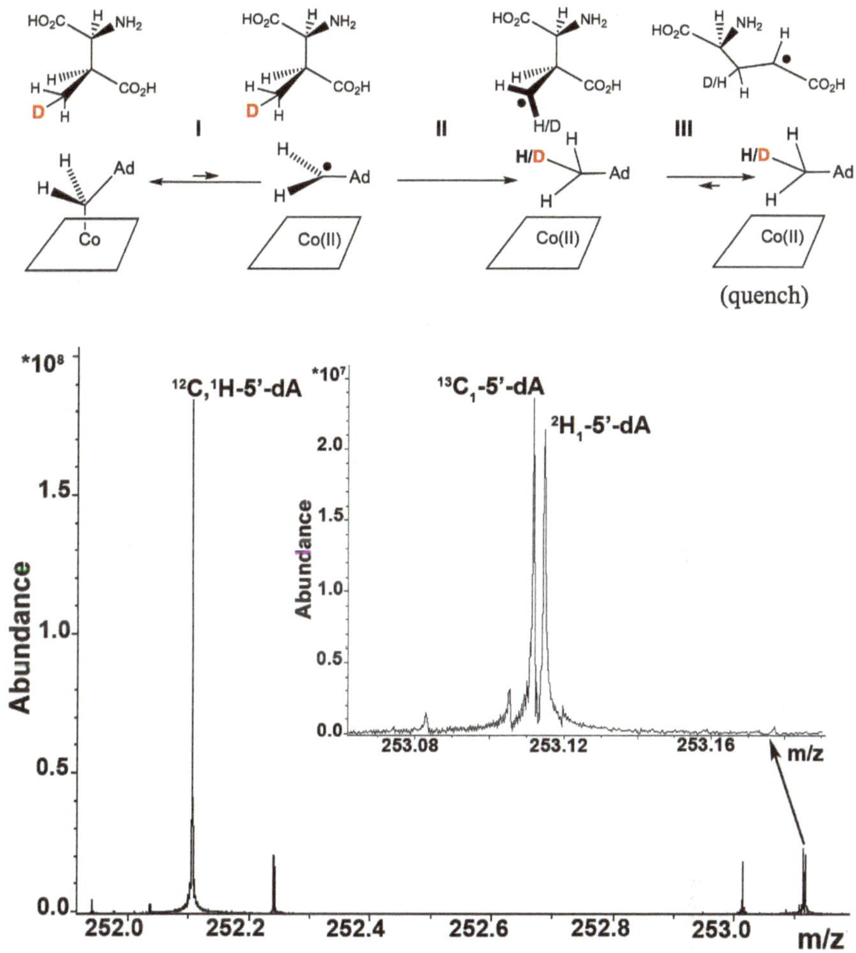

Fig. 6 *Top*: Scheme illustrating the internal competition between hydrogen and ^2H in the methyl group of methylaspartate for transfer to 5′-dAH. *Bottom*: Mass spectrum from an experiment in which 5′-dAH generated from the reaction of glutamate mutase with ^2H-methylaspartate was analyzed by electrospray FT-ICR-MS in positive ion mode. The $[M+H]^+$ peaks due to naturally occurring ^{13}C- and ^2H-labeled 5′-dAH molecules are baseline resolved and are of similar abundance.

2.0 mM d_1-methylaspartate in the same buffer (concentrations before mixing) using rapid quench flow apparatus maintained at 10°C. The solutions are allowed to age for various times between 20 and 80 ms before quenching with 2.5% trifluoroacetic acid. The 5′-dAH formed in the reaction is recovered by reverse-phase HPLC and the deuterium content analyzed by MS.

In this example, the deuterium content of the 5′-dAH was determined by ultra-high-resolution Fourier transform ion cyclotron mass spectrometry (FT-ICR-MS), although the analysis could also be performed with more widely available electrospray ToF instruments that have lower mass resolution. The 5′-dAH isolated from the reaction contains a mixture of isotopomers, with contributions from natural abundance ^{13}C, ^{15}N ^{17}O and ^{2}H isotopes in addition to any deuterium from the substrate. Normally, the presence of ^{13}C complicates the analysis of deuterium KIEs because the peaks due to ^{13}C-containing molecules overlap with deuterated molecular ion. However, an FT-ICR-MS instrument has sufficiently high resolution that the peaks due to ^{13}C-, ^{2}H- and ^{15}N-containing molecules are baseline resolved. This ultra-high resolution allows the ^{13}C-peak to be used as an internal standard to measure the relative abundance of the peak due to ^{2}H-labeled 5′-dAH (Fig. 6).

The intrinsic KIE is simply calculated by integrating the peak areas for ^{12}C-5′-dAH and ^{12}C,^{2}H-5′dA and comparing their ratio (Yoon et al., 2007). Because there are two protium atoms and only one deuterium atom in the methyl group of methylaspartate, the KIE must be corrected by a factor of 2. Alternatively, because the natural abundance of ^{13}C is known to great precision, the intrinsic KIE can also be calculated by comparing the ratio of ^{13}C-5′-dAH to ^{12}C,^{2}H-5′dA, and correcting the result for the natural abundance, n, using Eq. (4) below:

$$KIE = \frac{1}{2} \times \frac{[^{13}C - 5\text{'}dA]}{[^{2}H - 5\text{'}dA]} \times \frac{(1-n)}{n} \tag{4}$$

An important point to note is that, similarly to the above examples, the measurements need to be made under single turnover conditions; otherwise, if multiple turnovers occur, the apparent isotope effect will be diminished due to multiple transits of deuterium through the enzyme. This problem may be diagnosed, as some di-deuterated material will be evident at longer times. For glutamate mutase, the KIE for hydrogen transfer from methylaspartate to 5′-dAH calculated from reactions quenched at times ranging from 25 to 80 ms are very similar, indicating the single turnover condition is met; the average value of the KIE is 4.1 ± 0.2. The KIE may be calculated from either the ratio of ^{12}C-5′-dAH to ^{2}H-5′dA or ^{13}C-5′-dAH to ^{2}H-5′-dAH and in this experiment are, as expected, the same within experimental error.

These experiments resolved the ambiguity surrounding the magnitude of the intrinsic deuterium isotope effects in glutamate mutase. It had been generally assumed that the KIE would be much larger, based on KIE measurements on AdoCbl-dependent enzymes such as methylmalonyl-CoA mutase (Chowdhury & Banerjee, 2000; Padmakumar & Banerjee, 1997) and non-enzymatic model studies. Indeed, early measurements in our laboratory had produced values ranging as high as 35 for glutamate mutase (Marsh & Ballou, 1998), although we later demonstrated that the stopped-flow methodology we employed gave rise to an interesting kinetic phenomenon that led to the mis-interpretation of these data (Cheng & Marsh, 2005).

We subsequently extended these measurements on glutamate mutase to examine the temperature dependence of the deuterium KIE and thereby diagnose the contribution that hydrogen tunneling makes to this reaction (Yoon et al., 2010). This experiment provides one example of the utility of this method; however, it should be evident that the same experimental design can be adapted to measure intrinsic deuterium or tritium KIEs for a wide range of enzyme reactions that involve hydrogen transfer reactions at methyl groups. Indeed, if the reaction is irreversible and hydrogen transfer occurs between substrate and product, rather than substrate and coenzyme, the constraint of making measurements under single turnover conditions is removed, making the experiment technically much easier.

4. Conclusion

Kinetic isotope effect measurements have proved a valuable tool to decipher the mechanisms of AdoCbl-dependent enzymes. The use of rapid chemical quench methodology to measure isotope effects has the disadvantage of requiring much more protein and is more technically challenging than measuring steady state kinetic isotope effects. However, this approach allows individual steps in the mechanism to be interrogated and intrinsic (as opposed to apparent) isotope effects to be measured. Intrinsic KIEs are highly informative because they report directly on reaction transition state, which may be quite different in enzyme-catalyzed reactions compared to the non-enzymatic reaction. Going forward, the increasing availability of highly sensitive mass spectrometry methods should simplify the measurement of isotope effects in AdoCbl-dependent, and other enzymes, by eliminating the need to use radiolabeled substrates.

Acknowledgments

Research in my laboratory on B_{12} enzymes has been supported by NIH grant GM 093088. I am pleased to acknowledge the following members of my laboratory who, over several decades, contributed to the various experiments described here: Daniel Holloway,

Hao-Ping Chen, Prashanti Madhavapeddi, Hung-Wei Chih, Ipsita Roymoulik, Marja Huhta, Li Xia, Mou-Chi Cheng, Anjali Patwardhan, Huang Yeol Lee, Miri Yoon and Gabriel Roman-Melendez. I thank Craig Townsend and Erica Sinner for critically reading the manuscript.

References

Abeles, R. H., & Zagalak, B. (1966). Nature of hydrogen transfer in dimethylbenzimidazolylcobamide coenzyme-catalyzed conversion of 1,2-propanediol to propionaldehyde. *Journal of Biological Chemistry*, *241*(5), 1245.

Barker, H. A., Iodice, A. A., Rooze, V., & Suzuki, F. (1964). Glutamate mutase system—Assays + properties. *Journal of Biological Chemistry*, *239*(10), 3260.

Broderick, J. B., Duffus, B. R., Duschene, K. S., & Shepard, E. M. (2014). Radical S-adenosylmethionine enzymes. *Chemical Reviews*, *114*(8), 4229–4317. https://doi.org/10.1021/cr4004709.

Buckel, W., & Golding, B. T. (1996). Glutamate and 2-methyleneglutarate mutase: From microbial curiosities to paradigms for coenzyme B-12-dependent enzymes. *Chemical Society Reviews*, *25*(5), 329. https://doi.org/10.1039/cs9962500329.

Chen, H. P., & Marsh, E. N. G. (1997). Adenosylcobalamin-dependent glutamate mutase: Examination of substrate and coenzyme binding in an engineered fusion protein possessing simplified subunit structure and kinetic properties. *Biochemistry*, *36*(48), 14939–14945.

Cheng, M. C., & Marsh, E. N. G. (2004). Pre-steady-state measurement of intrinsic secondary tritium isotope effects associated with the homolysis of adenosylcobalamin and the formation of 5'-deoxyadensosine in glutamate mutase. *Biochemistry*, *43*(8), 2155–2158. https://doi.org/10.1021/bi036122w.

Cheng, M. C., & Marsh, E. N. G. (2005). Isotope effects for deuterium transfer between substrate and coenzyme in adenosylcobalamin-dependent glutamate mutase. *Biochemistry*, *44*(7), 2686–2691. https://doi.org/10.1021/bi047662b.

Cheng, M. C., & Marsh, E. N. G. (2007). Evidence for coupled motion and hydrogen tunneling of the reaction catalyzed by glutamate mutase. *Biochemistry*, *46*(3), 883–889.

Chih, H. W., & Marsh, E. N. G. (2000). Mechanism of glutamate mutase: Identification and kinetic competence of acrylate and glycyl radical as intermediates in the rearrangement of glutamate to methylaspartate. *Journal of the American Chemical Society*, *122*(43), 10732–10733. https://doi.org/10.1021/ja002488+.

Chih, H. W., & Marsh, E. N. G. (2001). Tritium partitioning and isotope effects in adenosylcobalamin-dependent glutamate mutase. *Biochemistry*, *40*(43), 13060–13067. https://doi.org/10.1021/bi011298o.

Chih, H. W., Roymoulik, I., Huhta, M. S., Madhavapeddi, P., & Marsh, E. N. G. (2002). Adenosylcobalamin-dependent glutamate mutase: Pre-steady-state kinetic methods for investigating reaction mechanism. *Enzyme Kinetics and Mechanism, Pt F: Detection and Characterization of Enzyme Reaction Intermediates*, *354*, 380–399.

Chowdhury, S., & Banerjee, R. (2000). Evidence for quantum mechanical tunneling in the coupled cobalt-carbon bond homolysis-substrate radical generation reaction catalyzed by methylmalonyl-CoA mutase. *Journal of the American Chemical Society*, *122*(22), 5417–5418.

Essenberg, M. K., Frey, P. A., & Abeles, R. H. (1971). Studies on mechanism of hydrogen transfer in coenzyme B12 dependent dioldehydrase reaction-Ii. *Journal of the American Chemical Society*, *93*(5), 1242.

Frey, P. A., & Abeles, R. H. (1966). Role of B12 coenzyme in conversion of 1,2-propanediol to propionaldehyde. *Journal of Biological Chemistry*, *241*(11), 2732.

Frey, P. A., Essenber, M. K., & Abeles, R. H. (1967). Studies on mechanism of hydrogen transfer in cobamide coenzyme-dependent dioldehydrase reaction. *Journal of Biological Chemistry*, *242*(22), 5369.

Gruber, K., & Kratky, C. (2002). Coenzyme B-12 dependent glutamate mutase. *Current Opinion in Chemical Biology*, *6*(5), 598–603.

Gruber, K., Reitzer, R., & Kratky, C. (2001). Radical shuttling in a protein: Ribose pseudorotation controls alkyl-radical transfer in the coenzyme B-12 dependent enzyme glutamate mutase. *Angewandte Chemie-International Edition*, *40*(18), 3377.

Holloway, D. E., & Marsh, E. N. G. (1993). Cloning and sequencing of glutamate mutase component-E from clostridium-tetanomorphum—Organization of the mut genes. *FEBS Letters*, *317*(1–2), 44–48. https://doi.org/10.1016/0014-5793(93)81488-L.

Holloway, D. E., & Marsh, E. N. G. (1994). Adenosylcobalamin-dependent glutamate mutase from clostridium-tetanomorphyum—Overexpression in *Escherichia coli*, purification, and characterization of the recombinant enzyme. *Journal of Biological Chemistry*, *269*(32), 20425–20430.

Lee, H. Y., Yoon, M., & Marsh, E. N. G. (2007). Synthesis of mono- and di-deuterated (2S,3S)-3-methylaspartic acids to facilitate measurement of intrinsic kinetic isotope effects in enzymes. *Tetrahedron*, *63*(22), 4663–4668. https://doi.org/10.1016/j.tet.2007.03.107.

Madhavapeddi, P., Ballou, D. P., & Marsh, E. N. G. (2002). Pre-steady-state kinetic studies on the Glu171Gln active site mutant of adenosylcobalamin-dependent glutamate mutase. *Biochemistry*, *41*(52), 15803–15809. https://doi.org/10.1021/bi020596y.

Marsh, E. N. G. (1995). Tritium isotope effects in adenosylcobalamin-dependent glutamate mutase—Implications for the mechanism. *Biochemistry*, *34*(22), 7542–7547. https://doi.org/10.1021/bi00022a030.

Marsh, E. N. G. (2000). Coenzyme-B-12-dependent glutamate mutase. *Bioorganic Chemistry*, *28*(3), 176–189. https://doi.org/10.1006/bioo.2000.1168.

Marsh, E. N. G., & Ballou, D. P. (1998). Coupling of cobalt-carbon bond homolysis and hydrogen atom abstraction in adenosylcobalamin-dependent glutamate mutase. *Biochemistry*, *37*(34), 11864–11872.

Marsh, E. N. G., & Holloway, D. E. (1992). Cloning and sequencing of glutamate mutase component-S from clostridium-tetanomorphum—Homologies with other cobalamin-dependent enzymes. *FEBS Letters*, *310*(2), 167–170. https://doi.org/10.1016/0014-5793(92)81321-C.

Marsh, E. N. G., & Melendez, G. D. R. (2012). Adenosylcobalamin enzymes: Theory and experiment begin to converge. *Biochimica Et Biophysica Acta-Proteins and Proteomics*, *1824*(11), 1154–1164.

Meier, T. W., Thoma, N. H., & Leadlay, P. F. (1996). Tritium isotope effects in adenosylcobalamin-dependent methylmalonyl-CoA mutase. *Biochemistry*, *35*(36), 11791–11796. https://doi.org/10.1021/bi961250o.

Padmakumar, R., & Banerjee, R. (1997). Evidence that cobalt-carbon bond homolysis is coupled to hydrogen atom abstraction from substrate in methylmalonyl-CoA mutase. *Biochemistry*, *36*(12), 3713–3718.

Reitzer, R., Gruber, K., Jogl, G., Wagner, U. G., Bothe, H., Buckel, W., et al. (1999). Glutamate mutase from *Clostridium cochlearium*: The structure of a coenzyme B-12-dependent enzyme provides new mechanistic insights. *Structure*, 7(8), 891–902. https://doi.org/10.1016/S0969-2126(99)80116-6.

Roman-Melendez, G. D., von Glehn, P., Harvey, J. N., Mulholland, A. J., & Marsh, E. N. G. (2014). Role of active site residues in promoting cobalt-carbon bond homolysis in adenosylcobalamin-dependent mutases revealed through experiment and computation. *Biochemistry*, *53*(1), 169–177. https://doi.org/10.1021/bi4012644.

Schramm, V. L. (2011). Enzymatic transition states, transition-state analogs, dynamics, thermodynamics, and lifetimes. In R. D. Kornberg, C. R. H. Raetz, J. E. Rothman, & J. W. Thorner (Eds.), *Vol. 80. Annual review of biochemistry* (pp. 703–732).

Schramm, V. L. (2015). Transition states and transition state analogue interactions with enzymes. *Accounts of Chemical Research*, *48*(4), 1032–1039. https://doi.org/10.1021/acs.accounts.5b00002.

Wang, J. R., Woldring, R. P., Roman-Melendez, G. D., McClain, A. M., Alzua, B. R., & Marsh, E. N. G. (2014). Recent advances in radical SAM enzymology: New structures and mechanisms. *ACS Chemical Biology*, *9*(9), 1929–1938. https://doi.org/10.1021/cb5004674.

Yoon, M., Kalli, A., Lee, H. Y., Hakansson, K., & Marsh, E. N. G. (2007). Intrinsic deuterium kinetic isotope effects in glutamate mutase measured by an intramolecular competition experiment. *Angewandte Chemie-International Edition*, *46*(44), 8455–8459. https://doi.org/10.1002/anie.200702448.

Yoon, M., Song, H. T., Hakansson, K., & Marsh, E. N. G. (2010). Hydrogen tunneling in adenosylcobalamin-dependent glutamate mutase: Evidence from intrinsic kinetic isotope effects measured by intramolecular competition. *Biochemistry*, *49*(14), 3168–3173. https://doi.org/10.1021/bi1001695.

CHAPTER EIGHT

Steady-state and pre-steady state kinetic analysis of ornithine 4,5-aminomutase

Amanda L. Darbyshire, Caitlyn Makins, and Kirsten R. Wolthers*
Department of Chemistry, University of British Columbia - Okanagan Campus, Kelowna, BC, Canada
*Corresponding author: e-mail address: kirsten.wolthers@ubc.ca

Contents

Abstract

Ornithine 4,5-aminomutase (4,5-OAM) is a pyridoxal 5′-phosphate and adenosylcobalamin-dependent enzyme that catalyzes a 1,2-rearrangement of the terminal amine of D-ornithine to form (2*R*, 4*S*)-diaminopentanoate. The gene encoding ornithine 4,5-aminomutase is clustered with other genes that function in the oxidative L-ornithine metabolic pathway present in a number of anaerobic bacteria. This chapter discusses the methodology for measuring 4,5-OAM activity using NAD^+-dependent diaminopentanoate dehydrogenase, which functions downstream of 4,5-OAM in the L-ornithine metabolic pathway. The use of ornithine racemace, which functions upstream of 4,5-OAM, for the synthesis of D,L-ornithine-3,3,4,4,5,5-d_6 is also presented. Finally, this chapter describes the anaerobic stopped-flow spectrophotometric analysis of 4,5-OAM. Information is provided on the integration of a stopped-flow system in the anaerobically-maintained glove, the preparation of anaerobic solutions, and the experimental approach.

Methods in Enzymology, Volume 669
ISSN 0076-6879
https://doi.org/10.1016/bs.mie.2021.11.016

1. Introduction

Ornithine 4,5-aminomutase (4,5-OAM; EC 5.4.3.5) and its homolog, lysine 5,6-aminomutase (5,6-LAM) are the only known adenosylcobalamin (AdoCbl)-dependent enzymes that also utilize pyridoxal 5′-phosphate (PLP) to catalyze a 1,2-interchange reaction (Baker, van der Drift, & Stadtman, 1973; Somack & Costilow, 1973b). 4,5-OAM and 5,6-LAM operate in anaerobic bacteria (e.g., *Clostridia*) where they participate in metabolic pathways for the degradation of L-ornithine and L-lysine, respectively.(Barker, 1981; Dyer & Costilow, 1968). For 4,5-OAM, the catalytic cycle initiates with formation of an external aldimine between the δ-NH_2 of D-ornithine and PLP (Fig. 1). Transimination triggers homolytic rupture of the AdoCbl Co—C bond, generating the highly reactive 5′-deoxyadenosyl radical (dAdo$^•$) and cob(*II*)alamin (Wolthers, Rigby, & Scrutton, 2008). Abstraction of the C4 hydrogen atom from the D-ornithinyl-PLP aldimine by dAdo$^•$ generates a substrate radical that undergoes internal addition to the imine N to form an aziridylcarbinyl-PLP radical adduct (Frey & Magnusson, 2003; Wu et al., 2000). Ring opening leads to formation of a product-like radical intermediate, which then re-abstracts a hydrogen atom from 5′-deoxyadenosine (AdoH). Recombination between dAdo$^•$ and cob(*II*)alamin reforms AdoCbl and transimination releases the product, (2*R*, 4*S*)-diaminopentanoate.

One of the more intriguing features of AdoCbl-dependent isomerases is their ability to control—to varying degrees—the productive trajectory of highly reactive radical intermediates during the course of the catalytic cycle. We sought to explore this aspect of AdoCbl enzymology by studying the kinetic properties of 4,5-OAM, as studies by our group and Frey's revealed that 5,6-LAM is remarkably less proficient at radical mediated chemistry compared to 4,5-OAM (Tang, Chang, & Frey, 2001; Wolthers et al., 2008). For example, during anaerobic turnover with L-β-lysine, 5,6-LAM undergoes substrate-induced suicide inactivation at a rate constant that is 300-fold higher than that observed for 4,5-OAM during anaerobic turnover with D-ornithine. Substrate suicide inactivation results from loss of the AdoH intermediate and oxidation of cob(*II*)alamin (Tang et al., 2001). 5,6-LAM also elicits a higher degree of substrate promiscuity as it can perform 1,2 amino shifts on L-β-lysine, the physiological substrate, as well has D- and L-lysine (Tang et al., 2001). A comparison of the substrate-bound complexes for 4,5-OAM and 5,6-LAM (generated through molecular modeling) reveals

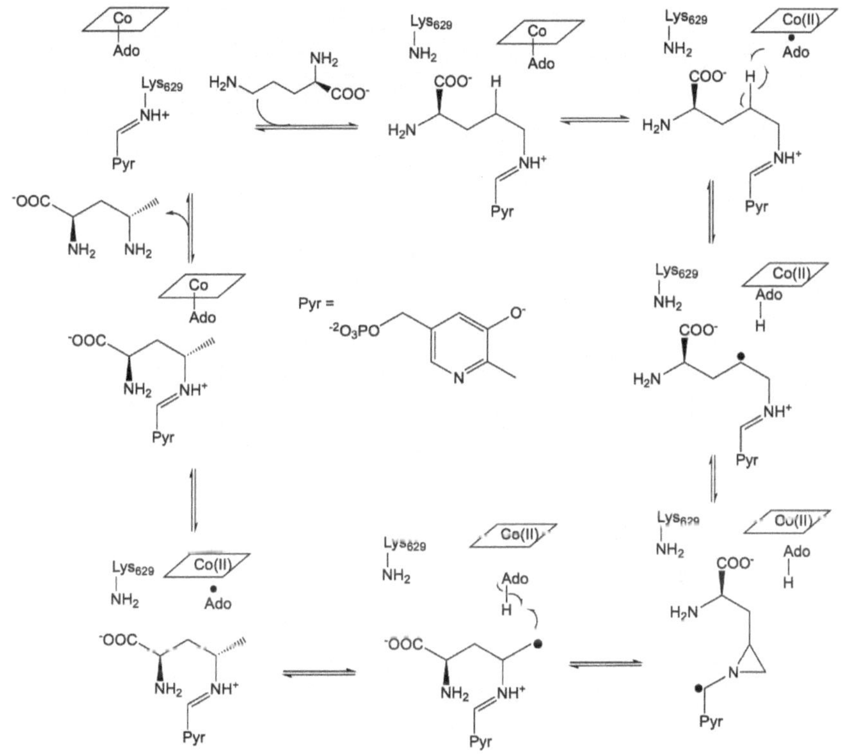

Fig. 1 Catalytic mechanism of 4,5-OAM. Transimination with D-ornithine triggers AdoCbl Co—C bond homolysis. Hydrogen abstraction by dAdo• generates a substrate-like radical that undergoes internal addition to a produce an aziridylcarbinyl-PLP radical adduct. Ring opening generates a product radical that re-abstracts a hydrogen from 5′-deoxyadenosine to reform dAdo•. (2*R*, 4*S*)-diaminopentanoate is released following transimination and recombination between dAdo• and cob(*II*)alamin.

more non-covalent interactions with the substrate in the former enzyme compared to the latter (Makins et al., 2017; Wolthers, Levy, Scrutton, & Leys, 2010).

We posit that more electrostatic interactions between D-ornithine and 4,5-OAM accounts for its narrower substrate scope and more controlled radical chemistry. This hypothesis was tested through kinetic isotope effects on steady state kinetic parameters and pre-steady state kinetic analysis of 4,5-OAM, both the wild type and several active site variants (Makins, Miros, Scrutton, & Wolthers, 2012; Makins et al., 2017; Makins, Pickering, Mariani, & Wolthers, 2013; Makins, Whitelaw, Mu, Walsby, & Wolthers, 2014). In this chapter, we present our methods for determining the steady-state kinetic parameters of 4,5-OAM using a coupled spectrophotometric assay. We also describe the preparation of deuterated D,L-ornithine and the measurement of deuterium kinetic isotope effects for k_{cat} and k_{cat}/K_M. Finally, we described the anaerobic stopped-flow system and the methodology we used to determine the observed rate constants for transimination and AdoCbl homolysis.

2. Steady state kinetic analysis of 4,5-OAM

The genes encoding 4,5-OAM (*oraE* and *oraS*) and 5,6-LAM (*kamD* and *kamE*) are located in putative operons for the anaerobic fermentation of L-ornithine and L-lysine, respectively (Barker, Kahn, & Hedrick, 1982; Fonknechten et al., 2009). The operons and metabolic pathways presented in Fig. 2 highlight the biological function of the aminomutases and the presence of downstream dehydrogenases that catalyze the oxidative deamination of the products of 4,5-OAM and 5,6-LAM catalysis. Given that the dehydrogenases are dependent on $NAD(P)^+$, they can be used to measure 4,5-OAM and 5,6-LAM turnover, as has been demonstrated for the former enzyme (Makins et al., 2012). Ornithine racemace, present in the L-ornithine oxidative pathway, can also be used to synthesize a racemic deuterated D,L-ornithine for measurements of kinetic isotope effects of the steady-state kinetic parameters.

2.1 Ornithine and lysine fermentation pathway

The oxidative L-ornithine fermentation pathway is one of many Stickland reactions present in several nonpathogenic *Clostridium* species (e.g., *C. sporogenes* and *C. sticklandii*) (Stickland, 1934; Woods, 1936). The Stickland reaction involves the oxidation of one amino acid coupled to the reduction of another amino acid (Barker, 1981). The L-ornithine oxidative pathway initiates with the conversion of L-ornithine to D-ornithine by ornithine racemase (EC 5.1.1.12) (Fig. 1) (Chen, Lin, Lee, Tsay, & Wu, 2000). As mentioned above, 4,5-OAM catalyzes the chemically difficult second step, the isomerization of D-ornithine to (2*R*, 4*S*)-diaminopentanoate. An NAD^+ -dependent diaminopentanoate dehydrogenase (DAPDH; EC 1.4.1.12) catalyzes the oxidative deamination of (2*R*, 4*S*)-diaminopentanoate generating 2-amino-4-ketopentanoate, ammonia, and NADH (Buckel & Golding, 2006; Fonknechten et al., 2009; Somack & Costilow, 1973a). The final step of the pathway is catalyzed by 2-amino-4-ketopentanoate thiolase (AKPT), a PLP-dependent enzyme that performs a thiolytic cleavage with coenzyme A

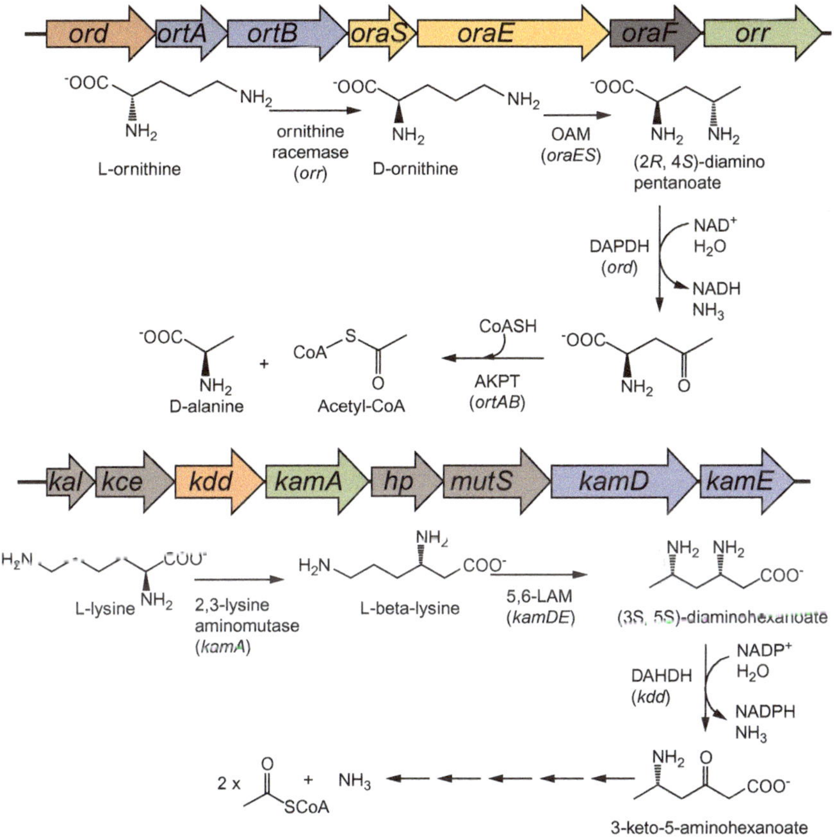

Fig. 2 Anaerobic fermentation of L-ornithine and L-lysine. (A) Genetic locus of the oxidative L-ornithine metabolic pathway and enzymatic steps for the degradation of L-ornithine. (B) Genetic locus for L-lysine metabolic pathway and the first three enzymatic steps involved in anaerobic lysine fermentation.

to form acetyl-CoA and D-alanine (Jeng, Somack, & Barker, 1974). The genes encoding the enzymes are clustered in the putative *ora* operon, which is also present in the nosocomial pathogen *Clostridium difficile*. The function of the gene product of the final gene in the *ora* operon, *ora*F, has not been biochemically determined. However, it may be involved in the ATP-dependent reactivation of the oxidized cobalamin cofactor of 4,5-OAM. Its primary sequence contains an HSP70-type ATP-binding motif that has also been detected in other putative reactivating enzymes of AdoCbl-dependent isomerases (Holloway & Marsh, 1993; Zelder, Beatrix, & Buckel, 1994). The genes encoding the reactivating enzymes often have a strong operonic association with gene(s) encoding AdoCbl-dependent isomerases (Dobson et al., 2002; Tobimatsu, Kajiura, Yunoki, Azuma, & Toraya, 1999).

The first three enzymatic steps of the lysine fermentation pathway are also illustrated to demonstrate that the activity of 5,6-LAM can also be measured through an analogous coupled spectrophotometric assay. The anaerobic lysine metabolic pathway initiates with two 1,2-amino shift reactions. The first is catalyzed by lysine 2,3-aminomutase, a PLP-dependent enzyme that employs an [4Fe-4S] center and S-adenosylmethionine to generate dAdo$^{\bullet}$ for the isomerization of L-lysine to L-β-lysine (Cosper, Booker, Ruzicka, Frey, & Scott, 2000). 5,6-LAM functions in the second step, conversion of L-β-lysine to *erythro*-3,5-diaminohexanoate. The first oxidative deamination step of the pathway is performed by *erythro*-3,5-diaminohexanoate dehydrogenase (DAHDH) which produces 5-amino-3-oxohexanoate and NH_3 from *erythro*-3,5-diaminohexanoate (Baker, Jeng, & Barker, 1972; Liu et al., 2021). Given that DAHDH transfers reducing equivalents from $NADP^+$ to NADPH, the enzyme could also be employed in a coupled spectrophotometric assay to measure 5,6-LAM turnover. This would avoid the labor-intensive process of quenching the enzymatic reaction at different time points, derivatizing the reaction with phenylisothiocyanate and then separating the derivatized components of the reaction by reverse phase high performance liquid chromatography (Tang, Casarez, Wu, & Frey, 2003).

2.2 Preparation of enzymes, 4,5-OAM, OR and DAPDH

The genes encoding 4,5-OAM (*oraE* and *oraS*), OR (*orr*) and DAPDH (*ord*) can be PCR amplified from (ATCC 12662) genomic DNA with primers designed to insert the PCR product into a pET expression vector for insertion of a hexahisidine tag onto the C-terminus of the protein (e.g., pET41a, pET23d). The genes encoding the two subunits of 4,5-OAM are amplified as one PCR product to allow for co-expression. A *C. sticklandii* ribosomal binding site is located upstream of the initiation codon of OraE and is embedded in the coding sequence for OraS. The presence of this embedded *C. sticklandii* ribosomal binding facilitates co-expression of both proteins in *Escherichia coli*. Recombinant expression of the large OraE subunit on its own results in insoluble protein (Chen, Wu, Lin, Chen, & Tsay, 2001). The three enzymes can then be co-expressed in *E. coli* Rosetta(DE3)pLysS using standard growth and induction conditions. The presence of the pLysS plasmid supplies endogenous lysozyme which assists in cell lysis during the protein purification procedure. Typically, we inoculate Terrific Broth medium (supplemented with the appropriate antibiotics) with a 1:250 dilution from an overnight culture, grow the cells at 30 °C, induce transcription with

100 μM of isopropyl α-thiogalactoside at OD_{600} of 0.6 to 0.8, and continue to grow the cells for 18 h and 25 °C. The cells are harvested and frozen at −70 °C.

All subsequent purification steps are performed at 4 °C or on crushed ice. The cell pellet is typically resuspended in lysis buffer (50 mM Tris-HCl, pH 7.5, 0.15 M NaCl, 10 mM imidazole with 1 mM phenylmethylsulfonyl fluoride and 0.5 mM benzamidine) at a ratio of 1 g of wet cell weight to 5 mL of lysis buffer. Following sonification and centrifugation to remove cell debris, the recombinant 4,5-OAM/DAPDH/OR enzymes are then purified from the resulting supernatant using Ni-nitrilotriacetic acid (Ni-NTA) affinity chromatography. The proteins are dialyzed against 50 mM Tris-HCl, 1 mM EDTA and 2 mM β-mercaptoethanol. 4,5-OAM and DAPDH are dialyzed for 16 h, but OR is dialyzed for 2 h; prolonged dialysis of OR results in protein precipitation. 4,5-OAM/DAPDH/OR proteins are then further purified by ion-exchange chromatography using Q-Sepharose high performance resin with a linear gradient from 0 to 0.5 M NaCl in 50 mM Tris-HCl pH 7.5. 4,5-OAM is then dialysed against 50 mM NH_4^+-EPPS pH 8.5 with 2 mM β-mercaptoethanol to remove Tris, which interferes with the formation of the external aldimine with D-ornithine. Activity assays for 5,6-LAM were also performed in NH_4^+-EPPS, pH 8.5 (Tang, Harms, & Frey, 2002). The purity of 4,5-OAM is determined via SDS-PAGE using 13% (v/v) of acrylamide, which ensures that both the OraE subunit (MW = 87 kDa) and the OraS subunit (MW = 13 kDa) are resolved (Fig. 3). The purity of DADPH (MW = 38 kDa) and OR (MW = 39 kDa) is assessed by 10% SDS-PAGE (Fig. 3). The protein concentrations are determined by the Lowry method using bovine serum albumin as a standard. 4,5-OAM/DAPDH/OR are concentrated to 10–20 mg/mL using a centrifugal filter unit with a 30 kDa cut-off filter, flash frozen in liquid N_2, and stored at −70 °C. Purified recombinant OR is yellow and the absorbance spectrum elicits a peak at 410 nm, confirming the presence of the PLP cofactor (Fig. 4A). Surprisingly, 4,5-OAM isolated from *E. coli* is colorless indicating that it is in the apoform but its activity is restored by reconstitution with PLP and AdoCbl. The spectrum of the holoenzyme following equimolar addition of both cofactors in shown in Fig. 4B.

2.3 Coupled spectroscopic assay with OAM and DAPDH

We used purified DAPDH to monitor the activity of 4,5-OAM as it catalyzes the NAD^+-dependent oxidative deamination of (2*R*, 4*S*)-diaminopentanoate to 2-amino-4-ketopentanoate. The formation of NADH is

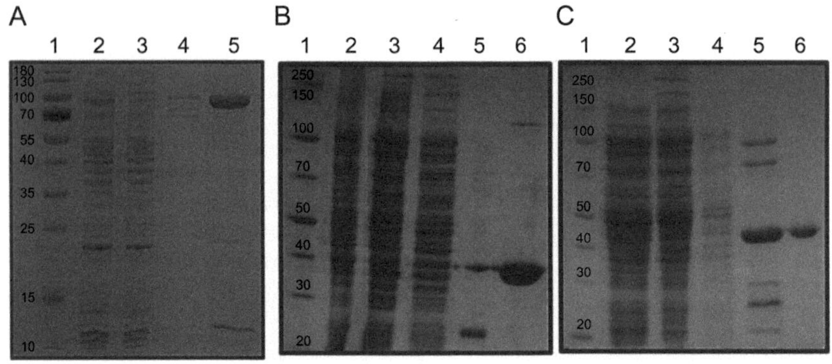

Fig. 3 (A) SDS 13% polyacrylamide gel showing purification of 4,5-OAM by Ni-NTA affinity and anion exchange chromatography Lane 1, protein molecular weight markers in kDa; Lane 2, crude extract from 4,5-OAM overexpressing Rosetta(DE3)pLysS cells. Lane 3. Flow-through from Ni-NTA column; Lane 4, wash with 50 mM Tris pH 7.5, 0.5 M NaCl, 20 mM imidazole Lane 5, Q-sepharose eluate from second purification step. (B) SDS 10% polyacrylamide gel showing purification of DAPDH by Ni-NTA affinity and anion exchange chromatography. Lane 1, protein molecular weight markers in kDa; Lane 2, crude extract from DADPH over-expressing Rosetta(DE3)pLysS cells; Lane 3. Flow-through from Ni-NTA column; Lane 4, wash with 50 mM Tris pH 7.5, 0.5 M NaCl, 20 mM imidazole; Lane 5, Ni-NTA eluate from first purification step; Lane 6, Q-sepharose eluate from second purification step. (C) SDS 10% polyacrylamide gel showing purification of OR by Ni-NTA affinity chromatography followed by Q-sepharose anion-exchange chromatography. Lane 1, protein molecular weight markers in kDa; Lane 2, crude extract from Rossetta(DE3)pLysS cells overexpressing OR; Lane 3, flow-through from initial Ni-NTA column; Lane 4, wash with 50 mM Tris pH 7.5, 0.5 M NaCl, 20 mM imidazole; Lane 5, Ni-NTA eluate from first purification step; Lane 6, Q-sepharose eluate from second purification step.

monitored at 340 nm using a UV–visible spectrophotometer. DAPDH had previously been shown to elicit >1000-fold higher k_{cat}/K_M for NAD^+ compared to $NADP^+$, consequently the former pyridine nucleotide is used as a cofactor (Fonknechten et al., 2009). A dual beam spectrophotometer with a deuterium lamp is preferred over a xenon lamp, as prolonged exposure to the latter can cause photolysis of AdoCbl, compromising kinetic analysis. The kinetic assays are performed in a 1.0 mL volume at 25 °C using a 1 cm pathlength quartz cuvette. The temperature of the cuvette holder is maintained a 25 °C using a circulating water bath and a thermostatted cell holder. The reaction mixtures contain 50 mM NH_4^+ -EPPS, pH 8.5, 100 nM 4,5-OAM, 100 nM of DAPDH, 10 μM PLP, 10 μM AdoCbl, 0.5 mM NAD^+, and variable concentrations of substrate. Under these conditions, the rate of NADH formation is proportional to the amount of 4,5-OAM added to the reaction mixture. Reactions are performed under low ambient light conditions and are initiated with the addition of substrate.

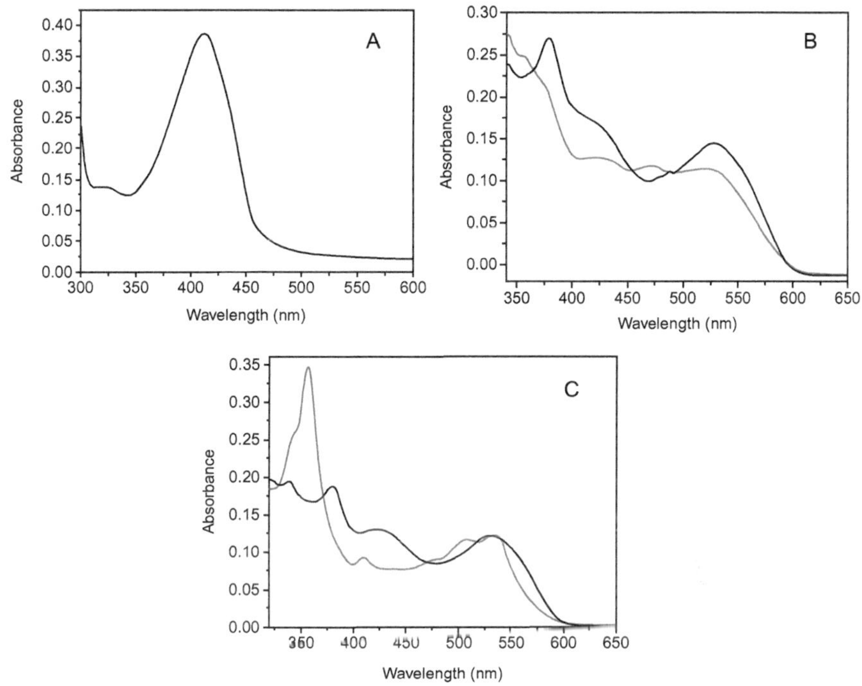

Fig. 4 Absorption spectra of cofactor bound from of OR and 4,5-OAM (A) UV–visible spectrum of 50 μM OR showing the prominent absorbance peak of the PLP cofactor at 410 nm. (B) UV–visible absorbance spectrum of an anaerobic solution of 15 μM 4,5-OAM reconstituted with an equimolar amount of PLP and AdoCbl before (black line) and 1 min after the addition of 2.5 mM 2,4-diaminobutyrate (gray line). The decrease in absorbance at 416 and 528 nm reflect formation of the external aldimine and homolysis of the Co—C bond, respectively. Accumulation of the cob(*II*)alamin species is denoted by the peak at 470 nm. (C) UV–visible absorbance spectrum of an aerobic solution of 12 μM 4,5-OAM reconstituted with an equimolar amount of PLP and AdoCbl before (black line) and (1 min) after the addition of 2.5 mM 2,4-diaminobutyrate (gray line). The prominent peak at 358 nm and the smaller double peaks at 505 and 524 nm denote the hydroxycob(*III*)alamin species.

The kinetic assays can be performed under aerobic conditions as during steady state turnover with D-ornithine, radical intermediates do not accumulate to sufficient levels to react with O_2.

2.3.1 Equipment

Dual beam UV–visible spectrophotometer with a thermostatted cell holder
Circulating water bath
1 mL quartz cuvette
Analytical balance

2.3.2 Chemicals and reagents

Purified OAM
Purified DAPDH
PLP (Millipore Sigma)
AdoCbl (Millipore Sigma)
NAD^+ (Millipore Sigma)
D-ornithine (Millipore Sigma)
50 mM NH_4^+-EPPS, pH 8.5

2.3.3 Procedure

1. Turn on the UV–visible spectrophotometer and the circulating water bath, and set temperature to 25 °C
2. Determine the concentration of apoOAM using the ratio of 1 OD_{280} = 1.2 mg/mL. Prepare a 20 μM stock in 50 mM NH_4^+-EPPS. The molecular mass of the combined OraS and OraE subunit is 97 kDa
3. Determine the concentration of DAPDH using the ratio of 1 OD_{280} = 3.1 mg/mL. Prepare a 20 μM stock in 50 mM NH_4^+-EPPS. The molecular mass of DAPDH is 38 kDa
4. Open up the bottle of AdoCbl in a very dimly lit room and add ~2–5 mg of the cofactor to an amber Eppendorf tube. Dissolve the cofactor in 0.5 mL 50 mM NH_4^+ EPPS, pH 8.5. Prepare a 1/200 dilution of the cofactor in a total volume of 1 mL and determine the concentration by the absorbance value at 525 nm ($\varepsilon = 8.3\,mM^{-1}\,cm^{-1}$). Prepare a 2 mM stock in 50 mM NH_4^+ EPPS, pH 8.5 buffer
5. Dissolve 2–5 mg of PLP in 0.5 mL 0.1 M NaOH. Prepare a 1/200 dilution of the cofactor in a total volume of 1 mL and determine the concentration by the absorbance value at 390 nm ($\varepsilon = 6.6\,mM^{-1}\,cm^{-1}$). Prepare a 2 mM stock in 50 mM NH_4^+ EPPS, pH 8.5 buffer
6. Dissolve 2–5 mg of NAD^+ in 0.5 mL of 50 mM NH_4^+ EPPS, pH 8.5 and determine the concentration of a 1/200 dilution by measuring the absorbance at 260 nm ($\Delta\varepsilon = 17.8\,mM^{-1}\,cm^{-1}$). Make a 0.1 M stock
7. Weigh out a known mass of D-ornithine using an analytical balance and dissolve in 50 mM NH_4^+-EPPS to achieve a concentration of 0.5 M
8. Under low ambient light conditions add 100 nM 4,5-OAM (apoform), 100 nM of DAPDH, 10 μM PLP, 0.5 mM NAD^+, and 10 μM AdoCbl to the quartz cuvette in 50 mM NH_4^+ EPPS
9. Invert the quartz cuvette four times and insert into the cuvette holder
10. After 3 min, initiate the reaction by adding D-ornithine. Invert the quartz cuvette four times again and record absorbance at 340 nm

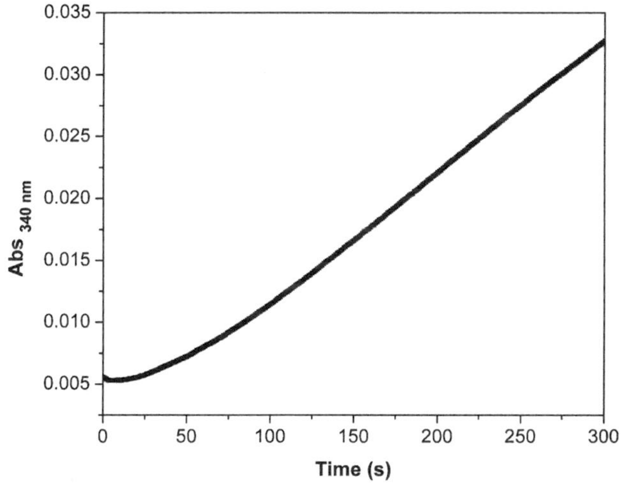

Fig. 5 Coupled-enzyme spectrophotometric assay for monitoring 4,5-OAM turnover. Sample absorbance data collected at 340 nm for wild type OAM with 2.5 mM D-ornithine during coupled enzyme assay with DAPDH. The initial velocity of the reaction was determined from the linear portion of the curve between 150 and 210 s, following the initial lag phase.

The absorbance trace at 340 nm elicits a lag time due to the coupled nature of the kinetic assay (Fig. 5). The initial velocity of the reaction is determined from the linear portion of the absorbance trace. The changes in absorbance over a select time domain along with the extinction coefficient of NADH ($\varepsilon_{340\ nm} = 6.2\,mM^{-1}\,cm^{-1}$) is used to determine the initial velocity of the reaction. The initial velocity data as a function of D-ornithine concentration are hyperbolic and consequently fitted to the Michaelis-Menten using nonlinear least squares analysis to extract values for k_{cat} and K_M for L-ornithine.

2.4 Preparation of DL-ornithine and DL-ornithine-3,3,4,4,5,5-d_6

The above coupled spectroscopic assay was also used to measure deuterium kinetic isotope effects of k_{cat} and k_{cat}/K_M using deuterated ornithine. The only deuterated form of ornithine that was commercially available to us was the L-isomer (e.g., L-ornithine-3,3,4,4,5,5-d_6 from CND isotopes). Accordingly, ornithine racemase from *C. sticklandii* was used to convert L-ornithine-3,3,4,4,5,5-d_6 to a racemic mixture of D,L-ornithine-3,3,4,4,5,5-d_6 which could be used as a substrate for 4,5-OAM. Prior studies revealed that maximal activity of ornithine racemace occurred at pH 8.5 (Chen et al., 2000). Therefore the following protocol was used to prepare D,L-ornithine-3,3,4,4,5,5-d_6. A 10 mL mixture containing 100 mM L-ornithine-3,3,4,4,5,5-d_6 in water was adjusted to a pH of 8.5. OR (100 μM) was added to the solution, and then the solution was stirred gently for 18 h at 20 °C. Aluminum foil was wrapped around the reaction flask to minimize exposure of the PLP cofactor to ambient light, and the pH of the mixture was monitored periodically. The reaction mixture was passed through a 30 kDa MW cut-off filter of a centrifugal filter unit to remove the enzyme. The products were then dried under reduced pressure. The resulting solid was washed with 10 mL of 95% ethanol and then dried under pressure, leaving behind a fine white powder. ^{1}H NMR was used to confirm the purity of the ornithine product. A racemic mixture of unlabelled D,L-ornithine was also synthesized from L-ornithine using the same protocol.

2.5 Measuring deuterium kinetic isotope effects

As shown in Table 1, a higher apparent K_M was obtained with D,L-ornithine compared to enantiomerically pure D-ornithine, while the k_{cat} values are comparable. These data indicate that L-ornithine present in the racemic mixture is acting as a competitive inhibitor. Thus, to obtain $^Dk_{cat}$ and $^Dk_{cat}/K_M$, the coupled spectrophotometric assay was performed with varying concentrations of D,L-ornithine and D,L-ornithine-3,3,4,4,5,5-d_6. The initial velocity data were fitted to the Michealis-Menten equation to obtain k_{cat} and k_{cat}/K_M values for the protected and deuterated substrate. The ratio of k_{cat} for D,L-ornithine over k_{cat} for D,L-ornithine-3,3,4,4,5,5-d_6 was used to obtain $^Dk_{cat}$, while the ratio of k_{cat}/K_M for D,L-ornithine over k_{cat} for D,L-ornithine-3,3,4,4,5,5-d_6 was used to obtain $^Dk_{cat}/K_M$.

Table 1 Summary of steady-state kinetic parameters for wild type OAM with substrates D-ornithine, D,L-ornithine and D,L-ornithine-3,3,4,4,5,5-d_6.

Substrate	k_{cat} (s^{-1})	K_M (μM)	k_{cat}/K_M (× 10^3 M^{-1} s^{-1})
D-ornithine	2.9 ± 0.1[a]	190 ± 13	15 ± 1.2
D,L-ornithine	2.9 ± 0.1	570 ± 45	5.1 ± 0.41
D,L-ornithine-d_6	0.38 ± 0.01	190 ± 11	2.0 ± 1.7

[a]From reference Makins et al. (2014).

3. Pre-steady-state kinetic analysis of 4,5-OAM

Pre-steady state analysis of AdoCbl-dependent eliminases and mutases with their physiological substrate reveals significant accumulation of the cob(*II*)alamin intermediate (Bandarian & Reed, 2000; Marsh & Ballou, 1998; Padmakumar, Padmakumar, & Banerjee, 1997; Toraya, Krodel, Mildvan, & Abeles, 1979; Zelder, Beatrix, Leutbecher, & Buckel, 1994). This is observed through a rapid absorbance decrease at 528 nm which tracks AdoCbl Co—C bond homolysis and an absorbance increase at 470 nm which denotes formation of the cob(*II*)alamin intermediate. Accumulation of radical intermediates is also observed in electron paramagnetic resonance (EPR) spectra of the AdoCbl-dependent isomerases in the presence of saturating amounts of substrate (Bandarian & Reed, 2002; Ke, 2003; Ke, Torrent, Museav, Morokuma, & Warncke, 1999; Padmakumar & Banerjee, 1995; Zhao, Abend, Kunz, Such, & Retey, 1994). In contrast, the addition of D-ornithine to 4,5-OAM does not lead to significant accumulation of radical intermediates to allow for detection through EPR or UV–visible spectroscopy. However, the addition of the substrate analog 2,4-diaminobutyrate does lead to accumulation of cob(*II*)alamin, because abstraction of the C4 hydrogen of the diaminobutryl-PLP adduct by dAdo$^{\bullet}$ leads to formation of a carbon-centered radical that is stabilized by delocalization through the Schiff base and pyridine ring of the cofactor (Wolthers et al., 2008). Given that 2,4-diaminobutyrate elicits a sizable absorbance change at 528 nm, it is used to measure the observed rate constants associated with Co—C bond homolysis (Fig. 4B). Stopped-flow absorbance spectrophotometry can also be used to measure the observed rate constants associated with formation of the external aldimine; this catalytic step is monitored by absorbance changes at 416 nm, the absorbance maximum of the PLP cofactor.

Pre-steady state kinetic analysis of 4,5-OAM must be performed under anaerobic conditions as the cob(*II*)alamin intermediate reacts rapidly with O_2 to form hydroxycob(*III*)alamin, which renders the enzyme inactive. The rapid formation of hydroxycob(*III*)alamin is denoted by a large absorbance peak at 358 nm and smaller but distinctive double peaks at 505 and 524 nm (Wolthers et al., 2008) (Fig. 4C). The spectral changes obscure expected absorbance changes at 525 and 470 nm. The AdoCbl cofactor is sensitive to photolysis, which further necessitates that the reactions are performed under low ambient light conditions. Below, we describe our experimental system for performing stopped-flow spectroscopy under anaerobic conditions.

3.1 Anaerobic glove box

A rigid modular glove box with three glove ports from Belle Technology (UK) is used by us to perform pre-steady state kinetic analysis of enzymatic reactions that are sensitive to O_2. The glove box has sufficient space to fit a small UV–visible spectrophotometer and the sample handling unit (SHU) of a stopped-flow system (Fig. 6). The system comes with an external gas purifying recirculator which sits on the floor below the glove box (Fig. 6B) and a catalyst regeneration system (Fig. 6C) that removes O_2 by heating the catalyst and flushing it with a 90% N_2, 10% H_2 gas mixture. Accessories include an oxygen meter to monitor O_2 levels during the course of the experiment, four electrical outlets, and a gas cooling unit (Fig. 6D). The box is outfitted with a large and small port for introduction of samples and reagents.

3.2 Stopped-flow system and UV–visible spectrophotometer

The SX20 from Applied Photophysics and the SF-61DX2 stopped-flow apparatus from TgK Scientific have both been used to monitor the pre-steady state kinetic behavior of 4,5-OAM (Makins et al., 2013; Wolthers et al., 2008). The SF-61DX2 comprises a SHU, two electronic control units, and a collection of optics for excitation and detection (Fig. 7). The SHU is the only component of the stopped-flow apparatus that is housed in the glove box (Figs. 6F, 7A and C). An access port was installed in the Belle Technology glove box to allow cables and tubing to be connected to the SHU from external sources (e.g., N_2 gas supply, electronic control units, optical detectors, circulating water bath). The SHU contains four syringes. Single mixing events (i.e., substrate and enzyme) uses syringes C and D, and double mixing or sequential mixing events uses all four syringes (A–D). Two pneumatic drive cylinders, one positioned under syringes C and D and one under syringes A and B, provide the tractive effort to drive syringes and mix solutions. A tank containing high purity N_2 gas (99.998%) is positioned on the left side of the box and provides air pressure of 100 psi to the drive syringes.

The CU-61 control unit, situated on the bench outside the glove box, contains circuit boards for the main, reference, and auxillary photomultiplier tubes (PMT) which plug into sockets in the back of the unit (Fig. 7). The stepper support unit (SSU-60) houses the control circuitry for the SHU and also powers the two stepper motors within the SHU. The lamp housing and excitation optics are carried by an optical rail which is located on top of the CU-61 control unit. A fiber optic cable couples the excitation light through an access port of the glove box to the observation cell in the SHU (Fig. 7).

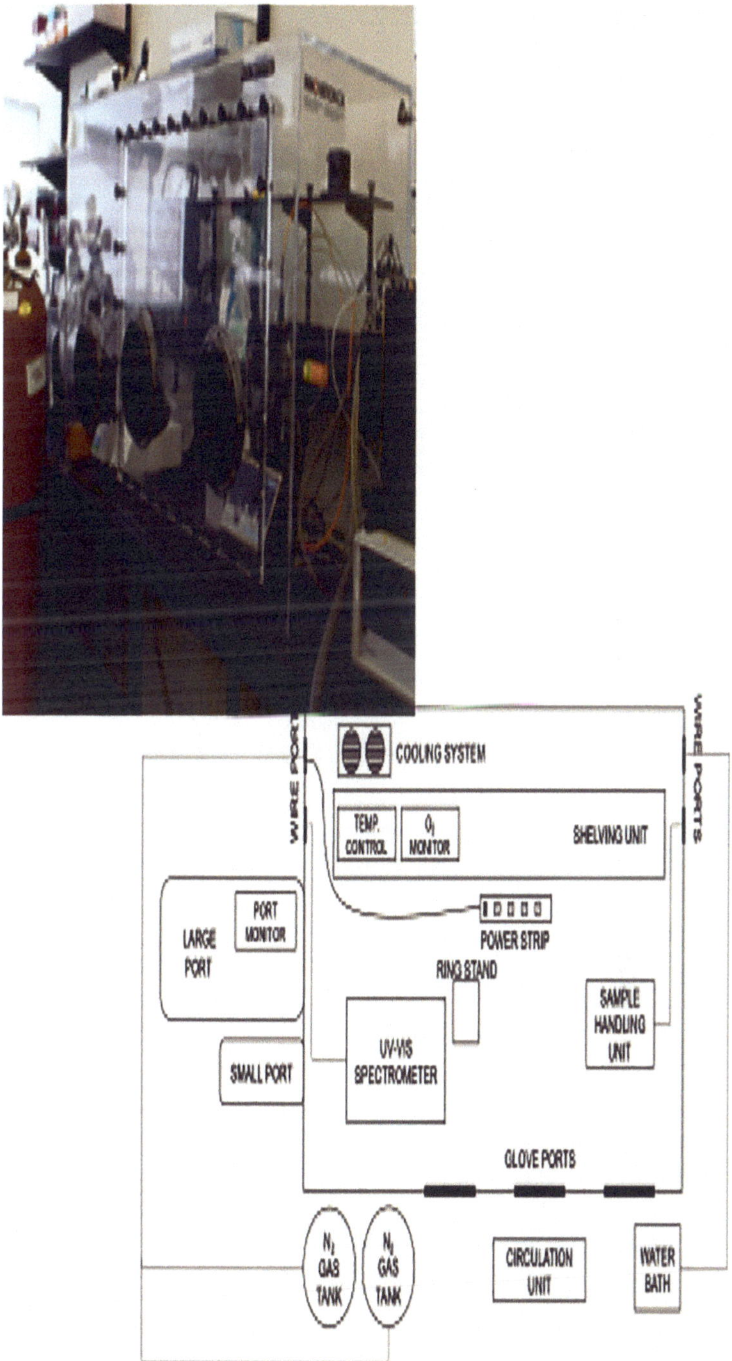

Fig. 6 Anaerobic glove box from Belle Technology with a UV–visible spectrophotometer and sample handling unit. The photographs counter clockwise are of (A) the exterior of the glove box, (B) the recirculator unit and water bath on the floor below the glove box, (C) cooling fan and O_2 meter located on the shelf in the glove box, (D) catalyst regenerator located in the fume hood, (E) small and large port, and (F) diagram of the system viewed from above.

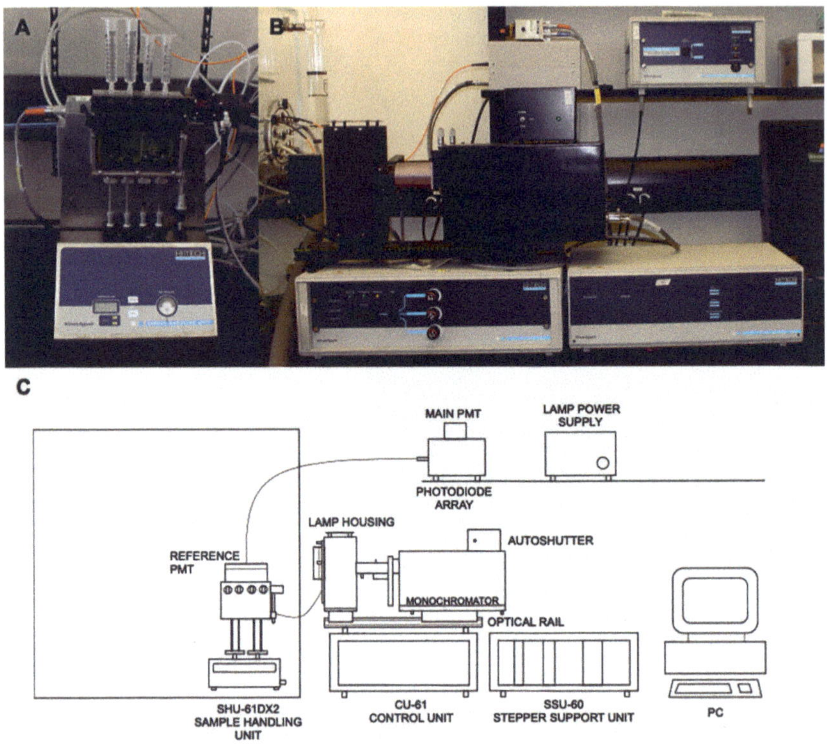

Fig. 7 SF-61DX2 stopped flow apparatus from TgK Scientific. (A) Photograph of the sample handling unit in the glove box and (B) photograph of the optical system and electronics. (C) Diagram of the SF-61DX2 stopped flow apparatus showing all of the components. *Adapted from the TgK Scientific manual.*

For our anaerobic set up, the main PMT is outside the glove box and sits above the photodiode array detector (Fig. 7). A second fiber optic light guide is used to transmit light from the optical cell through an access port to the main PMT. This light guide can easily be disconnected from the main PMT to the photodiode array detector allowing users to easily switch between collecting single wavelength and multiwavelength time resolved data in an anaerobic glove box. A circulating water bath (Fig. 6B) sits below the bench and tubing, insulated with foam is passed through the access port and is connected to the back of the SHU (Fig. 7).

3.3 Stopped-flow spectrophotometry of 4,5-OAM

3.3.1 Equipment

Stopped-flow system
Glove box

Circulating water bath
10 mL desalting column
1 mL quartz cuvette
Analytical balance
UV–visible spectrophotometer
Two gas cylinders with high purity N_2 (99.998%)

3.3.2 Chemicals and reagents

AdoCbl (Sigma Millipore)
PLP (Sigma Millipore)
4,5-OAM
2,4-Diaminobutyrate (Sigma Millipore)
50 mM NH_4^+ EPPS, pH 8.5

3.3.3 Procedure

3.3.3.1 Day 1

1. Prepare 50 mM NH_4^+ EPPS, pH 8.5 in a total volume of 200 mL. The bottle containing the buffer is positioned in the fume hood on a stir plate and a needle is inserted into a silicon lined cap. The buffer is then purged with high purity N_2 (99.998%) for 2 h while stirring. After 2 h, the needle is removed and the buffer in placed inside the glove box. The cap is removed and the solution is allowed to equilibrate in the anaerobic environment for 16 h. The same procedure is performed for 100 mL of 0.1 M NaOH.
2. Introduce non-perishable items into the glove box via the large port that will be required for the experiment (pipette tips, 10 mL desalting column, pipette tips, pipettes, Eppendorf tubes, quartz cuvette, waste container).

3.3.3.2 Day 2

1. The lamp of the stopped flow is ignited first and allowed to warm up for at least 30 min. The remaining electronics (CU-61 and SSU-60) are then turned on along with the computer and water bath.
2. AdoCbl (2–5 mg), PLP (2–5 mg), apo-OAM (2 mL of 20 mg/mL) and 2,4-diaminopentantoate (pre-weighed amount) in Eppendorf tubes are introduced into the glove box via the small port.
3. AdoCbl is dissolved in 1 mL of anaerobic 50 mM NH_4^+ EPPS, pH 8.5 buffer. The concentration of the cofactor is determined by the absorbance value at 525 nm of a 1/200 diluted sample ($\varepsilon = 8.3\ mM^{-1}\ cm^{-1}$).

4. PLP is dissolved in 0.5 mL of anaerobic 0.1 M NaOH. The solution is then diluted 1/200 in anaerobic 50 mM NH_4^+ EPPS pH 8.5 buffer and the concentration of the cofactor is determined by measuring the absorbance at 390 nm ($\varepsilon = 6.6\,mM^{-1}\,cm^{-1}$).
5. A 0.5 M stock solution of 2,4-diaminobutyrate is prepared by adding an appropriate volume of anaerobic 50 mM NH_4^+EPPS, pH 8.5 buffer to a known mass of the substrate analog.
6. A 2 mL solution of apoOAM (20 mg/mL) is applied to a 10 mL desalting column (e.g., Econo-Pac® 10DG Desalting column) equilibrated with three column volumes of anaerobic NH_4^+EPPS pH 8.5 buffer. Following absorption of apoOAM into the resin of the column by gravity flow, 20 mL of 50 mM NH_4^+EPPS pH 8.5 is applied to the column. The concentration of apoOAM, which elutes in the 20th to 30th drop, is determined by the absorbance at 280 nm using a calculated ratio of 1 OD = 1.2 mg/mL.
7. The PLP and AdoCbl cofactors are added in equimolar concentration to apoOAM to achieve a final holoOAM concentration of 50 μM. The protein is stored in amber colored tubes to protect it from ambient light.
8. The N_2 supply to the SHU is turned on; the internal pressure should read 0.4 MPa on the drive air pressure gauge.
9. The stopped-flow is typically left in water, so it must be flushed with anaerobic buffer. The syringes are emptied by switching valves of C and D to the FILL position. The drive syringes are pushed up and water is removed with 3 mL plastic transfer pipettes.
10. Buffer is added to reservoirs C and D and drive syringes are slowly lowered to the push plate. The valves of C and D are then switched to the DRIVE position.
11. With the SF-61DX2 model, buffer can be flushed through the sample flow circuit by manually acquiring "shots.". This is done by toggling the lever of the STOP/WASTE valve to the WASTE position and then pushing up on the stop syringe until it stops. The valve is then switched to the DRIVE position and pneumatic pressure is released by pressing the DRIVE PUSH button on the front of the SHU. This process is repeated 30 times to thoroughly flush the system with anaerobic buffer.
12. For single wavelength analysis, the fiber optic cable from the SHU observation cell is connected to the main PMT and the DUAL BEAM MODE is selected on the CU-61 control unit.
13. The shutters on the reference and main PMTs are opened.
14. The Kinetic Studio software is opened and the PHOTOMULTIPLIER is selected from the ACQUIRE menu.

15. A blank is acquired by selecting SPECTROPHOTOMETER and then SETUP AND SCAN BLANK. This option will allow data acquisition at several wavelengths (i.e., 416 and 528 nm).
16. Optimize the signal by adjusting the voltage of the main PMT and reference PMT using the control knobs on CU-61 control unit. The voltage should be at least 250 V for a linear response. The monochromator slits widths can be adjusted as necessary and the lamp position with respect to the monochromator can be changed by sliding it along the rail.
17. Remove buffer from syringes C and D and add 50 μM holoOAM to one syringe and 5 mM 2,4-diaminobutyrate to the other syringe. Put the valves back to the DRIVE position. Acquire three manual shots as described above.
18. With the Kinetic Studio, select a wavelength (i.e., 416 or 528 nm) and then a time base.
19. Acquire 5–10 absorbance traces and average them within Kinetic Studio.
20. Switch to a second wavelength and acquire 5–10 absorbance traces and average them within Kinetic Studio.
21. The absorbance traces are monophasic and observed rate constants can be obtained by fitting the traces to a single exponential equation (Fig. 8). This can be done under TOOLS > ANALYSIS > CONTROL PANEL.
22. After data acquisition, the flow circuitry is extensively washed with deionized water.

As shown in Fig. 8, the observed rate constants obtained from the absorbance traces at 416 nm ($464 \pm 36\,s^{-1}$) and 528 nm ($489 \pm 40\,s^{-1}$) are comparable, which indicates that formation of the external aldimine is slower than the subsequent homolysis step. Thus, AdoCbl Co—C bond homolysis in 4,5 OAM occurs with a rate constant $>500\,s^{-1}$ at 25 °C. For AdoCbl-dependent methylmalonyl CoA mutase and glutamate mutase, homolysis is tightly coupled to hydrogen abstraction by dAdo• from the substrate, and as a consequence a kinetic isotope effect is observed on the rate constant for Co—C bond homolysis in the presence of deuterated substrate (Marsh & Ballou, 1998; Padmakumar et al., 1997). However, we do not observe a kinetic isotope effect on the observed rate constant for Co—C bond homolysis with D,L-ornithine-3,3,4,4,5,5-d_6 as this catalytic step is fast and is "gated" by the preceding transimination step.

3.3.3.3 Notes

1. AdoCbl is light sensitive and the powder is dispensed into an amber Eppendorf tube in a dimly lit room

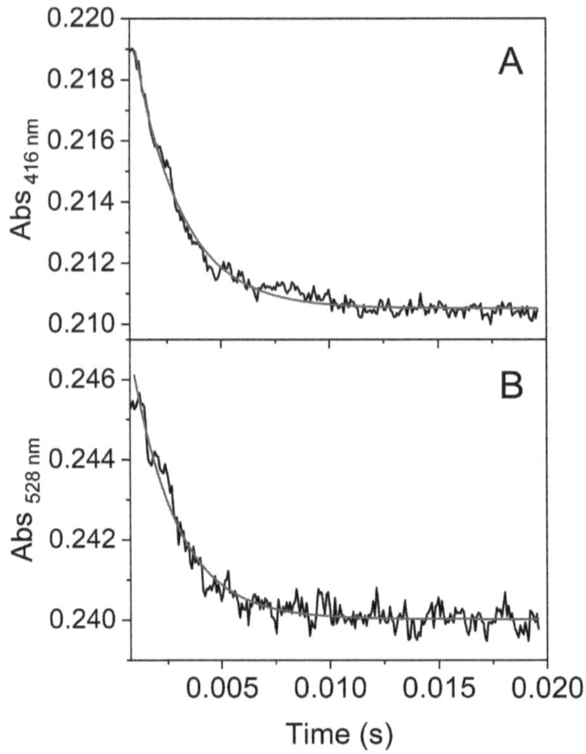

Fig. 8 Stopped flow absorbance traces following (A) transimination at 416nm and (B) AdoCbl Co—C bond homolysis at 528nm over 20ms following rapid-mixing of 50 μM wild type holoOAM with 5 mM 2,4-diaminobutyrate at 25 °C. The black line in each panel shows the average of five absorbance traces which were fitted to a single exponential equation to generate observed rate constants of $464 \pm 36\,s^{-1}$ for transimination and $489 \pm 40\,s^{-1}$ for homolysis.

2. The Eppendorf tubes containing the enzyme, cofactors, and substrate are left open in the small port while purging with N_2. Tin foil is loosely placed over the top of the Eppendorf to protect the AdoCbl cofactor from ambient light
3. Tin foil is also placed over the glass plate of the thermostated jacket of the SHU to protect holoOAM from photolysis during data acquisition

4. Conclusions

Steady-state and pre-steady state kinetic analysis of 4,5-OAM have provided valuable insight into the catalytic mechanism of the enzyme and the importance of key active site residues in supporting radical-mediated chemistry. Additional metabolic enzymes in the L-ornithine fermentation pathway have proved useful for developing a simple coupled spectrophotometric assay and in the synthesis of deuterated substrate. Inspection of the anaerobic lysine fermentation pathway reveals a $NADP^+$-dependent dehydrogenase that could also prove useful for measuring the activity of 5,6-LAM. Finally, we detailed an anaerobic stopped-flow system that is used to measure Co—C bond homolysis and transimination. This system can also be used to measure the pre-steady state kinetic properties of other enzymatic process that are sensitive to O_2.

Acknowledgment

Research reported in this publication is supported by the Natural Sciences and Engineering Council of Canada (NSERC) Discovery Grants (RGPIN-2015-06130).

References

Baker, J. J., Jeng, I., & Barker, H. A. (1972). Purification and properties of L-erythro-3,5-diaminohexanoate dehydrogenase from a lysine-fermenting Clostridium. *Journal of Biological Chemistry*, *247*(23), 7724–7734.

Baker, J. J., van der Drift, C., & Stadtman, T. C. (1973). Purification and properties of -lysine mutase, a pyridoxal phosphate and B 12 coenzyme dependent enzyme. *Biochemistry*, *12*(6), 1054–1063.

Bandarian, V., & Reed, G. H. (2000). Isotope effects in the transient phases of the reaction catalyzed by ethanolamine ammonia-lyase: Determination of the number of exchangeable hydrogens in the enzyme-cofactor complex. *Biochemistry*, *39*(39), 12069–12075.

Bandarian, V., & Reed, G. H. (2002). Analysis of the electron paramagnetic resonance spectrum of a radical intermediate in the coenzyme B(12)-dependent ethanolamine ammonia-lyase catalyzed reaction of S-2-aminopropanol. *Biochemistry*, *41*(27), 8580–8588.

Barker, H. A. (1981). Amino acid degradation by anaerobic bacteria. *Annual Review of Biochemistry*, *50*, 23–40.

Barker, H. A., Kahn, J. M., & Hedrick, L. (1982). Pathway of lysine degradation in fusobacterium nucleatum. *Journal of Bacteriology*, *152*(1), 201–207.

Buckel, W., & Golding, B. T. (2006). Radical enzymes in anaerobes. *Annual Review of Microbiology*, *60*, 27–49.

Chen, H. P., Lin, C. F., Lee, Y. J., Tsay, S. S., & Wu, S. H. (2000). Purification and properties of ornithine racemase from Clostridium sticklandii. *Journal of Bacteriology*, *182*(7), 2052–2054.

Chen, H. P., Wu, S. H., Lin, Y. L., Chen, C. M., & Tsay, S. S. (2001). Cloning, sequencing, heterologous expression, purification, and characterization of adenosylcobalamin-dependent D-ornithine aminomutase from Clostridium sticklandii. *Journal of Biological Chemistry*, *276*(48), 44744–44750.

Cosper, N. J., Booker, S. J., Ruzicka, F., Frey, P. A., & Scott, R. A. (2000). Direct FeS cluster involvement in generation of a radical in lysine 2,3-aminomutase. *Biochemistry*, *39*(51), 15668–15673.

Dobson, C. M., Wai, T., Leclerc, D., Wilson, A., Wu, X., Dore, C., et al. (2002). Identification of the gene responsible for the cblA complementation group of vitamin B12-responsive methylmalonic acidemia based on analysis of prokaryotic gene arrangements. *Proceedings of the National Academy of Sciences of the United States of America*, *99*(24), 15554–15559.

Dyer, J. K., & Costilow, R. N. (1968). Fermentation of ornithine by Clostridium sticklandii. *Journal of Bacteriology, 96*(5), 1617–1622.

Fonknechten, N., Perret, A., Perchat, N., Tricot, S., Lechaplais, C., Vallenet, D., et al. (2009). A conserved gene cluster rules anaerobic oxidative degradation of L-ornithine. *Journal of Bacteriology, 191*(9), 3162–3167.

Frey, P. A., & Magnusson, O. T. (2003). S-Adenosylmethionine: A wolf in sheep's clothing, or a rich man's adenosylcobalamin? *Chemical Reviews, 103*(6), 2129–2148.

Holloway, D. E., & Marsh, E. N. G. (1993). Cloning and sequencing of glutamate mutase component E from Clostridium tetanomorphum. Organization of the Mut genes. *FEBS Letters, 317*(1–2), 44–48.

Jeng, I. M., Somack, R., & Barker, H. A. (1974). Ornithine degradation in Clostridium sticklandii; pyridoxal phosphate and coenzyme a dependent thiolytic cleavage of 2-amino-4-ketopentanoate to alanine and acetyl coenzyme a. *Biochemistry, 13*(14), 2898–2903.

Ke, S. C. (2003). Spin-spin interaction in ethanolamine deaminase. *Biochimica et Biophysica Acta, 1620*(1–3), 267–272.

Ke, S. C., Torrent, M., Museav, D. G., Morokuma, K., & Warncke, K. (1999). Identification of dimethylbenzimidazole axial coordination and characterization of (14)N super-hyperfine and nuclear quadrupole coupling in cob(II)alamin bound to ethanolamine deaminase in a catalytically-engaged substrate radical-cobalt(II) biradical state. *Biochemistry, 38*(39), 12681–12689.

Liu, N., Wu, L., Feng, J. H., Sheng, X., Li, J., Chen, X., et al. (2021). Crystal structures and catalytic mechanism of l-erythro-3,5-diaminohexanoate dehydrogenase and rational engineering for asymmetric synthesis of beta-amino acids. *Angew. Chem. Int. Ed., 60*(18), 10203–10210.

Makins, C., Miros, F. N., Scrutton, N. S., & Wolthers, K. R. (2012). Role of histidine 225 in adenosylcobalamin-dependent ornithine 4,5-aminomutase. *Bioorganic Chemistry, 40*, 39–47.

Makins, C., Pickering, A. V., Mariani, C., & Wolthers, K. R. (2013). Mutagenesis of a conserved glutamate reveals the contribution of electrostatic energy to adenosylcobalamin co-C bond homolysis in ornithine 4,5-aminomutase and methylmalonyl-CoA mutase. *Biochemistry, 52*(5), 878–888.

Makins, C., Whitelaw, D. A., McGregor, M., Petit, A., Mothersole, R. G., Prosser, K. E., et al. (2017). Optimal electrostatic interactions between substrate and protein are essential for radical chemistry in ornithine 4,5-aminomutase. *Biochimica et Biophysica Acta - Proteins and Proteomics, 1865*(8), 1077–1084.

Makins, C., Whitelaw, D. A., Mu, C. H., Walsby, C. J., & Wolthers, K. R. (2014). Isotope effects for deuterium transfer and mutagenesis of Tyr187 provide insight into controlled radical chemistry in adenosylcobalamin-dependent ornithine 4,5-aminomutase. *Biochemistry, 53*(33), 5432–5443.

Marsh, E. N. G., & Ballou, D. P. (1998). Coupling of cobalt-carbon bond homolysis and hydrogen atom abstraction in adenosylcobalamin-dependent glutamate mutase. *Biochemistry, 37*(34), 11864–11872.

Padmakumar, R., & Banerjee, R. (1995). Evidence from electron paramagnetic resonance spectroscopy of the participation of radical intermediates in the reaction catalyzed by methylmalonyl-coenzyme a mutase. *Journal of Biological Chemistry, 270*(16), 9295–9300.

Padmakumar, R., Padmakumar, R., & Banerjee, R. (1997). Evidence that cobalt-carbon bond homolysis is coupled to hydrogen atom abstraction from substrate in methylmalonyl-CoA mutase. *Biochemistry, 36*(12), 3713–3718.

Somack, R., & Costilow, R. N. (1973a). 2,4-diaminopentanoic acid C 4 dehydrogenase. Purification and properties of the protein. *Journal of Biological Chemistry, 248*(2), 385–388.

Somack, R., & Costilow, R. N. (1973b). Purification and properties of a pyridoxal phosphate and coenzyme B 12 dependent D-ornithine 5,4-aminomutase. *Biochemistry*, *12*(14), 2597–2604.

Stickland, L. H. (1934). Studies in the metabolism of the strict anaerobes (genus Clostridium): The chemical reactions by which cl. Sporogenes obtains its energy. *The Biochemical Journal*, *28*(5), 1746–1759.

Tang, K. H., Casarez, A. D., Wu, W., & Frey, P. A. (2003). Kinetic and biochemical analysis of the mechanism of action of lysine 5,6-aminomutase. *Archives of Biochemistry and Biophysics*, *418*(1), 49–54.

Tang, K. H., Chang, C. H., & Frey, P. A. (2001). Electron transfer in the substrate-dependent suicide inactivation of lysine 5,6-aminomutase. *Biochemistry*, *40*(17), 5190–5199.

Tang, K. H., Harms, A., & Frey, P. A. (2002). Identification of a novel pyridoxal 5′-phosphate binding site in adenosylcobalamin-dependent lysine 5,6-aminomutase from *Porphyromonas gingivalis*. *Biochemistry*, *41*(27), 8767–8776.

Tobimatsu, T., Kajiura, H., Yunoki, M., Azuma, M., & Toraya, T. (1999). Identification and expression of the genes encoding a reactivating factor for adenosylcobalamin-dependent glycerol dehydratase. *Journal of Bacteriology*, *181*(13), 4110–4113.

Toraya, T., Krodel, E., Mildvan, A. S., & Abeles, R. H. (1979). Role of peripheral side chains of vitamin B12 coenzymes in the reaction catalyzed by dioldehydrase. *Biochemistry*, *18*(3), 417–426

Wolthers, K. R., Levy, C., Scrutton, N. S., & Leys, D. (2010). Large-scale domain dynamics and adenosylcobalamin reorientation orchestrate radical catalysis in ornithine 4,5-aminomutase. *Journal of Biological Chemistry*, *285*(18), 13942 13950.

Wolthers, K. R., Rigby, S. E., & Scrutton, N. S. (2008). Mechanism of radical-based catalysis in the reaction catalyzed by adenosylcobalamin-dependent ornithine 4,5-aminomutase. *Journal of Biological Chemistry*, *283*(50), 34615–34625.

Woods, D. D. (1936). Studies in the metabolism of the strict anaerobes (genus Clostridium): Further experiments on the coupled reactions between pairs of amino-acids induced by cl. Sporogenes. *The Biochemical Journal*, *30*(10), 1934–1946.

Wu, W., Booker, S., Lieder, K. W., Bandarian, V., Reed, G. H., & Frey, P. A. (2000). Lysine 2,3-aminomutase and trans-4,5-dehydrolysine: Characterization of an allylic analogue of a substrate-based radical in the catalytic mechanism. *Biochemistry*, *39*(31), 9561–9570.

Zelder, O., Beatrix, B., & Buckel, W. (1994). Cloning, sequencing and expression in Escherichia coli of the gene encoding component S of the coenzyme B12-dependent glutamate mutase from *Clostridium cochlearium*. *FEMS Microbiology Letters*, *118*(1–2), 15–21.

Zelder, O., Beatrix, B., Leutbecher, U., & Buckel, W. (1994). Characterization of the coenzyme-B12-dependent glutamate mutase from *Clostridium cochlearium* produced in *Escherichia coli*. *European Journal of Biochemistry*, *226*(2), 577–585.

Zhao, Y., Abend, A., Kunz, M., Such, P., & Retey, J. (1994). Electron paramagnetic resonance studies of the methylmalonyl-CoA mutase reaction. Evidence for radical intermediates using natural and artificial substrates as well as the competitive inhibitor 3-carboxypropyl-CoA. *European Journal of Biochemistry*, *225*(3), 891–896.

CHAPTER NINE

Exploring the mechanism of action of lysine 5,6-aminomutase using EPR and ENDOR spectroscopies

Amarendra Nath Maity, Jun-Ru Chen, and Shyue-Chu Ke*
Department of Physics, National Dong Hwa University, Hualien, Taiwan
*Corresponding author: e-mail address: ke@gms.ndhu.edu.tw

Contents

Abstract

Radical enzymes orchestrate challenging chemical transformations by devising strategies to tame the highly reactive radical intermediates. Electron paramagnetic resonance (EPR) spectroscopy is the most suitable technique to study various aspects of the radical enzymes. Lysine 5,6-aminomutase (5,6-LAM) is one such radical enzyme and employs coenzyme B_{12} and pyridoxal 5′-phosphate (PLP) to catalyze the 1,2-amino shift reaction through a radical mechanism. 5,6-LAM accepts either D-lysine or L-β-lysine as the substrate. EPR and electron nuclear double resonance (ENDOR) spectroscopies have played major roles in deciphering the mechanism of action of 5,6-LAM, while density functional theoretical (DFT) computation and synthetic isotopologues have played supporting

Methods in Enzymology, Volume 669
ISSN 0076-6879
https://doi.org/10.1016/bs.mie.2021.12.021

roles. This comprehensive toolkit has revealed that 5,6-LAM undergoes large-scale conformational movement to bring PLP and coenzyme B_{12} close together, which allows the reaction to progress. The conformational change also closes the active site, which protects the radical intermediates and enables their transformation to product without unwanted side reactions. The substrate-related radical (**S˙**), which is spin-coupled with Co^{2+} generated from homolysis of the Co—C bond in coenzyme B_{12}, was unequivocally characterized when a substrate analog, 4-thia-L-lysine, and isotopologues of it were reacted with 5,6-LAM. Studies with substrate analogs revealed a unique "odd-even" correlation with opening of the closed state. Moreover, mutagenesis studies identified the contributions that conserved residues in 5,6-LAM make toward binding of the substrate. Further studies with a cofactor analog, PLP-*N*-oxide, have shed light on various aspects of the mechanism of action of 5,6-LAM.

1. Introduction

Coenzyme B_{12} is one of the biologically active forms of vitamin B_{12} and is known as the nature's "free radical reservoir" (Banerjee, 1999; Dolphin, 1982). It facilitates many enzymatic reactions in living organisms by playing an active role as a cofactor (Banerjee, 2003; Brown, 2005; Frey, Hegeman, & Reed, 2006; Toraya, 2003). Lysine 5,6-aminomutase (5,6-LAM) is a fascinating bacterial enzyme that harnesses the capabilities of coenzyme B_{12} (5′-deoxyadenosylcobalamin, AdoCbl) (Frey, 2010; Frey & Chang, 1999; Frey & Reed, 2011; Maity, Chen, & Ke, 2014). The most powerful tool to investigate radical enzymes is electron paramagnetic resonance (EPR) spectroscopy. EPR and electron nuclear double resonance (ENDOR) spectroscopies together with synthetic isotopologues and density functional theoretical (DFT) computation constitute a comprehensive toolkit for investigating radical enzymes. In addition to coenzyme B_{12}, 5,6-LAM requires pyridoxal 5′-phosphate (PLP) as a cofactor to catalyze an otherwise challenging 1,2-amino amino shift reaction. 5,6-LAM possesses several interesting features: flexibility toward substrate as well as cofactor (Fig. 1), which makes it a promising candidate for biocatalytic applications; large-scale conformational movement; suicide inactivation (Tang, Chang, & Frey, 2001); magnetic field effects (Chen & Ke, 2018). The proposed mechanism of 5,6-LAM is illustrated in Fig. 2. In the resting open state of 5,6-LAM, PLP forms an internal aldimine with Lys144β residue such that PLP and AdoCbl are positioned ~25 Å apart. External aldimine formation by the substrate (D-lysine), with PLP releasing Lys144β, elicits a large scale conformational movement to bring AdoCbl close to the active site and to form the catalytic closed state. During this event, simultaneous rupture of Co—C bond of AdoCbl generates cob(*II*)

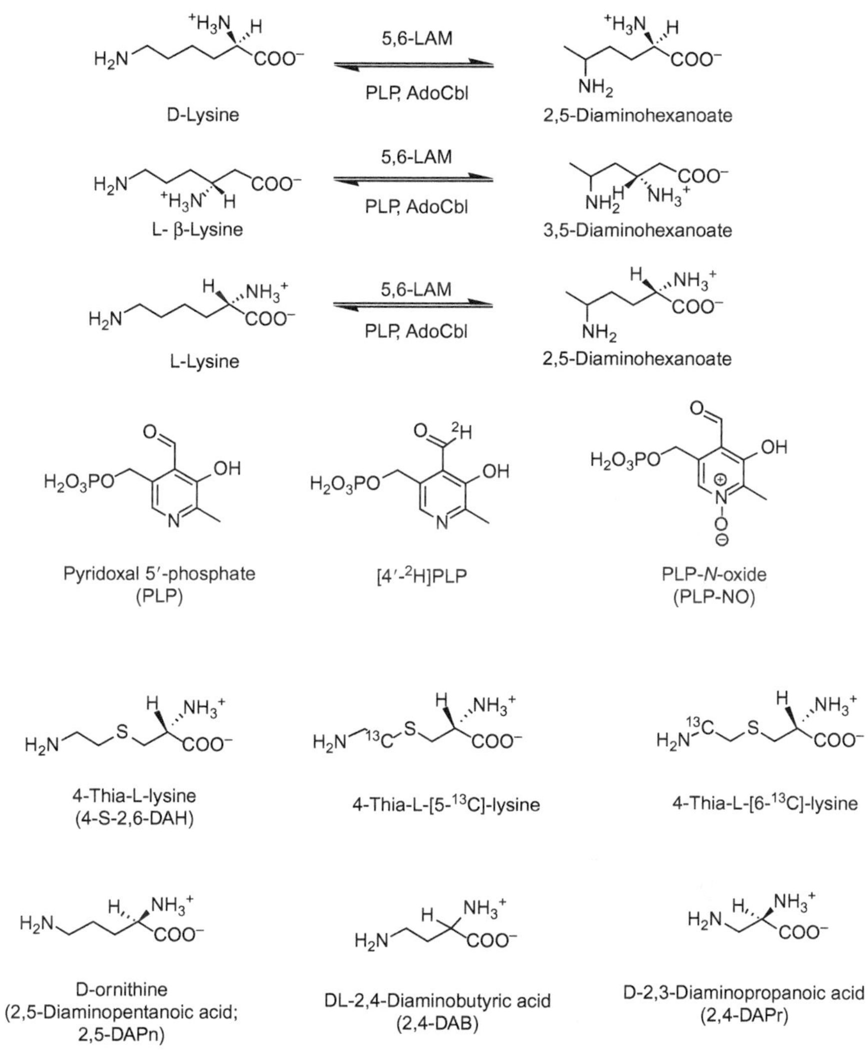

Fig. 1 Chemical structures of substrates, cofactor and relevant analogs in the reaction of 5,6-LAM.

alamin and 5′-deoxyadenosyl radical (Ado•) which then abstracts a hydrogen atom attached to C5 of lysine to produce the substrate related radical (**S**•). Intramolecular rearrangement of this radical results in the product related radical (**P**•) through a three membered cyclic intermediate radical, the azacyclopropylcarbinyl radical intermediate (**I**•). Subsequently, **P**• re-abstracts a hydrogen atom from 5′-deoxyadenosine (AdoH) to furnish product-PLP complex. Transaldimination of PLP with lysine144β releases

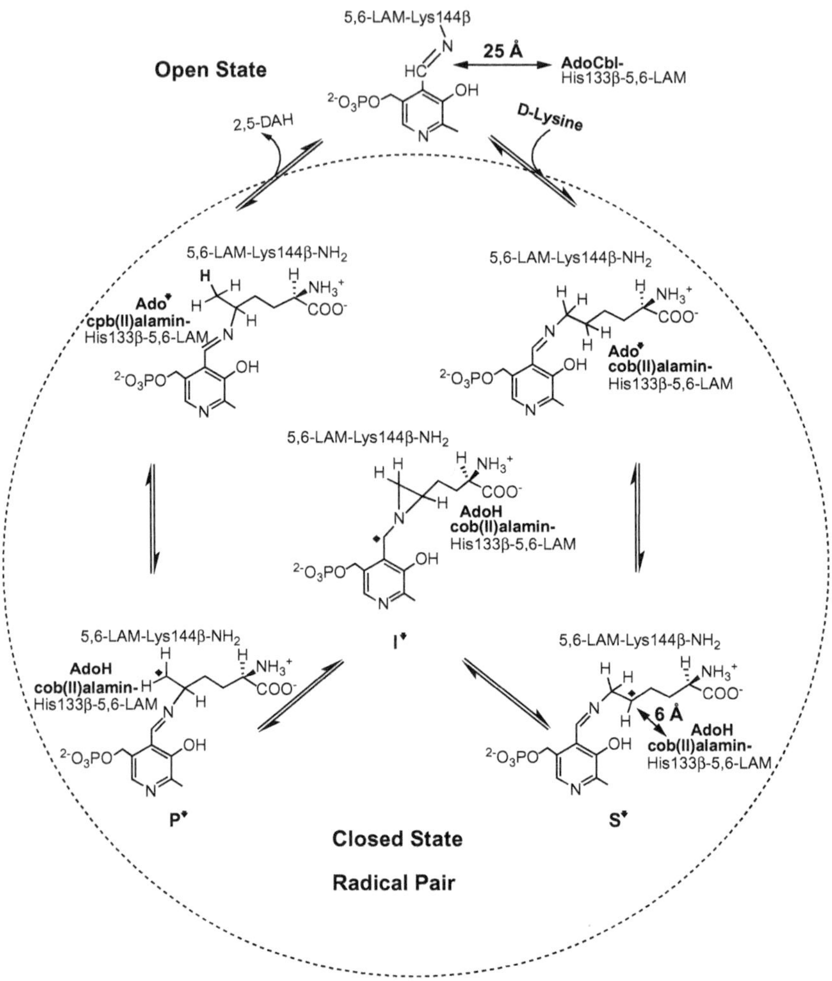

Fig. 2 The mechanism of action of 5,6-LAM.

the product, 2,5-diaaminohaxanoic acid (2,5-DAH) and takes 5,6-LAM back to the open state with concomitant reformation of AdoCbl by recombination of Ado• and cob(*II*)alamin. 5,6-LAM belongs to the family of radical aminomutases that includes ornithine 4,5-aminomutase (4,5-OAM) (Wolthers, Rigby, & Scrutton, 2008), lysine 2,3-aminomutase (2,3-LAM) (Frey & Magnusson, 2003; Lepore, Ruzicka, Frey, & Ringe, 2005) and glutamate 2,3-aminomutase (2,3-GAM) (Frey & Chang, 1999; Frey & Reed, 2011; Maity et al., 2014; Ruzicka & Frey, 2007). The case of 4,5-OAM, which is most closely related, is discussed in a separate chapter in this volume. Although in the case of 2,3-LAM and 2,3-GAM the source of Ado• is *S*-adenosyl-L-methionine (SAM), the latter part of the mechanism similarly involves PLP. 5,6-LAM was discovered in the 1970s by Professor Thressa C. Stadtman and coworkers (Baker & Stadtman, 1982; Baker, van der Drift, & Stadtman, 1973; Morley & Stadtman, 1970). Nonetheless, the research on 5,6-LAM did not progress much further because of the presence of the inactivated AdoCbl bound to the purified protein. After a long hiatus, with the advent of recombinant technology, research on 5,6-LAM was revived in the beginning of this century by Professor Perry A. Frey and coworkers. Subsequently, during the last two decades, a series of studies performed in Frey's and our laboratories have provided valuable insights about 5,6-LAM. Among the various radical intermediates mentioned above, corresponding **S**$^{\bullet}$ and **P**$^{\bullet}$ were previously identified in the reaction of 2,3-LAM (Ballinger, Frey, & Reed, 1992; Horitani et al., 2015; Lees et al., 2006; Wu et al., 2000; Wu, Lieder, Reed, & Frey, 1995). Besides, after evading detection for decades, Ado$^{\bullet}$ has very recently been detected after generating it by cryogenic photoinduced rupture of S-C5′ bond of SAM in pyruvate formate-lyase activating enzyme (PFL-AE) and characterized by Yang and coworkers using EPR and ENDOR (Yang et al., 2019). This has been followed by another observation of Ado$^{\bullet}$ in the radical SAM enzyme HydG with a non-native substrate by Sayler and coworkers (Sayler et al., 2019). In this chapter, we elucidate in detail how EPR and ENDOR have been playing crucial roles in unraveling the various aspects of mechanism of action of 5,6-LAM.

2. Protocol for the preparation of protein samples for EPR and ENDOR measurements

Quartz EPR tubes (Wilmad Lab Glass) having dimensions of either 4 mm outer diameter and 250 mm length, or 2 mm outer diameter and 160 mm length are used for X- and Q-band EPR measurements, respectively. The same tube can be used for ENDOR also. It is convenient to prepare the respective stock solutions of various reagents first. As PLP and AdoCbl are light sensitive, caution is needed to avoid exposure to the light and the handling of these must be done under dim red light. Therefore, the stock solutions of PLP and AdoCbl are stored in amberized Eppendorf microcentrifuge tubes. A liquid nitrogen chilled ethanol bath is also prepared: ~200 mL ethanol is poured into a Dewar flask and liquid nitrogen is slowly added to it by stirring with a teflon rod until the mixture becomes sticky viscous liquid. Two 1 mL syringes attached with long needles are arranged such that the needle can reach to the bottom of the EPR tubes. 5,6-LAM and its variants are purified using reported procedure

Table 1 Relative amounts of required ingredients for preparing the EPR tube.

	5,6-LAM	NH_4EPPS	DTT	AdoCbl	PLP	4-Thia-L-lysine	Total volume (μL)
Volume taken (μL)	200	12	2	10	5	4	233
Concentration of stock solution (mM)	0.2	2000	1000	5	10	1000	
Final concentration (mM)	0.172	103.004	8.584	0.215	0.215	17.167	

(Chang & Frey, 2000). Either 5,6-LAM or one of its variants (usually ~200 μM but must be higher than 400 μM for ENDOR) in 100 mM NH_4EPPS buffer (pH 8.5) is mixed with dithiothreitol (DTT) and incubated on ice for 45 min. PLP and AdoCbl solutions are prepared and kept on ice. The exact amounts of various reagents required for preparing an EPR tube are presented in Table 1.

The following operations are performed under dim red light. Holoenzyme is prepared by incubating 5,6-LAM along with PLP and AdoCbl at 37 °C for 5 min. Then holoenzyme is transferred into an EPR tube using a syringe. Substrate, or analog, is drawn up in another syringe and inserted into EPR tube and held just above the surface of holoenzyme. Substrate analog is mixed with holoenzyme and after the desired time has elapsed—usually between 8 s and 3 min, the EPR tube is dipped into liquid nitrogen chilled ethanol bath. It is immersed into liquid nitrogen bath after wiping thoroughly using tissue paper to remove the solvent. After this, all the operations can be performed under normal light. The EPR tube is stored in liquid nitrogen Dewar flask.

3. Typical spectrometer settings for measurements of EPR and ENDOR spectra

EPR measurements are performed using a Bruker EMX spectrometer equipped with either a TE102 or a ER5106QT cavity for X-band or Q-band, respectively. Acquisitions are accomplished using Bruker WinEPR program. For a quick EPR X-band measurement, a quartz liquid nitrogen finger Dewar (Wilmad Lab Glass) is used. For prolonged stable measurements the cryostat equipment of Bruker is used. ENDOR spectra are acquired with a Bruker DICE ENDOR assembly equipped with a high power radio frequency (RF) amplifier (ENI 3200L) to generate the continuous-wave (CW) B2 field in the cavity. ENDOR spectra are collected in frequency modulation mode. The representative instrument settings for measurements of X- and Q-band EPR, and X-band ENDOR are summarized in Tables 2 and 3, respectively.

Table 2 Instrument settings for X- and Q-band EPR.

	X-band	**Q-band**
Microwave frequency (GHz)	9.535	33.96
Microwave power (mW)	2	1.46
Modulation amplitude (G)	8	8
Modulation frequency (kHz)	100	100
Center field (G)	3250	12,000
Sweep width (G)	1200	4000
Static field (G)	2650	10,000
Temperature (K)	80	80
Receiver gain	1.00×10^5	6.32×10^4
Time constant (ms)	81.92	40.96
Conversion time (ms)	40.96	40.96
Sweep time (s)	167.77	167.77
Harmonic	1	1
Resolution in X	4096	4096
Number of scans	4	1

Table 3 Instrument settings for X-band ENDOR.

	X-band ENDOR
Magnetic field (G)[a]	3315[b]
Microwave frequency (GHz)	9.595
Microwave power (mW)	2
Radiofrequency power (W)	200
Modulation depth (kHz)	100
Number of scans	36
Temperature (K)	20
Scan range (MHz)	1–20

[a]The field is dependent on the sample.
[b]The magnetic field at which maximum EPR absorption occurs in the reaction of 5,6-LAM with 4-thia-L-lysine.

4. Simulation protocol of EPR and ENDOR spectra

The situation in 5,6-LAM gives rise to complex spectra, which are difficult to analyze, because those originate from various interactions between two spins. Simulation of the spectra becomes indispensable to extract the required information accurately. EasySpin (Stoll & Schweiger, 2006), a software package that works on MATLAB platform, is suitable for this purpose. EasySpin is available free of charge in the following website: https://www.easyspin.org/. We first use Bruker WinEPR program to correct the baseline of the spectra. The saved data, in ASCII format, are then used for the simulation. The microwave frequency measured with a Hewlett-Packard 5246L electronic counter is used for simulation. Many parameters are needed to be taken account: two sets of g values associated with two spin centers—one on the cobalt(II) atom of AdoCbl having low-spin d^7 electronic configuration and the other spin residing on PLP-substrate complex; hyperfine interactions of the unpaired electron spin residing on the cobalt(II) atom with the ^{59}Co nuclear spin ($I = 7/2$) and the axial ^{14}N nuclear spin ($I = 1$); the isotropic exchange coupling constant (J_{iso}) and zero-field splitting parameters (D and E) originating from inter-spin interactions; two sets of Euler's angles needed to describe the spatial arrangement of the spin centers. X- and Q-band EPR spectra were simulated using the "pepper" function of EasySpin whereas ENDOR spectra were simulated using the "salt" function.

5. Tracking conformational movement

EPR, in particular pulsed dipolar electron paramagnetic resonance spectroscopy, has become one of the favorite tools for obtaining crucial structural information of biomolecules (Abdullin & Schiemann, 2020). The dipolar coupling between two paramagnetic centers, which can be obtained from EPR spectra, provides information about the distance and relative orientation between those centers. Therefore, the presence of two or more paramagnetic centers is needed so that the dipolar coupling between these centers can be measured. In the absence of intrinsic paramagnetic centers within the biomolecules, these can be introduced either by replacing diamagnetic metals with paramagnetic ones or through site-directed spin labeling (SDSL) of the biomolecules (Berliner, 2002). Fortunately, the proposed mechanism of action of 5,6-LAM, as elucidated above, involves generation of two paramagnetic centers and EPR is the best tool to extract structural information during catalysis. The existence of radical intermediates as pairs during the catalytic cycle gives rise to two situations—first, the radicals are close enough to experience spin–spin interaction; second, the radicals reside far enough apart that spin–spin interaction is not feasible. These two situations result in two different types of the EPR spectra—first one is referred as spin-coupled spectrum while the second one referred as uncoupled spectrum. To our advantage, spin–spin interaction is dependent on relative positions of the radicals and an inter-spin distance can be estimated from the parameters obtained from the simulation of an experimental spin-coupled EPR spectrum (Bencini & Gatteschi, 1990; Ke, 2003; Reed & Mansoorabadi, 2003). The zero-field splitting parameter (D) originating from dipole–dipole interaction varies with distance according to the equation, $D = D_0/r^3$, where r (Å) is the distance between the spins. Distance can also be estimated from isotropic exchange coupling constant (J) using the empirical Eq. $J = J_0 \exp(-r)$. Nevertheless, the distance estimated from D provides better approximation as it depends explicitly on inter-spin distance. The inter-spin distance is usually used as a marker to track conformational dynamics in proteins containing SDSL. In the case of AdoCbl dependent enzymes, two paramagnetic centers—an organic radical and Co^{2+}—usually appears as a pair in the active site at some point during catalytic cycle. These two spin centers could be weakly or strongly coupled depending upon the distance between them. Weakly coupled spin systems have been observed in the case of ethanolamine ammonia-lyase (EAL) (Bandarian & Reed, 2002; Ke, 2003) and diol dehydratase (DDH) (Abend, Bandarian, Reed, & Frey, 2000) whereas strongly coupled systems have been reported for ribonucleotide reductase (Orme-Johnson, Beinert, & Blakley, 1974), glutamate mutase (Zelder, Beatrix, Leutbecher, & Buckel, 1994), methyleneglutarate mutase (Michel, Albracht, & Buckel, 1992), and methyl-malonyl-CoA mutase (Padmakumar & Banerjee, 1995; Zhao, Abend, Kunz, Such, & Retey, 1994). Weakly coupled spin system refers to an estimated inter-spin distance of 10–12 Å while in the strongly coupled system the two spins are 5–7 Å apart. A very strongly coupled (<4 Å interspin separation) system was observed in the reaction of DDH with 3′,4′-anhydroadenosylcobalamin, an analog of AdoCbl (Magnusson & Frey, 2002); this provided indirect evidence for Ado• that was elusive at that time. The variation of EPR spectral features with interspin distance was nicely illustrated in a review article by Reed and Mansoorabadi (Reed & Mansoorabadi, 2003). So, information about conformational states could be gathered by simple EPR measurements, without the requirement of spin labeling. Nonetheless, no radical intermediate could be detected during the reaction of the natural substrate D-lysine with 5,6-LAM. Luckily, there is a way to circumvent this obstacle, which is discussed in the following sections.

5.1 Importance of analogs

Analogs of substrates and cofactors play vital roles in the mechanistic investigations of enzymes. In the case of radical enzymes, those analogs are often used to trap the corresponding radical intermediates that do not accumulate during the reaction with the natural substrate or cofactor. For instance, 4-thialysine, in which C4 of lysine is replaced by a sulfur atom, allows the detection of substrate-related radical when reacted with 5,6-LAM as the sulfur atom adjacent to the radical center stabilizes the radical by delocalization of the spin (Maity et al., 2009; Tang, Mansoorabadi, Reed, & Frey, 2009). In another example, 2,4-diaminobutyric acid (2,4-DAB), a homolog of the substrate D-ornithine, served the same purpose in the case of 4,5-OAM (Wolthers et al., 2008). Detection of these radical intermediates provided the necessary evidence for the radical mechanism of action in 5,6-LAM and 4,5-OAM. Depending upon reaction time, two different radicals (Fig. 3) are observed in the reaction of 4-thialysine with 5,6-LAM—a transient radical at a shorter reaction time and a persistent radical at a longer reaction time. Tautomerization of the transient radical affords the highly stable persistent radical with time. Among the two enantiomers of 4-thialysine, 4-thia-D-lysine reacts, as expected, faster than 4-thia-L-lysine because D-lysine is a better substrate for 5,6-LAM than L-lysine. The transient radical in the case of D enantiomer, which appears at the reaction time of 2 s, can only be detected by rapid mixing freeze-quench EPR while the persistent radical appears at the reaction time of 10 s (Tang et al., 2009). On the other

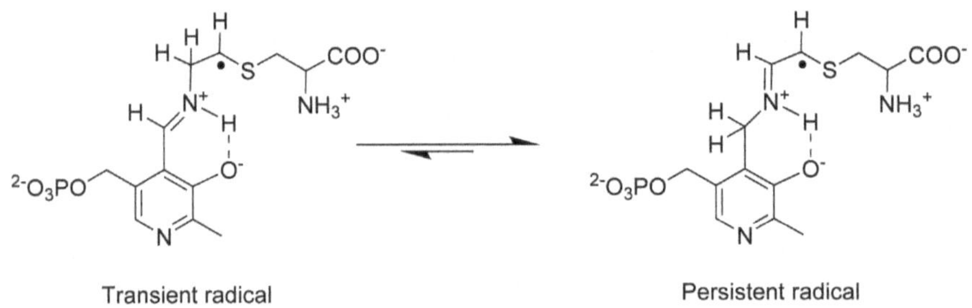

Fig. 3 Chemical structures of transient and persistent radical intermediates in the reaction of 4-thia-L-lysine with 5,6-LAM.

hand, in the case of 4-thia-L-lysine, the corresponding time periods for transient and persistent radicals are 8 s and 2 min, respectively; therefore, experiments can be performed manually without the need of assistance from sophisticated rapid mixing apparatus. Due to this experimental simplicity associated with 4-thia-L-lysine in comparison to 4-thia-D-lysine, further studies have been performed only on the L-isomer to gain deeper insights into the mechanism of action in 5,6-LAM.

In another study, the homologs of D-lysine revealed a unique "odd-even" dependence of the number of carbon atoms in the substrate or substrate analog on the opening of the closed conformational state (Chen, Maity, Frey, & Ke, 2013). The reactions of cofactor analogs have also been explored in 5,6-LAM. For example, DFT calculations suggested that PLP-*N*-oxide (PLP-NO), a synthetic analog of cofactor PLP, might assist in trapping the cyclic intermediate radical, $\mathbf{I}^{\bullet}$. But the experimental attempts to trap this radical with PLP-NO synthesized following the literature procedure (Ando & Emoto, 1975; Fukui, Ohishi, Nakai, & Shimizu, 1969) were unfortunately unsuccessful, demonstrating that the radical is not stabilized sufficiently such that it accumulates to the critical concentration needed for EPR detection (Maity, Lin, Chiang, Lo, & Ke, 2015). However, PLP-NO is functional as a cofactor in the place of PLP for 5,6-LAM, albeit at a slower rate, and, furthermore, displays a much smaller deuterium kinetic isotope effect relative to PLP. This means that in the case of PLP-NO a step other than the hydrogen atom transfer is partially rate limiting. Such a step could involve the azacyclopropylcarbinyl radical reaction.

5.2 Advantages of multifrequency EPR

The commonly used frequency for measuring EPR is 9 GHz (X-band), however, measurements at more than one frequency provide valuable information as some parameters are field dependent while others are not. For example, g anisotropy is field dependent whereas hyperfine interactions are not; higher frequency, for example, 35 GHz (Q-band), enhances the resolution related to g anisotropy. Thus, by measuring spectra at different frequencies spectral features associated with g anisotropy can be distinguished from other field independent interactions such as hyperfine interactions. An example of this is manifested in the spectral differences of the persistent radical in the reaction of 4-thia-L-lysine with 5,6-LAM obtained at X- and Q-band (Fig. 4), wherein the g-splittings are well resolved in the Q-band spectrum while J and A splittings remain unaltered. Simulation of both X- and Q-band spectra with

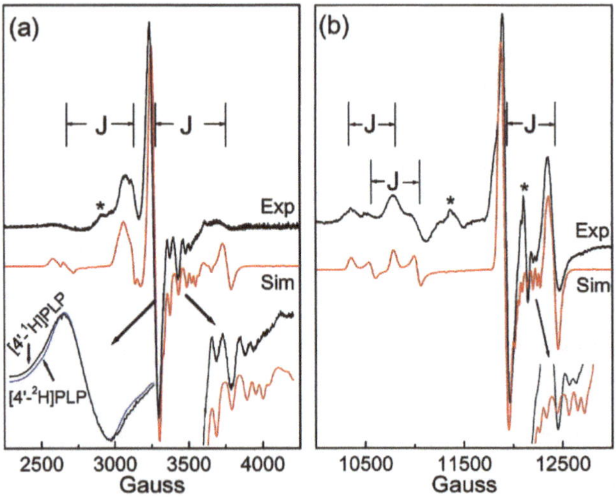

Fig. 4 (A) X-band and (B) Q-band EPR spectra of the radical intermediate in 5,6-LAM reaction with 4-thia-L-lysine. Experimental: microwave frequency, (A) 9.535, (B) 33.96 GHz; power, (A) 2, (B) 1.46 mW; modulation, 8 G at 100 kHz; $T=80$ K. The starred feature in part a is the g_{xy} signal of free Co^{2+}, and those in part b are contaminants present in variable amounts in the samples. Simulation: $g_{Co}=[2.285, 2.236, 2.01]$, $g_{radical}=[2.01, 2.0, 1.995]$, Euler angle = [60°, 10°, 120°]; $A_{Co}=[8, 5, 112]$ G; J = 500 G, D = −25 G, E = 0, Euler angle = [0°, 80°, 0°]; line width = [40, 40, 20] G. *Reprinted with permission from the Journal of Physical Chemistry B* ***2009,*** *113, 12,161–12,163; Copyright 2009, American Chemical Society.*

identical parameters confirmed the accuracy of those values (Maity et al., 2009). These parameters contain useful information about the characteristics of the radical intermediates; for instance, of great importance is the value of D, which, as mentioned above, encodes the distance between the spin centers and is discussed in detail in the following section.

The advantages of multifrequency EPR measurements are nicely documented in the case of the transient radical formed in the reaction of 5,6-LAM and are illustrated in Fig. 5. Initially, with only the X-band data available, the spectrum was satisfactorily simulated using J = 4000 G (Tang et al., 2009). Later, with the Q-band spectrum in hand, attempts to simulate both spectra demonstrate that the best value for J is 8000 G rather than 4000 G (Lo, Lin, Maity, & Ke, 2016). As shown in Fig. 5, the simulated spectra for the X-band spectrum with these two distinct J values are almost identical while those for Q-band are easily distinguishable, J = 8000 G being the closest to the experimental spectrum. Thus, to obtain the accurate simulation parameters it is pertinent to measure multifrequency EPR and to perform the subsequent simulation of both spectra with identical parameters.

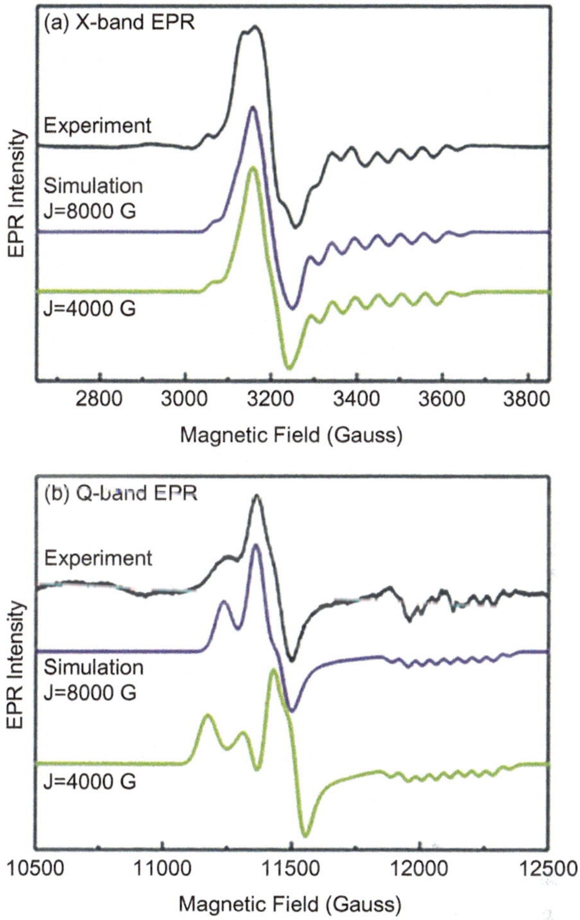

Fig. 5 The (A) X-band and (B) Q-band EPR spectra of the 8 s transient radical intermediate of the 5,6-LAM reaction with 4-thia-L-lysine. Microwave frequency, (A) 9.538 and (B) 33.94 GHz. $T = 77$ K. Simulations: $g_{Co} = [2.28, 2.235, 2.002]$, $g_{radical} = [2.01, 2.005, 2.0]$, Euler angle = [73°, 28°, 73°]; $A_{Co} = [10, 7, 112]$ G, $A_N = 19$ G; J = 8000 G (blue), J = 4000 G (green), D = −180 G, E = 0, Euler angle = [0°, 42°, 36°]; line width = 10 G. *Reproduced from Chemical Communications, 2016, 52, 6399–6402; with permission from The Royal Society of Chemistry.*

5.3 Correlation of spin-coupled/uncoupled spectra with closed/open states

A spin-coupled EPR spectrum, which originates from spin–spin interactions between PLP-substrate based radical and Co^{2+} of AdoCbl in 5,6-LAM, corresponds to the closed state while an uncoupled spectrum relates to the open state. The spin-coupled EPR spectra (Fig. 4) obtained from the reaction of 5,6-LAM with 4-thialysine provided the evidence for the presence of a putative closed state and thereby a large-scale conformational movement (Maity et al., 2009; Tang et al., 2009). The spin-coupled EPR spectra of the persistent radical derived from 4-thia-L-lysine can be simulated with the zero field splitting (D) value of −25 G (Fig. 4). Using the equation $D = 6.95 \times 10^3\ g1g2/R^3$, the inter-spin distance is estimated as 10.6 Å between Co^{2+} and the PLP-lysyl based radical center (Maity et al., 2009). Similarly, this inter-spin distance is estimated as ~10.8 Å when the simulation-derived J value of 500 G (Fig. 4) is used in the equation $|J| \leq 1.35 \times 107\ e^{-1.8R}$. In the substrate free crystal structure of 5,6-LAM in the open state, PLP and AdoCbl are 25 Å apart from each other (Berkovitch et al., 2004). Taken together, spin-coupled spectra relate to a different conformation, in which the distance between AdoCbl and PLP is significantly reduced in comparison to the open state and is referred as the closed state. This transition from open to closed state could be achieved by a large-scale conformational movement of 5,6-LAM upon binding of a substrate or a substrate analog. A similar large-scale conformation movement triggered by product release is needed to complete the catalytic cycle by bringing PLP and AdoCbl 25 Å apart as seen in the open state. If radical intermediates can be trapped in this state, the spin–spin interaction between Co^{2+} and the PLP-substrate analog-based radical will be lost as the distance between those are too long. Thus an uncoupled EPR spectrum observed in 5,6-LAM corresponds to the open state. Indeed, observation of uncoupled spectra with substrate homologs (Chen et al., 2013) (Fig. 6) or with the variant of 5,6-LAM (Fig. 7) demonstrated the opening of the closed state (Lo et al., 2016). An interesting correlation between substrate analogs containing odd or even number of carbon atoms and the open/closed state was observed. D-2,3-diaminopropanoic acid (2,3-DAPr) and D-ornithine (D-2,3-diaminopentanoic acid, D-2,5-DAPn) containing odd numbers of carbon atoms—three and five carbon atoms, respectively,—produced uncoupled spectra suggesting the open state. On the other hand, in the case of even numbers of carbon atoms, 2,4-DAB (which contains four carbon atoms) produced coupled spectra similar as 4-thia-L-lysine which represents D-lysine (which contains six carbon atoms). This unique odd-even dependence is explained by relative orientation at C2 and discussed in detail later in Section 6.3.

The importance of an aspartate residue in binding the substrate is illustrated in Fig. 7 (Lo et al., 2016). Mutation of the aspartate residue to alanine

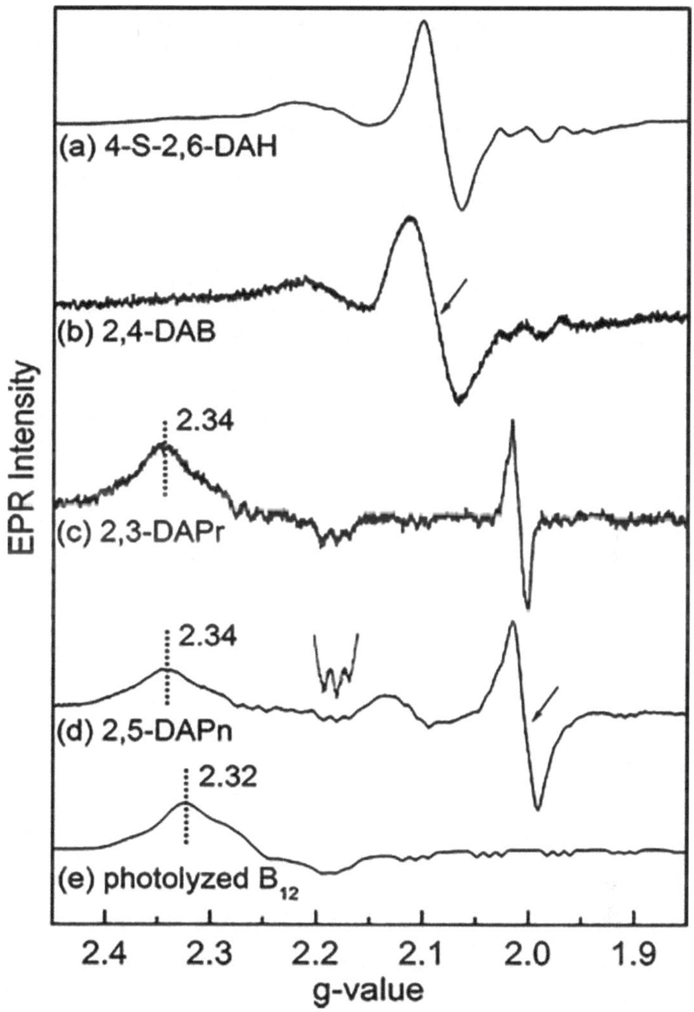

Fig. 6 77 K EPR spectra for radicals generated in the reaction of 5,6-LAM with 4-S-2,6-DAH (4-thia-L-lysine), 2,4-DAB, 2,3-DAPr, and D-2,5-DAPn (from top to bottom). 5,6-LAM at 0.3 mM was incubated with 100 mM NH_4EPPS buffer (pH 8.5), 5 mM dithiothreitol, 0.3 mM AdoCbl, and 0.3 mM PLP at 37 °C. The reaction was initiated by addition of 30 mM analogs and allowed to proceed for 3 min aerobically. Spectrum (E) is photolyzed AdoCbl in solution. Arrows mark the positions for ENDOR measurements. Instrument settings: microwave frequency, 9.6 GHz; microwave power, 20 mW; modulation amplitude, 4 G at 100 kHz. *Reprinted with permission from Journal of the American Chemical Society 2013, 135, 788–794; Copyright 2013, American Chemical Society.*

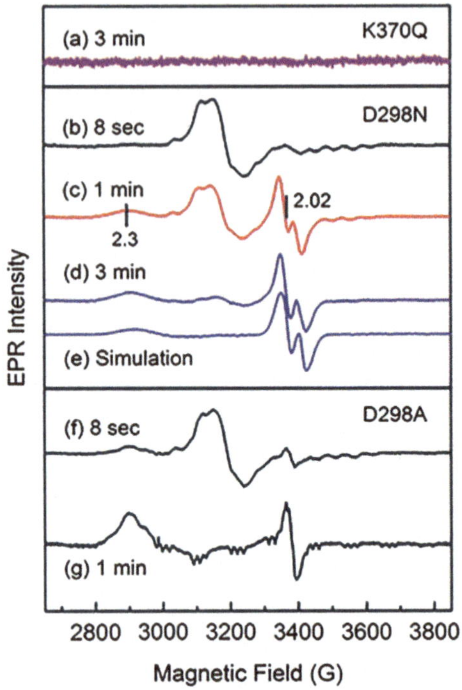

Fig. 7 The X-band EPR spectra of 4-thia-L-lysine in the reaction with the K370Q, D298N, and D298A variants at the indicated incubation times. Simulation: g_{Co} = [2.29, 2.24, 2.003], $g_{radical}$ = 2.003; A_{Co} = [10, 7, 112] G; A_N = 19 G; J = 46 G, D = 28 G, E = 0, Euler angle = [0°, 45°, 0°]; line width = [40, 40, 12] G. *Reproduced from Chemical Communications, 2016, 52, 6399–6402; with permission from The Royal Society of Chemistry.*

results in the opening of the active site within 1 min. On the other hand, when aspartate is replaced by asparagine, complete opening of active state is not achieved even after 3 min; although the strongly coupled spectrum is replaced by new features at 3 min, simulation revealed that it is still weakly coupled as it can be seen in middle panel of Fig. 7.

6. Identification and characterization of relevant radical intermediates

6.1 Utility of synthetic Isotopologues

EPR experiments with isotopologues of substrate, substrate analogs, and cofactor shed light on spin density distribution, which leads to the identification of radical intermediates. In particular, site-specific isotopologues produce the evidence for unambiguous characterization of the radical intermediates. Although the required isotopologues can sometimes be purchased from chemical vendors, often those are commercially unavailable, especially site-specifically labeled compounds, and must be synthesized. In the absence of a published synthesis of the specific isotopologue, the procedure for the synthesis of the natural abundance isotopologue is usually adopted using the corresponding isotopologue of the starting material or reagent. In rare cases, one needs to design new synthetic schemes if the corresponding isotopologue of the starting material is also commercially unavailable. In our experience, in order to synthesize a particular isotopologue, we found it beneficial to first standardize all the steps using natural abundance starting material as even the isotopologues of the starting material are often extremely expensive.

A great deal of information can be obtained by manipulating the nuclear spin of the atom that is presumed to carry a substantial amount of spin density. Of particular interest are the elements hydrogen, carbon, nitrogen and phosphorus. Natural abundance hydrogen (^{1}H), phosphorus (^{31}P), and nitrogen (^{14}N) have nuclear spins ½, ½, and 1, respectively, whereas natural abundance carbon, ^{12}C, does not possess any nuclear spin. The relevant stable isotopes of these elements having nuclear spin are ^{2}H ($I=1$), ^{13}C (I = 1/2), and ^{15}N (I = 1/2). An isotopologue enriched with ^{2}H usually ensues line narrowing in EPR spectra in comparison to natural abundance isotopologue while ^{13}C-isotopologue results in line broadening. Hyperfine coupling of a natural abundance ^{14}N gives rise to a triplet feature with 1:1:1 relative intensity. The replacement with ^{15}N changes the triplet feature to a doublet.

Taking advantage of this fact, [^{15}N-DMB]-AdoCbl, in which the axial ligand, dimethylbenzimidazole (DMB) moiety, is approximately 60% enriched with ^{15}N, was used to investigate the binding mode of AdoCbl in 5,6-LAM (Chang & Frey, 2000). The EPR spectrum obtained when 5,6-LAM reconstituted with [^{15}N-DMB]-AdoCbl is treated with methylhydrazine suggests that the axial ligation comes from a ^{14}N nucleus instead of a ^{15}N nucleus as the spectrum resembles that of photolyzed natural abundance AdoCbl rather than that of photolyzed [^{15}N-DMB]-AdoCbl (Fig. 8). This means that DMB is no longer ligated to the Co atom in 5,6-LAM; rather a histidine residue of 5,6-LAM is coordinated to the Co atom. This implies that AdoCbl binds with 5,6-LAM in "base off-His on" mode. Later, "base off-His on" mode of binding in 5,6-LAM was confirmed when it was found that the imidazole side chain from His-133β is the axial ligand to the Co atom in the crystal structure of 5,6-LAM holoenzyme (Berkovitch et al., 2004).

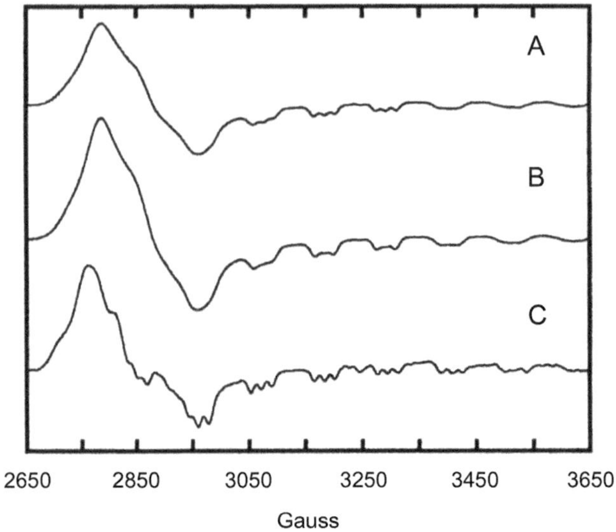

Fig. 8 AdoCbl binding mode. (A) Photolyzed natural abundance AdoCbl in solution; (B) photolyzed 60% enriched [^{15}N-DMB]AdoCbl in solution; (C) 5,6-LAM reconstituted with [^{15}N-DMB]AdoCbl and treated with methylhydrazine. *Reproduced with permission from Journal of Biological Chemistry **2000**, 275, 106–114.*

AdoCbl binds with enzymes in two distinct modes—"base off-His on" as in 5,6-LAM, 4,5-OAM (Wolthers, Levy, Scrutton, & Leys, 2010), glutamate mutase (Chen & Marsh, 1997), methylmalonyl-coenzyme A mutase (Mancia et al., 1996) and "base on-His off" as in ethanolamine ammonia lyase (Abend et al., 1999), ribonucleotide reductase (Lawrence et al., 1999), diol dehydratase (Yamanishi et al., 1998). Besides, methionine synthase binds methylcobalamin in "base off-His on" mode (Jarrett et al., 1996).

The effect of isotopic substitution is evident in the spectra of both the transient and the persistent radical in the reaction of 5,6-LAM with various synthetic isotopologues as reported by Tang and coworkers (Tang et al., 2009). Fig. 9 illustrates the obvious effects—line narrowing and broadening for labeling by deuterium and ^{13}C, respectively—that provide critical clues to the location of radical center. These effects are noticeable when the label is positioned at C5 and C6, while labeling at other positions reproduced spectra identical to the unlabeled substrate analog. These experiments show that the radical center could be at either C5 or C6. But this could not confirm whether it is C5 or not.

Two site specific isotopologues—4-thia-L-[5-^{13}C]lysine and 4-thia-L-[6-^{13}C]lysine—are required to characterize the radical intermediate unequivocally. We synthesized 4-thia-[5-^{13}C]lysine (Maity & Ke, 2011)

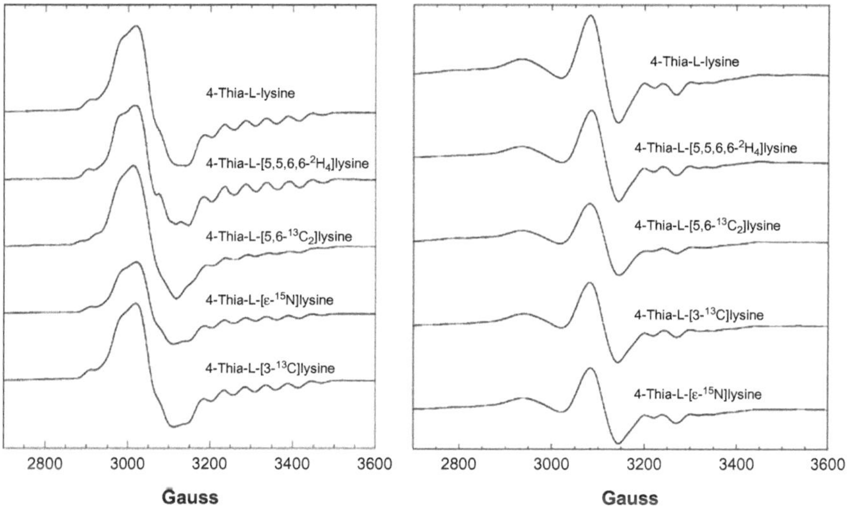

Fig. 9 Left: Isotope-edited EPR spectra of the transient radical with 4-thia-L-lysine. EPR spectra at 9.1 GHz of the transient radical with 100 μM activated 5,6-LAM, 100 μM AdoCbl, and 100 μM PLP reacting with 20 mM 4-thia-L-lysine without and with ^{2}H, ^{13}C, or ^{15}N labeling and freeze quenched at 10 s. In this spectrum, $g=2.0$ corresponds to 3261 G. Right: Isotope-edited EPR spectra of the persistent radical with 4-thia-L-lysine. Spectra at 9.1 GHz were obtained as in left panel but after freeze quenching at 10 min. In this spectrum, $g=2.0$ corresponds to 3261 G. *Reprinted with permission from Biochemistry* ***2009****, 48, 8151–8160; Copyright 2009, American Chemical Society.*

in two steps following a literature procedure, while we needed to design a synthetic scheme to synthesize 4-thia-L-[6-^{13}C]lysine due to the unavailability of required labeled precursor. As shown in Fig. 10, the synthesis of 4-thia-L-[6-^{13}C]lysine was achieved by starting from commercially available [2-^{13}C]glycine (Maity et al., 2012).

These two site-specific isotopologues of 4-thia-L-lysine helped to characterize the radical intermediate unambiguously as depicted in Fig. 11 (Chen, Maity, Pan, Frey, & Ke, 2011). The spectrum with 4-thia-L-[5-^{13}C]lysine displays line broadening in comparison to that with natural abundance 4-thia-L-lysine whereas the spectrum with 4-thia-L-[6-^{13}C]lysine is almost identical to that of natural abundance 4-thia-L-lysine. In the case of the transient radical that is presumed to be the substrate-related radical, which has the radical center at C5, the effect, as shown in Fig. 11, is much more pronounced on the spectral features between $g=2.08$ and 1.88; these features originate from cobalt hyperfine interaction. In addition, the singlet feature at $g=2.24$ in the natural abundance 4-thia-L-lysine spectrum splits into a doublet with

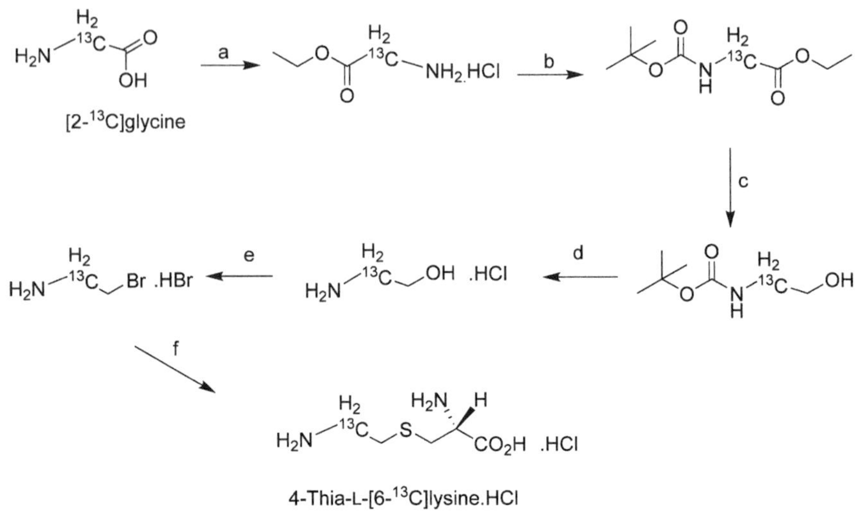

Reagents and conditions: a $SOCl_2$, EtOH; b Boc_2O, Et_3N, CH_3OH; c $NaBH_4$, LiCl, MeOH; d HCl gas, CH_2Cl_2; e 48% HBr; f (1) L-Cysteine, KOH, H_2O, (2) HCl

Fig. 10 Synthesis of 4-thia-L-[6-^{13}C]lysine.

the splitting of ~40 G in the case of the spectrum with 4-thia-[5-^{13}C]-L-lysine. These observations with site-specific isotopologues unambiguously confirm that the transient radical is C5-based; and, therefore, it is indeed the corresponding substrate related radical, **S$^•$**. Similar effects were also observed in the case of the persistent radical as shown in the bottom panel of Fig. 11.

[4′-^{2}H]PLP, an essential isotopologue of PLP, synthesized following a literature procedure (Fu & Silverman, 1999) was used to understand the nature of the persistent radical. An obvious line narrowing can be observed in the EPR spectrum of [4′-^{2}H]PLP with respect to that of the unlabeled PLP, as illustrated in Fig. 4, suggesting a substantial hyperfine interaction with 4′-^{2}H of [4′-^{2}H]PLP or 4′-^{1}H of PLP (Maity et al., 2009). In the following section, we discuss how the ENDOR spectra of these isotopologues revealed further insights about the radical intermediate.

6.2 Significance of ENDOR

ENDOR is an advanced technique able to extract information about electron–nuclear hyperfine interactions from complex EPR spectra in cases

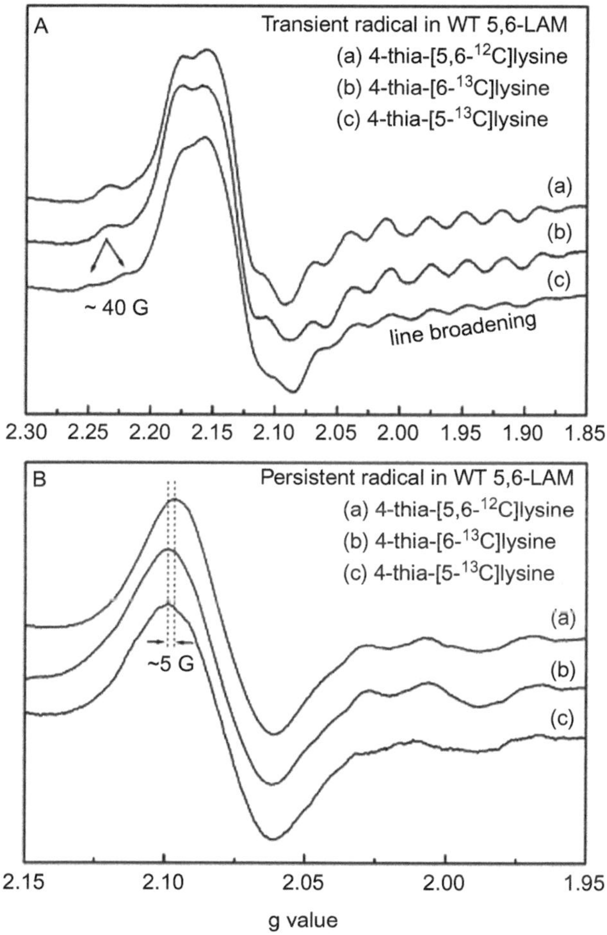

Fig. 11 EPR spectra of the radical regions of the transient (A) and persistent (B) radicals in the reaction of 5,6-LAM at 77 K in reaction with natural abundance and isotopically labeled 4-thia-L-lysine, as indicated. Instrument settings: microwave frequency, 9.54 GHz; microwave power, 2 mW; modulation amplitude, 4 G at 100 kHz. *Reprinted with permission from Journal of the American Chemical Society 2011, 133, 17152–17155; Copyright 2011, American Chemical Society.*

where these interactions are buried within the linewidth. ENDOR has been employed widely to gain insights into metalloproteins including B_{12}-dependent enzymes (Cutsail, Telser, & Hoffman, 2015; LoBrutto et al., 2001; Yang et al., 2019). We have used ^{2}H X-band ENDOR to

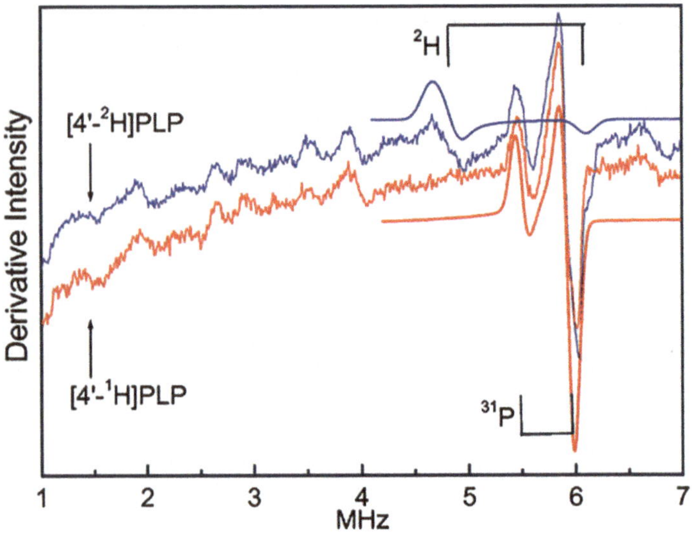

Fig. 12 X-band ENDOR spectra of the radical intermediate (persistent radical) in the reaction of 5,6-LAM with 4-thia-L-lysine. Conditions: B = 3315 G; ν = 9.595 GHz; microwave power, 2 mW; RF power, 200 W; modulation depth,100 kHz; T = 20 K. Simulation parameters: g_{2H} = 0.857; g_{31P} = 2.263; A_{2H} = [−5, −5.4, −7.9] MHz; A_{31P} = [−0.4, −0.5, 19.2] MHz; linewidth, 150 kHz. *Reprinted with permission from the Journal of Physical Chemistry B* ***2009,*** *113, 12161–12163; Copyright 2009, American Chemical Society.*

identify the radical intermediate in 5,6-LAM using [4′-^{2}H]PLP (Fig. 12) (Maity et al., 2009). Two pairs of lines are visible in the ENDOR spectrum for [4′-^{2}H]PLP, whereas only the pair of lines at 5.54 and 5.90 MHz mirrored about the Larmor precession frequency of ^{31}P (5.72 MHz) is present in the spectrum of unlabeled PLP. Therefore, the additional pair of lines at 4.8 and 6.1 MHz in the case of [4′-^{2}H]PLP is attributed to deuterium hyperfine coupling. By neglecting ^{2}H nuclear quadrupole contribution (normally <0.5 MHz), the broad derivative feature at 4.8 MHz and the trough at 6.1 MHz can be assigned to perpendicular and parallel transitions, respectively. The matching low frequency peaks expected at ~0.5 and ~1.8 MHz—i.e., to mirror about half the hyperfine coupling and split by twice the free ^{2}H Larmor frequency (2.16 MHz)—were not observed. Simulation of the spectra revealed a deuterium hyperfine coupling constant (HFCC), $|A_{iso}(^2H)|$, of ~6 MHz that corresponds to a proton HFCC, $|A_{iso}(^1H)|$, of ~40 MHz as the gyromagnetic ratio of ^{1}H (γ_{1H}) is 6.51 times larger than γ_{2H}. The observed transition related to ^{31}P, mentioned above, can originate from either the phosphate group of PLP or phosphodiester of AdoCbl. Nevertheless, the possibility of phosphodiester group of AdoCbl is ruled out because: first, these features are absent in the region where the Co^{2+} EPR signal occurs; and second, in the holoenzyme the phosphodiester of AdoCbl is positioned away from the corrin ring, making it improbable that the hyperfine interaction is between this phosphorus and the spin center. Thus, the transition related to ^{31}P is attributed to the phosphate of PLP. The two derivative shaped features are assumed to be the perpendicular singularities of a dipolar powder pattern and simulated nicely with $A_{31P} = [-0.4, -0.5, 19.2]$.

Similarly, X-band ENDOR spectra with substrate homologs confirmed that the radicals are still attached to PLP (Fig. 13) (Chen et al., 2013).

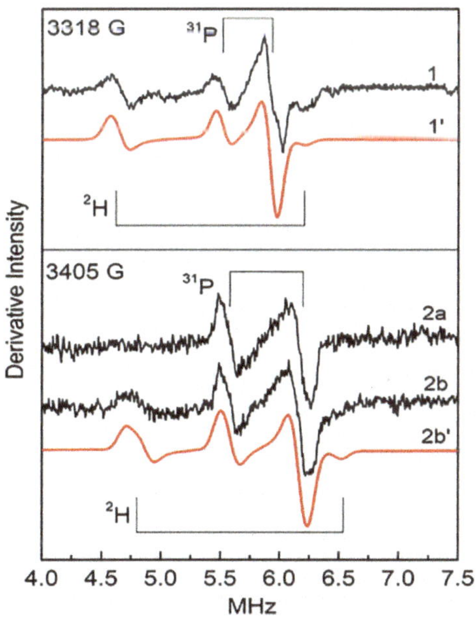

Fig. 13 X-band ENDOR spectra (1, 2a, 2b) along with powder spectra simulation (1′, 2b′) of radical intermediates trapped at 3 min in the reaction of 5,6-LAM with 2,4-DAB (top panel) and with D-2,5-DAPn (bottom panel). Coenzyme [4′-^{2}H]PLP is used for the data collection of spectra 1 and 2b; [4′-^{1}H]PLP for spectrum 2a. The spectra were taken at the maximum EPR absorption as indicated by arrows in Fig. 6. Instrument settings: microwave power, 2 mW; RF power, 200 W; modulation depth, 100 kHz; T, 20 K; magnetic fields are as indicated. *Reprinted with permission from Journal of the American Chemical Society* ***2013,*** *135, 788–794; Copyright 2013, American Chemical Society.*

As illustrated in Fig. 13, the spectra can be accounted for by two different transitions. As mentioned above, these pair of transitions—one pair of transitions at 4.6 and 6.2 MHz and the other at 5.53 and 5.9 MHz—can be assigned to ^{2}H, and ^{31}P nuclei of [4′-^{2}H]PLP, respectively.

6.3 Assistance from DFT calculations

DFT calculations provide many valuable insights for the characterization of the putative radical intermediates. For example, the feasibility of detecting a particular radical intermediate by EPR can be predicted by comparing the relative energies, computed using DFT, of various probable radical intermediates in radical aminomutases (Maity & Ke, 2013, 2015; Maity et al., 2015; Pang, Scrutton, & Sutcliffe, 2014; Sandala, Smith, & Radom, 2006, 2010; Wetmore, Smith, & Radom, 2001). If the radical intermediate is not stable enough and does not accumulate sufficiently, it evades the detection by EPR. The radical intermediate with lowest relative energy is usually the one detected. The other information obtained from DFT calculations are the values of HFCCs, which are of paramount importance in the case of radical intermediates. These computed values can reliably be used to determine the right structure of a radical intermediate from the pool of the various putative structures. In the case of 5,6-LAM, DFT calculations have been performed using the Gaussian suite of programs (Frisch et al., 2004) following the strategies established by Leo Radom and coworkers (Sandala et al., 2006; Wetmore et al., 2001). In this strategy, geometry optimizations are performed at the level of B3LYP/6-31G(d,P) while for single point energy calculations RMP2/G3MP2large level is employed and corrected for zero point energy. B3-LYP/6-311G-(2d,p) level of computation is used to calculate HFCCs. Moreover, the methyl group and the phosphate handle of PLP are modeled with hydrogen to save computation time, as these groups are known to a play negligible role in the catalytic cycle. DFT calculations validated proposed mechanism for radical aminomutases, in which the formation and subsequent opening of the three membered ring of **I**$^{\bullet}$ are the most crucial events (Maity & Ke, 2015; Pang et al., 2014; Sandala et al., 2006; Wetmore et al., 2001).

DFT calculations have been helpful to identify the structure of the persistent radical in the reaction of 4-thia-L-lysine with 5,6-LAM as a tautomer of the substrate radical. The observation of a second radical intermediate, which is different from corresponding **S**• and appeared at longer reaction time was at first jubilantly construed, albeit erroneously, as the putative **I**• radical. But the computed relative energies from DFT revealed that the observed radical could not be the **I**• radical, instead, the radical is actually the tautomer radical. This assignment was based on the computed relative energies and comparison of computed ^{1}H and ^{31}P HFCCs of a pool of probable radical candidates including **I**• with experimentally determined values (Maity et al., 2009; Tang et al., 2009).

In another case, the relative energy computed from DFT predicted the probable trapping of elusive **I**• radical if the reaction is performed with PLP analog PLP-NO. However, attempts to detect it failed, suggesting that the accumulation of **I**• could not reach to the required level for the detection by EPR as the stabilization was not adequate (Maity et al., 2015). In another instance, comparison of computed HFCCs of multiple putative radical candidates with the experimental HFCCs obtained from ENDOR revealed that the radical intermediates detected in the reaction of homologs, D-2,5-DAPn and D-2,4-diaminobutaric acid (D-2,4-DAB), are C5 and C4 based, respectively (Chen et al., 2013). As it can be seen from Table 4, computed ^{2}H HFCCs of these are in agreement with the values obtained from ENDOR whereas computed ^{2}H HFCCs of other putative candidates are very different.

The correspondence between substrate and substrate analogs containing odd or even numbers of carbon atoms and the open/closed state of 5,6-LAM is interpreted by correlating relative orientations at C2 among the analogs, substrate, and product on the basis of their relative energies obtained from DFT calculations, which is illustrated in Fig. 14. The similar variation of energies among the rotamers of the substrate (D-lysine that is abbreviated as D-2,6-DAH), and D-2,4-DAB, which contain even number of carbons, is evident and the corresponding radical intermediates are trapped in the closed state. On the other hand, homologs containing an odd number of carbon atoms—D-2,5-DAPn and D-2,3-DAPr—mimic the energy landscape of the rotamers of the product, D-2,5-DAH, and corresponding radical intermediates are found to be present in the open state. Thus, it is hypothesized that the conformational change at C2 plays a crucial role in the opening of the closed state and thereby triggering the product release.

Table 4 ^{2}H and ^{31}P hyperfine coupling parameters.

	Experiment						Theory				
	$A(^{31}P)_{ENDOR}$ (MHz)			$A(^{2}H)_{ENDOR}$ (MHz)			Radical	Atom	$A(^{2}H)_{DFT}$ (MHz)		
Analog	A_{xx}	A_{yy}	A_{zz}	A_{xx}	A_{yy}	A_{zz}			A_{xx}	A_{yy}	A_{zz}
D-2,5-DAPn	−0.75	−0.55	14.0	−4.95	−5.30	−8.60	C5	C4'($^{2}H_{\alpha}$)	−2.09	−5.40	−7.58
							C4	C4'(^{2}H)	−0.34	−0.26	0.29
							C4 based tautomer	C4'($^{2}H_{\beta1}$)	11.46	11.57	12.85
								C4'($^{2}H_{\beta2}$)	3.10	3.62	4.48
D-2,4-DAB	−0.45	−0.45	19.0	−4.80	−5.00	−8.10	C4	C4'($^{2}H_{\alpha}$)	−2.02	−5.37	−7.53
							C3	C4'(^{2}H)	−0.29	−0.15	0.46
							C3 based tautomer	C4'($^{2}H_{\beta1}$)	11.00	11.22	12.38
								C4'($^{2}H_{\beta2}$)	4.35	4.88	5.75

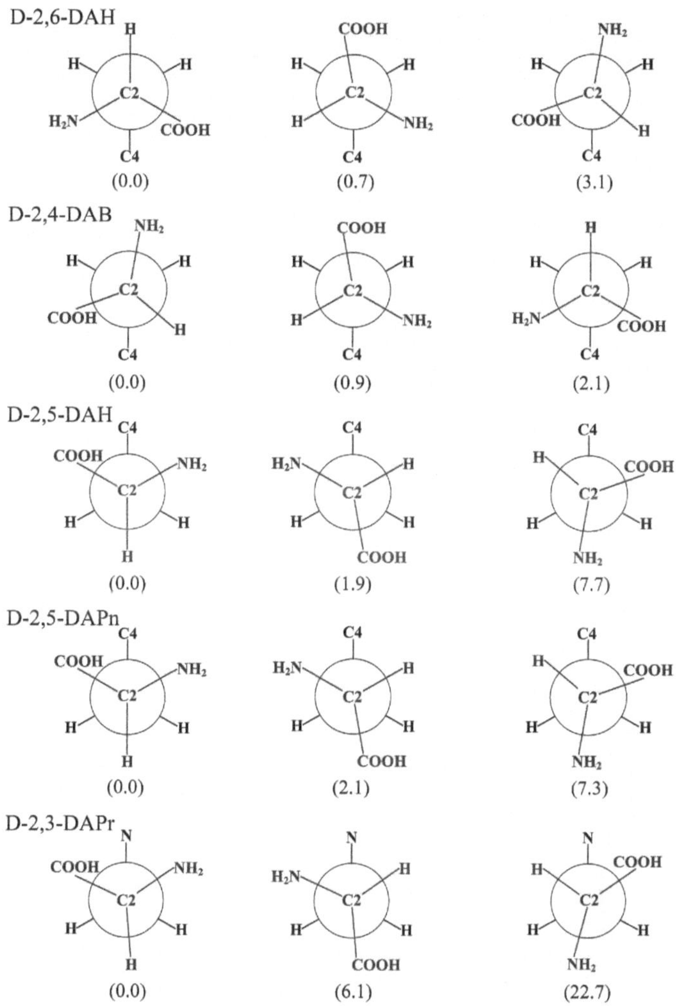

Fig. 14 Relative energies in kJ mol^{-1} of optimized structures of substrate-, product-, homolog-PLP complexes displayed by Newman projections along C2—C3 bond. *Reprinted with permission from Journal of the American Chemical Society* ***2013****, 135, 788–794; Copyright 2013, American Chemical Society.*

7. Conclusions

The indispensable roles of EPR and ENDOR in studying the radical enzymes are presented in this chapter by taking 5,6-LAM, which uses coenzyme B_{12} and PLP as cofactors, as a representative example. EPR and ENDOR together with synthetic isotopologues and DFT computation established the radical mechanism of action by structurally characterizing the important radical intermediates. In doing so, it also provided the evidence for the putative closed state that protects the highly reactive radical intermediates generated during the catalytic cycle and was proposed on the basis of location of PLP and AdoCbl which are away from each other in the resting open state, as observed in the crystal structure of 5,6-LAM. In future, EPR and ENDOR are definitely going to play prominent roles in detection and characterization of the elusive azacyclopropylcarbinyl radical that is central to the proposed mechanism of action of the radical aminomutases.

Acknowledgments

Research described in this chapter was funded by Ministry of Science and Technology, Taiwan.

References

Abdullin, D., & Schiemann, O. (2020). Pulsed dipolar EPR spectroscopy and metal ions: Methodology and biological applications. *ChemPlusChem*, *85*(2), 353–372. https://doi.org/10.1002/cplu.201900705.

Abend, A., Bandarian, V., Nitsche, R., Stupperich, E., Retey, J., & Reed, G. H. (1999). Ethanolamine ammonia-lyase has a "base-on" binding mode for coenzyme B_{12}. *Archives of Biochemistry and Biophysics*, *370*(1), 138–141. https://doi.org/10.1006/abbi.1999.1382.

Abend, A., Bandarian, V., Reed, G. H., & Frey, P. A. (2000). Identification of cis-ethanesemidione as the organic radical derived from glycolaldehyde in the suicide inactivation of dioldehydrase and of ethanolamine ammonia-lyase. *Biochemistry*, *39*(20), 6250–6257. https://doi.org/10.1021/bi992963k.

Ando, M., & Emoto, S. (1975). Catalytic activities of pyridoxal analogues. I. the catalytic effects of pyridoxal, pyridoxal Methochloride, and pyridoxal *N*-oxide on the racemization of L-glutamic acid. *Bulletin of the Chemical Society of Japan*, *48*(5), 1655–1656. https://doi.org/10.1246/bcsj.48.1655.

Baker, J. J., & Stadtman, T. C. (1982). Amino mutases. In D. Dolphin (Ed.), *Vol. 2. B_{12}* (pp. 203–232). New York: John Wiley & Sons.

Baker, J. J., van der Drift, C., & Stadtman, T. C. (1973). Purification and properties of β-lysine mutase, a pyridoxal phosphate and B_{12} coenzyme dependent enzyme. *Biochemistry*, *12*(6), 1054–1063. https://doi.org/10.1021/bi00730a006.

Ballinger, M. D., Frey, P. A., & Reed, G. H. (1992). Structure of a substrate radical intermediate in the reaction of lysine 2,3-aminomutase. *Biochemistry*, *31*(44), 10782–10789. https://doi.org/10.1021/bi00159a020.

Bandarian, V., & Reed, G. H. (2002). Analysis of the electron paramagnetic resonance spectrum of a radical intermediate in the coenzyme B_{12}-dependent ethanolamine ammonia-lyase catalyzed reaction of *S*-2-aminopropanol. *Biochemistry*, *41*(27), 8580–8588. https://doi.org/10.1021/bi0201217.

Banerjee, R. (Ed.). (1999). *Chemistry and biochemistry of B_{12}*. New York: Wiley.

Banerjee, R. (2003). Radical carbon skeleton rearrangements: Catalysis by coenzyme B_{12}-dependent mutases. *Chemical Reviews*, *103*(6), 2083–2094. https://doi.org/10.1021/cr0204395.

Bencini, A., & Gatteschi, D. (1990). *EPR of exchange coupled systems*. New York: Springer-Verlag.

Berkovitch, F., Behshad, E., Tang, K.-H., Enns, E. A., Frey, P. A., & Drennan, C. L. (2004). A locking mechanism preventing radical damage in the absence of substrate, as revealed by the x-ray structure of lysine 5,6-aminomutase. *Proceedings of the National Academy of Sciences of the United States of America*, *101*, 15870–15875. https://doi.org/10.1073/pnas.0407074101.

Berliner, L. J. (Ed.). (2002). *Spin labeling the next millennium*. New York: Springer US.

Brown, K. L. (2005). Chemistry and enzymology of vitamin B_{12}. *Chemical Reviews*, *105*(6), 2075–2150. https://doi.org/10.1021/cr030720z.

Chang, C. H., & Frey, P. A. (2000). Cloning, sequencing, heterologous expression, purification, and characterization of adenosylcobalamin-dependent D-lysine 5,6-aminomutase from *clostridium sticklandii*. *Journal of Biological Chemistry*, *275*(1), 106–114. https://doi.org/10.1074/jbc.275.1.106.

Chen, J. R., & Ke, S. C. (2018). Magnetic field effects on coenzyme B_{12}- and B_6-dependent lysine 5,6-aminomutase: Switching of the J-resonance through a kinetically competent radical-pair intermediate. *Physical Chemistry Chemical Physics*, *20*(18), 13068–13074. https://doi.org/10.1039/C8CP01497C.

Chen, Y. H., Maity, A. N., Frey, P. A., & Ke, S. C. (2013). Mechanism-based inhibition reveals transitions between two conformational states in the action of lysine 5,6-aminomutase: A combination of electron paramagnetic resonance spectroscopy, electron nuclear double resonance spectroscopy, and density functional theory study. *Journal of the American Chemical Society*, *135*(2), 788–794. https://doi.org/10.1021/ja309603a.

Chen, Y. H., Maity, A. N., Pan, Y. C., Frey, P. A., & Ke, S. C. (2011). Radical stabilization is crucial in the mechanism of action of lysine 5,6-aminomutase: Role of tyrosine-263α as revealed by electron paramagnetic resonance spectroscopy. *Journal of the American Chemical Society*, *133*(43), 17152–17155. https://doi.org/10.1021/ja207766c.

Chen, H. P., & Marsh, E. N. (1997). How enzymes control the reactivity of adenosylcobalamin: Effect on coenzyme binding and catalysis of mutations in the conserved histidine-aspartate pair of glutamate mutase. *Biochemistry*, *36*(25), 7884–7889. https://doi.org/10.1021/bi970169y.

Cutsail, G. E., III, Telser, J., & Hoffman, B. M. (2015). Advanced paramagnetic resonance spectroscopies of iron-sulfur proteins: Electron nuclear double resonance (ENDOR) and electron spin echo envelope modulation (ESEEM). *Biochimica et Biophysica Acta-Molecular Cell Research*, *1853*(6), 1370–1394. https://doi.org/10.1016/j.bbamcr.2015.01.025.

Dolphin, D. (Ed.). (1982). *B_{12} (Vol. 2)*. New York: John Wiley & Sons.

Frey, P. A. (2010). *Comprehensive natural products II: Chemistry and biology*. *Vol.* 7 (pp. 501–546). Oxford, U. K: Elsevier.

Frey, P. A., & Chang, C. H. (1999). Aminomutases. In R. Banerjee (Ed.), *Chemistry and biochemistry of B_{12}* (pp. 835–857). New York: John Wiley & Sons.

Frey, P. A., Hegeman, A. D., & Reed, G. H. (2006). Free radical mechanisms in enzymology. *Chemical Reviews*, *106*(8), 3302–3316. https://doi.org/10.1021/cr050292s.

Frey, P. A., & Magnusson, O. T. (2003). *S*-adenosylmethionine: A wolf in sheep's clothing, or a rich man's adenosylcobalamin? *Chemical Reviews*, *103*(6), 2129–2148. https://doi.org/10.1021/cr020422m.

Frey, P. A., & Reed, G. H. (2011). Pyridoxal-5′-phosphate as the catalyst for radical isomerization in reactions of PLP-dependent aminomutases. *Biochimica Et Biophysica Acta-Proteins and Proteomics*, *1814*(11), 1548–1557. https://doi.org/10.1016/j.bbapap.2011.03.005.

Frisch, M. J., Trucks, G. W., Schlegel, H. B., Scuseria, G. E., Robb, M. A., Cheeseman, J. R., et al. (2004). *Gaussian 03, Revision C.02*. Wallingford CT: Gaussian, Inc.

Fu, M., & Silverman, R. B. (1999). Isolation and characterization of the product of inactivation of gamma-aminobutyric acid aminotransferase by gabaculine. *Bioorganic & Medicinal Chemistry*, 7(8), 1581–1590. https://doi.org/10.1016/s0968-0896(99)00081-4.

Fukui, S., Ohishi, N., Nakai, Y., & Shimizu, S. (1969). Properties and catalytic activities of vitamin B_6 *N*-oxide. *Archives of Biochemistry and Biophysics*, *130*(1), 584–593. https://doi.org/10.1016/0003-9861(69)90074-5.

Horitani, M., Byer, A. S., Shisler, K. A., Chandra, T., Broderick, J. B., & Hoffman, B. M. (2015). Why nature uses radical SAM enzymes so widely: Electron nuclear double resonance studies of lysine 2,3-aminomutase show the 5′-dAdo· "free radical" is never free. *Journal of the American Chemical Society*, *137*(22), 7111–7121. https://doi.org/10.1021/jacs.5b00498.

Jarrett, J. T., Amaratunga, M., Drennan, C. L., Scholten, J. D., Sands, R. H., Ludwig, M. L., et al. (1996). Mutations in the B_{12}-binding region of methionine synthase: How the protein controls methylcobalamin reactivity. *Biochemistry*, *35*(7), 2464–2475. https://doi.org/10.1021/bi952389m.

Ke, S. C. (2003). Spin-spin interaction in ethanolamine deaminase. *Biochimica et Biophysica Acta-General Subjects*, *1620*(1–3), 267–272. https://doi.org/10.1016/S0304-4165(03)00006-0.

Lawrence, C. C., Gerfen, G. J., Samano, V., Nitsche, R., Robins, M. J., Retey, J., et al. (1999). Binding of cob(II)alamin to the adenosylcobalamin-dependent ribonucleotide reductase from *lactobacillus leichmannii*. Identification of dimethylbenzimidazole as the axial ligand. *Journal of Biological Chemistry*, *274*(11), 7039–7042. https://doi.org/10.1074/jbc.274.11.7039.

Lees, N. S., Chen, D., Walsby, C. J., Behshad, E., Frey, P. A., & Hoffman, B. M. (2006). How an enzyme tames reactive intermediates: Positioning of the active-site components of lysine 2,3-aminomutase during enzymatic turnover as determined by ENDOR spectroscopy. *Journal of the American Chemical Society*, *128*(31), 10145–10154. https://doi.org/10.1021/ja061282r.

Lepore, B. W., Ruzicka, F. J., Frey, P. A., & Ringe, D. (2005). The X-ray crystal structure of lysine-2,3-aminomutase from *Clostridium subterminale*. *Proceedings of the National Academy of Sciences of the United States of America*, *102*(39), 13819–13824. https://doi.org/10.1073/pnas.0505726102.

Lo, H. H., Lin, H. H., Maity, A. N., & Ke, S. C. (2016). The molecular mechanism of the open-closed protein conformational cycle transitions and coupled substrate binding, activation and product release events in lysine 5,6-aminomutase. *Chemical Communications*, *52*(38), 6399–6402. https://doi.org/10.1039/C6CC01888B.

LoBrutto, R., Bandarian, V., Magnusson, O. T., Chen, X., Schramm, V. L., & Reed, G. H. (2001). 5'-deoxyadenosine contacts the substrate radical intermediate in the active site of ethanolamine ammonia-lyase: ^{2}H and ^{13}C electron nuclear double resonance studies. *Biochemistry*, *40*(1), 9–14. https://doi.org/10.1021/bi001865s.

Magnusson, O. T., & Frey, P. A. (2002). Interactions of diol dehydrase and 3′,4′-anhydroadenosylcobalamin: Suicide inactivation by electron transfer. *Biochemistry*, *41*(5), 1695–1702. https://doi.org/10.1021/bi011947w.

Maity, A. N., Chen, Y. H., & Ke, S. C. (2014). Large-scale domain motions and pyridoxal-5′-phosphate assisted radical catalysis in coenzyme B_{12}-dependent aminomutases. *International Journal of Molecular Sciences*, *15*(2), 3064–3087. https://doi.org/10.3390/ijms15023064.

Maity, A. N., Hsieh, C. P., Huang, M. H., Chen, Y. H., Tang, K. H., Behshad, E., et al. (2009). Evidence for conformational movement and radical mechanism in the reaction of 4-thia-L-lysine with lysine 5,6-aminomutase. *The Journal of Physical Chemistry B*, *113*(36), 12161–12163. https://doi.org/10.1021/jp905357a.

Maity, A. N., & Ke, S. C. (2011). Synthesis of 4-thia[5-^{13}C]lysine. *Journal of Labelled Compounds & Radiopharmaceuticals*, *54*(9–10), 589–590. https://doi.org/10.1002/jlcr.1893.

Maity, A. N., & Ke, S. C. (2013). 5-Fluorolysine as alternative substrate of lysine 5,6-aminomutase: A computational study. *Computational and Theoretical Chemistry, 1022*, 1–5. https://doi.org/10.1016/j.comptc.2013.08.007.

Maity, A. N., & Ke, S. C. (2015). 4′-CyanoPLP presents better prospect for the experimental detection of elusive cyclic intermediate radical in the reaction of lysine 5,6-aminomutase. *Biochemical and Biophysical Research Communications, 457*(2), 161–164. https://doi.org/10.1016/j.bbrc.2014.12.076.

Maity, A. N., Lin, H. H., Chiang, H. S., Lo, H. H., & Ke, S. C. (2015). Reaction of pyridoxal-5′-phosphate-*N*-oxide with lysine 5,6-aminomutase: Enzyme flexibility toward cofactor analog. *ACS Catalysis, 5*(5), 3093–3099. https://doi.org/10.1021/acscatal.5b00671.

Maity, A. N., Shaikh, A. C., Srimurugan, S., Wu, C. J., Chen, C. P., & Ke, S. C. (2012). Synthesis of 4-thia-[6-^{13}C]lysine from [2-^{13}C]glycine: Access to site-directed isotopomers of 2-aminoethanol, 2-bromoethylamine and 4-thialysine. *Amino Acids, 42*(1), 309–315. https://doi.org/10.1007/s00726-010-0808-8.

Mancia, F., Keep, N. H., Nakagawa, A., Leadlay, P. F., McSweeney, S., Rasmussen, B., et al. (1996). How coenzyme B_{12} radicals are generated: The crystal structure of methylmalonyl-coenzyme a mutase at 2 angstrom resolution. *Structure, 4*(3), 339–350. https://doi.org/10.1016/s0969-2126(96)00037-8.

Michel, C., Albracht, S. P., & Buckel, W. (1992). Adenosylcobalamin and cob(II)alamin as prosthetic groups of 2-methyleneglutarate mutase from *Clostridium barkeri*. *European Journal of Biochemistry, 205*(2), 767–773. https://doi.org/10.1111/j.1432-1033.1992.tb16841.x.

Morley, C. G., & Stadtman, T. C. (1970). Studies on the fermentation of D-α-lysine. Purification and properties of an adenosine triphosphate regulated B_{12} -coenzyme-dependent D-α-lysine mutase complex from *clostridium sticklandii*. *Biochemistry, 9*(25), 4890–4900. https://doi.org/10.1021/bi00827a010.

Orme-Johnson, W. H., Beinert, H., & Blakley, R. L. (1974). Cobamides and ribonucleotide reduction. XII. The electron paramagnetic resonance spectrum of "active coenzyme B_{12}". *Journal of Biological Chemistry, 249*(8), 2338–2343. https://doi.org/10.1016/S0021-9258(19)42735-X.

Padmakumar, R., & Banerjee, R. (1995). Evidence from electron paramagnetic resonance spectroscopy of the participation of radical intermediates in the reaction catalyzed by methylmalonyl-coenzyme a mutase. *Journal of Biological Chemistry, 270*(16), 9295–9300. https://doi.org/10.1074/jbc.270.16.9295.

Pang, J., Scrutton, N. S., & Sutcliffe, M. J. (2014). Quantum mechanics/molecular mechanics studies on the mechanism of action of cofactor pyridoxal 5′-phosphate in ornithine 4,5-aminomutase. *Chemistry—a European Journal, 20*(36), 11390–11401. https://doi.org/10.1002/chem.201402759.

Reed, G. H., & Mansoorabadi, S. O. (2003). The positions of radical intermediates in the active sites of adenosylcobalamin-dependent enzymes. *Current Opinion in Structural Biology, 13*(6), 716–721. https://doi.org/10.1016/j.sbi.2003.10.011.

Ruzicka, F. J., & Frey, P. A. (2007). Glutamate 2,3-aminomutase: A new member of the radical SAM superfamily of enzymes. *Biochimica Et Biophysica Acta-Proteins and Proteomics, 1774*(2), 286–296. https://doi.org/10.1016/j.bbapap.2006.11.008.

Sandala, G. M., Smith, D. M., & Radom, L. (2006). In search of radical intermediates in the reactions catalyzed by lysine 2,3-aminomutase and lysine 5,6-aminomutase. *Journal of the American Chemical Society, 128*, 16004–16005. https://doi.org/10.1021/ja0668421.

Sandala, G. M., Smith, D. M., & Radom, L. (2010). Modeling the reactions catalyzed by coenzyme B_{12}-dependent enzymes. *Accounts of Chemical Research, 43*(5), 642–651. https://doi.org/10.1021/ar900260c.

Sayler, R. I., Stich, T. A., Joshi, S., Cooper, N., Shaw, J. T., Begley, T. P., et al. (2019). Trapping and electron paramagnetic resonance characterization of the 5'dAdo· Radical in a radical *S*-adenosyl methionine enzyme reaction with a non-native substrate. *ACS central. Science*, *5*(11), 1777–1785. https://doi.org/10.1021/acscentsci.9b00706.

Stoll, S., & Schweiger, A. (2006). EasySpin, a comprehensive software package for spectral simulation and analysis in EPR. *Journal of Magnetic Resonance*, *178*(1), 42–55. https://doi.org/10.1016/j.jmr.2005.08.013.

Tang, K. H., Chang, C. H., & Frey, P. A. (2001). Electron transfer in the substrate-dependent suicide inactivation of lysine 5,6-aminomutase. *Biochemistry*, *40*(17), 5190–5199. https://doi.org/10.1021/bi010157j.

Tang, K. H., Mansoorabadi, S. O., Reed, G. H., & Frey, P. A. (2009). Radical triplets and suicide inhibition in reactions of 4-thia-D- and 4-thia-L-lysine with lysine 5,6-aminomutase. *Biochemistry*, *48*(34), 8151–8160. https://doi.org/10.1021/bi900828f.

Toraya, T. (2003). Radical catalysis in coenzyme B_{12}-dependent isomerization (eliminating) reactions. *Chemical Reviews*, *103*(6), 2095–2127. https://doi.org/10.1021/cr020428b.

Wetmore, S. D., Smith, D. M., & Radom, L. (2001). Enzyme catalysis of 1,2-amino shifts: The cooperative action of B_6, B_{12}, and aminomutases. *Journal of the American Chemical Society*, *123*(36), 8678–8689. https://doi.org/10.1021/ja010211j.

Wolthers, K. R., Levy, C., Scrutton, N. S., & Leys, D. (2010). Large-scale domain dynamics and adenosylcobalamin reorientation orchestrate radical catalysis in ornithine 4,5-aminomutase. *Journal of Biological Chemistry*, *285*(18), 13942–13950. https://doi.org/10.1074/jbc.M109.068908.

Wolthers, K. R., Rigby, S. E. J., & Scrutton, N. S. (2008). Mechanism of radical-based catalysis in the reaction catalyzed by adenosylcobalamin-dependent ornithine 4,5-aminomutase. *Journal of Biological Chemistry*, *283*(50), 34615–34625. https://doi.org/10.1074/jbc.M807911200.

Wu, W., Booker, S., Lieder, K. W., Bandarian, V., Reed, G. H., & Frey, P. A. (2000). Lysine 2,3-aminomutase and trans-4,5-dehydrolysine: Characterization of an allylic analogue of a substrate-based radical in the catalytic mechanism. *Biochemistry*, *39*(31), 9561–9570. https://doi.org/10.1021/bi000658p.

Wu, W., Lieder, K. W., Reed, G. H., & Frey, P. A. (1995). Observation of a second substrate radical intermediate in the reaction of lysine 2,3-aminomutase: A radical centered on the β-carbon of the alternative substrate, 4-thia-L-lysine. *Biochemistry*, *34*(33), 10532–10537. https://doi.org/10.1021/bi00033a027.

Yamanishi, M., Yamada, S., Muguruma, H., Murakami, Y., Tobimatsu, T., Ishida, A., et al. (1998). Evidence for axial coordination of 5,6-dimethylbenzimidazole to the cobalt atom of adenosylcobalamin bound to diol dehydratase. *Biochemistry*, *37*(14), 4799–4803. https://doi.org/10.1021/bi972572a.

Yang, H., McDaniel, E. C., Impano, S., Byer, A. S., Jodts, R. J., Yokoyama, K., et al. (2019). The elusive 5′-deoxyadenosyl radical: Captured and characterized by electron paramagnetic resonance and electron nuclear double resonance spectroscopies. *Journal of the American Chemical Society*, *141*(30), 12139–12146. https://doi.org/10.1021/jacs.9b05926.

Zelder, O., Beatrix, B., Leutbecher, U., & Buckel, W. (1994). Characterization of the coenzyme-B_{12}–dependent glutamate mutase from *Clostridium cochlearium* produced in *Escherichia coli*. *European Journal of Biochemistry*, *226*(2), 577–585. https://doi.org/10.1111/j.1432-1033.1994.tb20083.x.

Zhao, Y., Abend, A., Kunz, M., Such, P., & Retey, J. (1994). Electron paramagnetic resonance studies of the methylmalonyl-CoA mutase reaction. Evidence for radical intermediates using natural and artificial substrates as well as the competitive inhibitor 3-carboxypropyl-CoA. *European Journal of Biochemistry*, *225*(3), 891–896. https://doi.org/10.1111/j.1432-1033.1994.0891b.x.

CHAPTER TEN

Resolution and characterization of contributions of select protein and coupled solvent configurational fluctuations to radical rearrangement catalysis in coenzyme B_{12}-dependent ethanolamine ammonia-lyase

Meghan Kohne, Wei Li, Alina Ionescu, Chen Zhu, and Kurt Warncke*
Department of Physics, Emory University, Atlanta, GA, United States
*Corresponding author: e-mail address: kwarncke@physics.emory.edu

Contents

Methods in Enzymology, Volume 669
ISSN 0076-6879
https://doi.org/10.1016/bs.mie.2021.12.017

Abstract

Coenzyme B_{12} (adenosylcobalamin) -dependent ethanolamine ammonia-lyase (EAL) is the signature enzyme in ethanolamine utilization metabolism associated with microbiome homeostasis and disease conditions in the human gut. The enzyme conducts a complex choreography of bond-making/bond-breaking steps that rearrange substrate to products through a radical mechanism, with themes common to other coenzyme B_{12}-dependent and radical enzymes. The methods presented are targeted to test the hypothesis that particular, select protein and coupled solvent configurational fluctuations contribute to enzyme function. The general approach is to correlate enzyme function with an introduced perturbation that alters the properties (for example, degree of concertedness, or collectiveness) of protein and coupled solvent dynamics. Methods for sample preparation and low-temperature kinetic measurements by using temperature-step reaction initiation and time-resolved, full-spectrum electron paramagnetic resonance spectroscopy are detailed. A framework for interpretation of results obtained in ensemble systems under conditions of statistical equilibrium within the reacting, globally unstable state is presented. The temperature-dependence of the first-order rate constants for decay of the cryotrapped paramagnetic substrate radical state in EAL, through the chemical step of radical rearrangement, displays a piecewise-continuous Arrhenius dependence from 203 to 295 K, punctuated by a kinetic bifurcation over 219–220 K. The results reveal the obligatory contribution of a class of select collective protein and coupled solvent fluctuations to the interconversion of two resolved, sequential configurational substates, on the decay time scale. The select class of collective fluctuations also contributes to the chemical step. The methods and analysis are generally applicable to other coenzyme B_{12}-dependent and related radical enzymes.

1. Introduction

The development of the mechanistic enzymology of the coenzyme B_{12} (adenosylcobalamin; AdoCbl)-dependent enzymes has advanced to the refined level of understanding of enzyme mechanisms in general (Richard, 2013). High-resolution protein structures of resting and intermediate states (Berman et al., 2000) provide the snapshots that anchor detailed physico-chemical descriptions of the transformations of the cofactor and reactants that are acquired by spectroscopic and kinetic experiments (Banerjee, 1999, 2003; Brown, 2005; Frey, 2010; Marsh & Melendez, 2012; Toraya, 2003). And, high-level ab initio, empirical valence bond, and multi-layer computations (Khoroshun, Warncke, Ke, Musaev, & Morokuma, 2003; Semialjac & Schwartz, 2003; Sharma, Chu, Olsson, & Warshel, 2007; Wetmore, Smith, Bennet, & Radom, 2002) have expanded the microscopic interpretations to the current limits on treatable system size. The accumulating, detailed knowledge of enzyme structure and mechanism has been shadowed by persistent questions about the complete understanding of the fundamental basis of enzyme catalysis (Agarwal, 2019; Herschlag & Natarajan, 2013; Jencks, 1969; Kraut, Carrol, & Herschlag, 2003). The role of protein dynamics emerged as an explanatory thrust, with intense activity in defining concepts and outcomes of experimental and theoretical studies (Henzler-Wildman & Kern, 2007; Kamerlin & Warshel, 2010; Nagel & Klinman, 2009). The clarification that emerged led to the "ensemble perspective," in which the enzyme's protein structure was appropriately recognized to occupy a hierarchical distribution of conformational states, in accord with the expectations of statistical equilibrium (Hammes, Benkovic, & Hammes-Schiffer, 2011). A protein structure, corresponding to a particular minimum-energy, canonical intermediate state, ergodically samples, or is in exchange with, conformations (configurations) that resemble the structures of the preceding and succeeding states (and transition states) on the progress coordinate (Boehr, Dyson, & Wright, 2006; Boehr, McElheny, Dyson, & Wright, 2006; Bruice, 2002; Ma & Nussinov, 2010). From this interpretive framework, the further, accessible understanding of mechanistic enzymology was attributed to the collection of additional protein structures distributed within the ensembles, allowing the reaction progress coordinate to be resolved in finer detail (Hammes et al., 2011). This direction has been realized by the high through-put, large data set sampling approach (Yabukarski et al., 2020).

This chapter defines concepts relevant to the contribution of protein and coupled solvent dynamics to the core, chemical steps of enzymes, and describes experimental and analytical methodologies, and an interpretive framework, for their identification and elucidation, in ensemble systems. The enzyme is the Class II, AdoCbl-dependent ethanolamine ammonia-lyase, EAL [EC 4.3.1.7; cobalamin (vitamin B_{12})-dependent enzyme superfamily] (Bandarian & Reed, 1999; Frey, 2010; Toraya, 2003) from *Salmonella enterica* Serovar typhimurium (Faust & Babior, 1992). EAL is the first in a sequence of enzymes that process ethanolamine (aminoethanol) in the ethanolamine utilization (Eut) metabolic pathway (Roof & Roth, 1988), that is associated with microbiome homeostasis (Banerjee, 2017), and *Salmonella enterica* and *Escherichia coli* –induced disease conditions, in the human gut (Anderson & Kendall, 2017; Staib & Fuchs, 2014; Thiennimitr et al., 2011). EAL catalyzes the deamination of aminoethanol and (*R*)- and (*S*)-2-aminopropanol through the multi-step reaction cycle depicted in Fig. 1. The central adiabatic reaction of radical rearrangement (step 3) is surrounded by chemical sub-steps, that include: Cobalt-carbon (Co—C) bond cleavage leading to radical pair formation (step 1), substrate radical-forming and product radical-quenching hydrogen atom transfer reactions (steps 2 and 4, respectively), and Co—C bond reformation and return to the diamagnetic state (step 5). Following Co—C bond cleavage, the low-spin Co(II) in cobalamin (d^7; electron spin, $S=1/2$; denoted cob(II)alamin) interacts with the radical ($S=1/2$) to form a radical pair (Gerfen, 1999). The cob(II)alamin-substrate radical pair state, formed by using the native substrate (Bender, Poyner, & Reed, 2008; Warncke, Schmidt, & Ke, 1999) or 2-aminopropanol (Babior, Moss, Orme-Johnson, & Beinert, 1974), accumulates under steady-state turnover as the only paramagnetic intermediate detectable by electron paramagnetic resonance (EPR) spectroscopy. Simulations of the radical pair EPR spectra yield the magnetic dipolar interaction coupling constant, from which electron-electron dipolar distances, R_{ee}, are obtained (Gerfen, 1999). The cryotrapped cob(II)alamin-substrate radical pair states in EAL display R_{ee} values of 9–11 Å (Bender et al., 2008; Canfield & Warncke, 2002; Ke, 2003). The simulation-derived R_{ee} values and radical, cofactor, and active site structure (Bandarian & Reed, 2002; Bender et al., 2008; Canfield & Warncke, 2002; LoBrutto et al., 2001; Sun, Groover, Canfield, & Warncke, 2008) are consistent with high-resolution X-ray crystallographic *Escherichia coli* (Shibata et al., 2010; Shibata, Sueyoshi, Higuchi, & Toraya, 2018) and homology-modeled *S. typhimurium* (Bovell & Warncke, 2013) structures of EAL. Fig. 2 shows the active site region in *S. typhimurium* EAL.

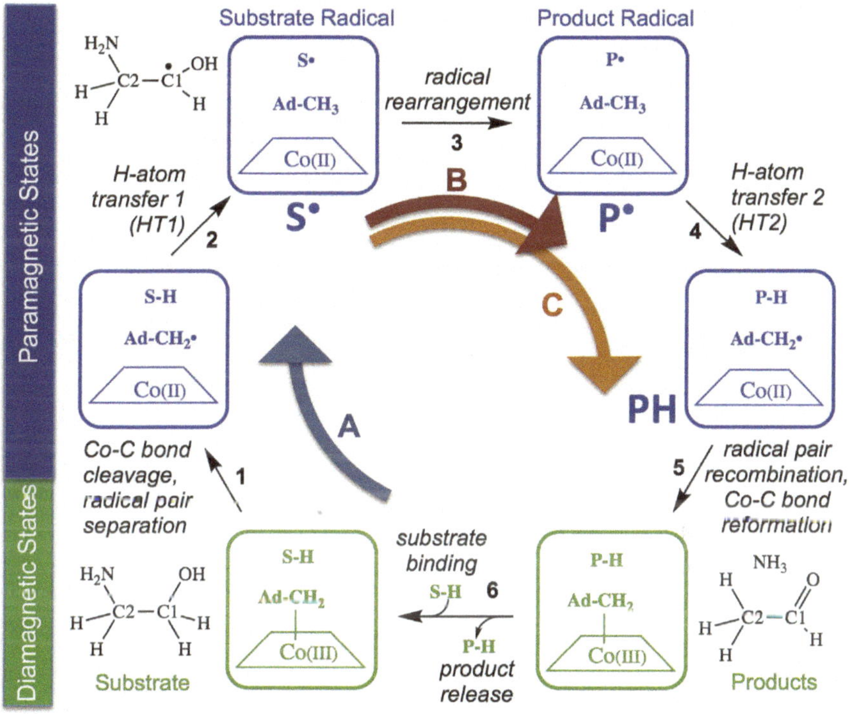

Fig. 1 Catalytic cycle of EAL, showing canonical intermediate states, including the **S**$^{\bullet}$, **P**$^{\bullet}$ and **PH** states, and steps. Highlighted steps A and B, and sequence C (with ^{2}H-substrate), are resolved at low-temperature, by using time-resolved, full-spectrum EPR spectroscopy. Labels: SH, substrate; S$^{\bullet}$, substrate radical; P$^{\bullet}$, product radical, PH, product; trapezoid with Co(III/II), cobalamin; Ad-CH_2-, AdCH$_2^{\bullet}$, Ad-CH_3, Co-bonded, radical, and 5′-methyl deoxyadenosyl species, respectively.

We have developed and applied low-temperature approaches to resolve individual steps in the sequence of radical reactions in EAL, as depicted by processes (A) (Wang & Warncke, 2008, 2013), (B) (Kohne, Zhu, & Warncke, 2017; Ucuncuoglu & Warncke, 2018; Zhu & Warncke, 2008), and (C) (Kohne, Li, Zhu, & Warncke, 2019) in Fig. 1 for kinetic analysis. Methods of time-resolved, full-spectrum EPR spectroscopy to determine the first-order rate constants for steps as a function of temperature over the wide range of 190 to 295 K have been described (Wang, Zhu, Kohne, & Warncke, 2015). Here, we focus on the core adiabatic radical rearrangement reaction of the aminoethanol substrate radical (state, **S**$^{\bullet}$) in EAL (Fig. 1, step 3), which forms the product radical (state, **P**$^{\bullet}$), which is followed by the hydrogen transfer reaction, HT2, that forms state, **PH**.

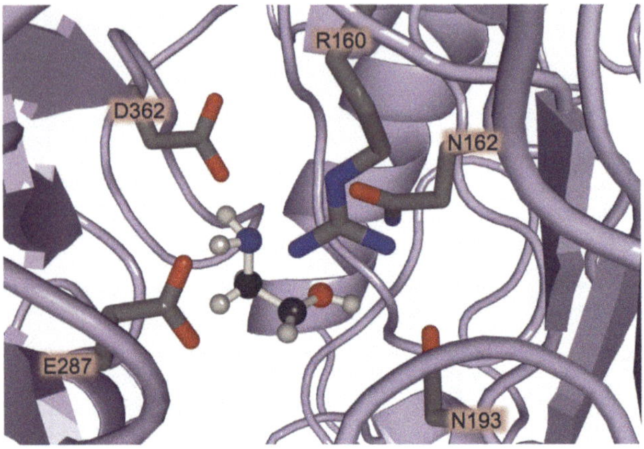

Fig. 2 Active site structure in *S. typhimurium* EAL from the homology model (Bovell & Warncke, 2013), based on the *Escherichia coli* EAL X-ray crystallographic structure (Shibata et al., 2010). Active site region in the EutB subunit is shown in the absence of the 5′-deoxyadenosyl group, with aminoethanol substrate radical modeled in substrate position, surrounded by interacting amino acid side chains.

We articulate the rationale for, and interpretation of, the experiments, in the context of the contribution of select protein and coupled solvent dynamics to the catalysis of chemical steps in EAL, and enzyme catalysis more generally.

2. Protein dynamics and contributions to enzyme catalysis: Definitions and concepts

2.1 Definitions of protein dynamics

Fundamental features of the motions (fluctuations) of protein and associated solvent among configurations, defined and distinguished by different positions of the constituent atoms, and their role in enzyme catalysis, have been examined (Glowacki, Harvey, & Mulholland, 2012; Kamerlin & Warshel, 2010). To provide context for our aims, results, and interpretations, we reiterate some of the basic features, here. The word, "dynamics," is intuitively used as synonymous with "motion," or in the case of molecules, "change" in atomic coordinates (Henzler-Wildman & Kern, 2007). Dynamics is the formalism that describes the causes of motion, which are governed by Newton's Laws: The acceleration (**a**; bold font indicates a vector quantity) imparted to a mass (m) by a force (**F**), or compactly, $\mathbf{F}=m\mathbf{a}$ (second Law), as observed from the initial state and subsequent trajectory of the mass. The motional behavior of enzymes is commonly studied in samples that include a large number of protein molecules, because this affords suitable signal-to-noise ratios. The requirements for *directly* observing reaction dynamics in proteins in such a statistical ensemble are stringent. An example is creation of an excited electronic state of a chromophore molecule bound in a protein, by absorption of a photon of a particular energy, in the photosynthetic reaction center protein (Vos et al., 1991). In an ensemble sample, this event positions molecules at a particular point on an excited-state, vibrational potential energy surface. The vibrationally coherent electronic state subsequently oscillates back and forth, within the harmonic potential, as the coherence decays and the states "thermalize" to eventually reach the statistical population distribution. The reaction of the coherent excited state by barrier crossing to a lower-energy, product state is periodic, and thus, the observed time evolution cannot be described succinctly by the Arrhenius or transition state theories. This is an example of a direct measurement of reaction "dynamics," in which the observed changes in motion (**a**) of the molecule ("*m*") is dictated by the restoring, Hooke's law force (**F**). The stringency arises from the need to create and maintain coherence, by using precise initiation (singular energy) and measurement on ultra-fast (sub-picosecond to few picoseconds) time scales. The potential energy surface is the same as it would be for the equilibrium decay process, but the reacting population is not thermalized. The inability of Arrhenius or transition state theories to describe the events is diagnostic of a direct measurement of reaction dynamics (Kamerlin & Warshel, 2010).

In typical measurements of the kinetics of transient processes in a sample containing a large number of enzymes in a globally unstable state, the reactant state or macrostate ensemble is thermalized—the population of configurational substates within the macrostate energy minimum, and microstates within the substate energy minima, is distributed in accord with equilibrium Boltzmann statistics. Significant changes in system geometry are conducted by configuration changes. Each configuration, defined by a particular position of the protein and solvent atoms, has its complement of $3N$-6 (N, number of atoms) comparable vibrational states. The probability, P_i, or relative population, of configurational microstate, i, is proportional to the factor, $\exp(-E_i/k_B T)$, where E_i is the energy, k_B is Boltzmann's constant, and T is absolute temperature. The relative probability, or population, of states, i and j, is described by the Boltzmann factor, $P_i/P_j = \exp[-(E_i - E_j)/k_B T]$.

The equilibrium or thermalized population of substates and microstates extends, by "conformational sampling," over the reaction progress coordinate accessible to the macrostate (Boehr, Dyson, & Wright, 2006; Boehr, McElheny, et al., 2006; Bruice, 2002; Ma & Nussinov, 2010).

2.2 Arrhenius rate law for reactions from protein configurations at statistical equilibrium

The kinetic expression for the first-order rate constant, k (s^{-1}), from the Arrhenius rate theory is:

$$k = A\ \exp\left(-E_a/RT\right) \tag{1}$$

where A is the phenomenological Arrhenius prefactor (s^{-1}), E_a is the activation energy, R is the gas constant, and T is the absolute temperature (Cornish-Bowden, 1981). The plot of natural logarithm of the rate constant versus inverse temperature is linear for T-independent E_a, with a slope of $-E_a/R$ and intercept of $\ln A$:

$$\ln k = \ln A - (E_a/R)(1/T) \tag{2}$$

Eq. (1) shows that the first-order rate constant for a single-step process is proportional to $\exp(-E_a/k_B T)$, which is analogous to the equilibrium Boltzmann factor, where $E_a = E_i - E_j$, and E_i and E_j correspond to the transition and energy minimum states, respectively. Note that the transition "state" is a maximum on the energy surface, not a true state, as defined by a local energy minimum. In the Arrhenius model, the energies of the states that compose the energy surface do not depend on temperature. Therefore, E_a is temperature-independent. The temperature dependence of the rate constant comes from the Boltzmann factor, $\exp(-E_a/k_B T)$. For example, as T increases, $\exp(-E_a/k_B T)$ becomes larger, which represents an increase in population at the transition state. The thermalized population in the reactant macrostate decays because the reactant macrostate is globally higher in energy than product macrostates. In EAL, the cryotrapped aminoethanol substrate radical state, $\mathbf{S}^\bullet$, is higher in energy than the subsequent diamagnetic product states.

2.3 Indistinguishability of protein and coupled solvent dynamics

Within the minimum and along the reaction progress coordinate on the energy surface, transitions among the configurational microstates of the system are driven randomly and incessantly by the available thermal energy. These fluctuations represent the Newtonian dynamics of the protein and coupled solvent, which are the responses (accelerations) to the forces imparted among the parts of the protein structure, owing to the random motion that arises from the thermal energy at a given temperature (½k_BT, per degree of freedom). However, the properties of these dynamics are not resolved by the standard ensemble, first-order kinetic measurement, because the probability of the population of any state, including the transition state, as given by $\exp(-E_a/k_BT)$, is independent of the mechanism of its population. The transition state may have been achieved by many cumulative incremental fluctuations among configurations of the protein and coupled solvent, or by a few collective configurational changes. Therefore, configurations and fluctuations distinguished by spatial (for example, local changes of amino acid side chains, or regional changes involving the polypeptide backbone, or changes that extend to larger protein domains and to coupled solvent) (Lewandowski, Halse, Blackledge, & Emsley, 2015) and associated temporal (for example, fluctuation frequencies from $\sim 10^{-13}$ to 10^{-3} s) (Henzler-Wildman & Kern, 2007) properties are not resolved by using standard temperature-dependence measurements of first-order rate constants in enzymes. For the standard approach of measuring the temperature dependence of the first-order rate constant, the constant energy surface corresponds to linear Arrhenius (ln$k$ versus $1/T$ plot) behavior.

3. Revealing contributions of select protein configurational dynamics to enzyme catalysis under ensemble conditions

3.1 Select and generic classes of protein and coupled solvent configurational dynamics

Select configurational dynamics is defined as a subset of fluctuations, and the configurations that they connect, out of the total, broad band of fluctuations of the protein and coupled solvent, that provide a preferential pathway through the configuration space along the reaction progress coordinate. The subset can be identified with distinguishing properties, such as the degree of collectiveness (gauged by the number of participating elements, or size of the network) in the configuration change, particular values of the fluctuation frequencies, and temperature dependence. As an example, coarse-grain distinction between types of configurational fluctuations is obtained from the description of motions in super-cooled fluids and glasses

(Angell, 1995; Stillinger, 1995). Local, single-solvent particle displacements, known as β-fluctuations, lead to incremental changes in system configuration, while collective, multiple-particle displacements, known as α-fluctuations, lead to more significant, non-local or extended changes in system configuration. A class of collective displacements of a small particle number has been identified, which undergoes a transition to incremental fluctuations of the cluster constituents (Johari-Goldstein β-fluctuation) (Goldstein, 2011; Johari & Goldstein, 1970; Liu, Fujita, Aji, Matsuura, & Chen, 2014). Characteristic temperature-dependences distinguish the different fluctuation types. The types of configurational fluctuations in isolated, single, hydrated proteins have some analogy to the classes in the glass-forming systems (Nakanishi & Sokolov, 2015; Panagopoulou, Kyritsis, Shinyashiki, & Pissis, 2012). In the materially more complex, isolated protein, the "particles" or fluctuating entities may be conceived of as internal or hydration-associated solvent molecules, amino acid side chains, backbone segments, and more extended domains (Lewandowski et al., 2015). Motions in proteins surrounded by a glass-forming solution are coupled to the solvent fluctuations (Fenimore, Frauenfelder, McMahon, & Young, 2004; Frauenfelder et al., 2009).

3.2 Distinguishing select protein and coupled solvent configurational fluctuations

Revealing the contributions of select protein configurational dynamics to an enzyme progress coordinate under ensemble conditions requires a selective perturbation of a subset of the fluctuations, and observation of the effect on the kinetics of the step, under the same conditions—a "dynamics-function" correlation. The collective classes of fluctuation have distinctive dynamical transitions, that are inducible by changes in temperature. The activation energy for α-fluctuations is temperature-dependent, increasing non-linearly with decreasing temperature (Angell, 1995). This has been attributed to an increase in size of the cooperative cluster (Adam & Gibbs, 1965; Dyre, Hechsher, & Niss, 2009), and leads to a precipitous drop in frequency, near the glass transition, at a temperature, T_g, which is defined as the temperature at which the mean α-fluctuation rate is $10^{-2}\ s^{-1}$ (a decrease of $>10^8$-fold, relative to room temperature) (Angell, 1995). The Johari-Goldstein β-fluctuations display a transition from a non-Arrhenius temperature dependence characteristic of the collective cluster to a linear dependence characteristic of incremental fluctuations of the cluster constituents, at a "fragile-to-strong dynamical crossover" temperature (Capaccioli, Ngai, & Shinyashiki, 2007; Chen et al., 2006; Mallamace et al., 2006). Below the glass and crossover transition temperatures, the collective motion dynamics are arrested, and, depending on the time scale of the observed protein function, they may be effectively eliminated. Thus, these collective fluctuations are potentially valuable for identification of dynamical selectivity in enzyme processes because the effect of their effective elimination on function can be measured.

3.3 Precedents in non-enzymic protein systems

Measurements of small molecule (CO, O_2) migration and heme-iron binding in the protein, myoglobin, have previously revealed roles for the collective and incremental fluctuation classes in non-enzymic processes (Agmon, 2004; Frauenfelder et al., 2009). The general approach was to measure the temperature-dependences of the frequencies of protein and coupled solvent fluctuations, and separately, the temperature dependence of the kinetics of CO rebinding to heme-iron following photolysis (the protein "function"), under the same conditions in global (whole sample) glass-forming solvents, and over an exceptional temperature range. Comparison of the forms of the logarithm of rate constant versus inverse temperature relations showed that entry and exit of CO were coupled to protein motions responsive to solvent α-fluctuations, while the protein-internal processes of migration and rebinding of CO to heme-iron were driven by a class of β-fluctuations coupled to fluctuations in the protein hydration layer (denoted β_h) (Frauenfelder et al., 2009).

The approach to identification of select coupling of protein and coupled solvent dynamics to the enzyme progress coordinate in EAL follows this general method of manipulation protein and coupled solvent dynamics, and correlation with function—in this case, the adiabatic, chemical reaction of the substrate radical. The method utilizes the effective elimination of the select protein and coupled fluctuations, on the time scale of the protein function (here, the lifetime of the $\mathbf{S}^{\bullet}$ states), and measurement of the impact on kinetics of a step or steps in the enzyme catalytic cycle. The method generally requires cryotrapping a canonical intermediate state in frozen, polycrystalline solution, followed by temperature-step initiation of the reaction of the intermediate, and measurement of the first-order kinetics over a range of low temperatures. For EAL, the range is ~200–230 K, for the aminoethanol substrate radical decay reactions, (Kohne et al., 2017, 2019; Zhu & Warncke, 2008) which extends to 250 K for the slower decay of the non-native substrate, 2-aminopropanol radical (Ucuncuoglu & Warncke, 2018). Integrity of protein structure and fidelity in reactivity is achieved over the low temperature range, by using small amounts of added cryosolvent in the frozen sample preparation.

4. Sample preparation and measurement of reaction kinetics

4.1 Background

The approach described here requires measurements over a broad temperature range, including values significantly below 273 K, while preserving native properties of the protein and its interaction with solvent. Three general types of methods have been developed to resolve intermediates and reaction steps in metalloproteins in solid state samples at cryogenic temperatures: (1) A sample containing a redox-poised metal center in the protein at equilibrium, is subsequently frozen, and then reduced by using γ-irradiation, at 77 K (cryoreduction method) (Blumenfeld, Davydov, Magonov, & Vilu, 1974; Davydov et al., 2002; Davydov & Hoffman, 2011). Subsequent graded-annealing, by using stepwise increases in temperature, thermally activates the reduced, out-of-equilibrium metal center and promotes relaxation through protein conformational changes, electron transfer, or chemical reaction. This technique has been primarily applied to heme and non-heme iron proteins. (2) The low-temperature, out-of-equilibrium state is created by photodissociation of protein-bound metal-ligand complexes in frozen solutions. The prototype is the optically-monitored migration and rebinding of CO or O_2 to the heme iron in myoglobin after photolysis of the carboxy- or oxy-heme state in frozen solutions at temperatures from <10 to 270 K (Agmon, 2004; Austin, Beeson, Eisenstein, Frauenfelder, & Gunsalus, 1975; Frauenfelder et al., 2009). This technique has been extended to other heme proteins and metalloproteins, including to trap the 5′-deoxyadenosyl radical in an enzyme (Yang et al., 2019) (3) A kinetically-unstable enzyme intermediate is created in solution, and then cryotrapped. The intermediate is subsequently promoted to relax by repeated temperature-step, elevated-temperature-decay, low-temperature-measurement cycles (Lukoyanov, Barney, Dean, Seefeldt, & Hoffman, 2007) or temperature-step followed by continuous measurement of the reaction (Zhu & Warncke, 2008). Here, we describe the latter method.

The cob(II)alamin-substrate radical pair state, formed by using the native EAL substrate, aminoethanol, or the non-native substrate, 2-aminopropanol, accumulates as the only detectable paramagnetic intermediate during steady-state turnover of EAL at room temperature (Babior et al., 1974; Bandarian & Reed, 2002; Bender et al., 2008; Warncke et al., 1999). The kinetically unstable substrate radical intermediate state formed from the aminoethanol substrate can be cryotrapped Warncke et al. (1999), and is stable under storage at 77 K. The forward reaction of the substrate radical ensemble is synchronously initiated at low temperature by a temperature step, and the evolution of the reaction is monitored by time-resolved, full-spectrum EPR spectroscopy. The methods have allowed the study of the radical rearrangement and second hydrogen atom transfer (Fig. 1, steps 3 and 4) for natural isotopic abundance and 1,1,2,2-$^2H_4$-labeled aminoethanol substrate over the range of 190–235 K (Kohne et al., 2017, 2019; Zhu & Warncke, 2008), with extension to 250 K for 2-aminopropanol (Ucuncuoglu & Warncke, 2018).

The frozen solution system consists of EAL protein, surrounded by successive protein-associated domain (PAD; akin to the hydration layer) and mesodomain solvent layers, that are bounded by polycrystalline ice, as depicted in Fig. 3 (Ionescu et al., 2021; Nforneh & Warncke, 2017, 2019). The formation of the system follows the phenomenon, that particles over an exceptionally broad size range, from several Ångstroms (molecules) to micrometers (polymer spheres) are excluded from the advancing ice fronts during the freezing of aqueous solutions, which leads to their collection in the freeze-concentrated liquid, or mesodomain, within the polycrystalline

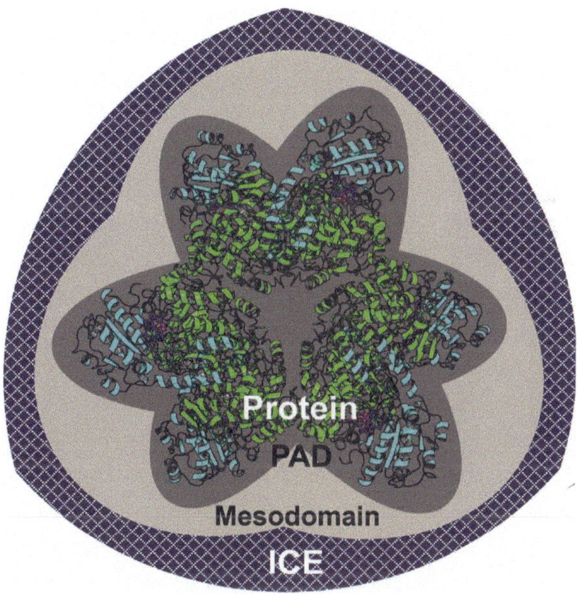

Fig. 3 Depiction of the protein and solvent environment in frozen aqueous samples containing small amounts of added cryosolvent (aminoethanol, dimethylsulfoxide) (Ionescu, Li, Nforneh, & Warncke, 2021). The protein is surrounded by successive layers of protein associated domain (PAD; hydration layer), aqueous-cosolvent mesodomain at the maximum freeze concentration (Mesodomain) and polycrystalline water ice (Ice).

frozen aqueous solution (Curtis et al., 2012; Hubei, Bischof, & Aksan, 2009; Körber, 1988). In the presence of small amounts of cryosolvent added to the samples prior to cryotrapping (for example, 0.5–4% dimethylsulfoxide, DMSO) (Nforneh & Warncke, 2017, 2019), the frozen mesodomain is composed of the aqueous-cryosolvent at the maximum freeze concentration (Franks, 2003), and maintains nominal fluidity through 200 K (Nforneh & Warncke, 2019). Cryosolvents are preferentially excluded from the PAD (Arakawa, Kita, & Timasheff, 2007; Timasheff, 2002). Aminoalkanols are cryoprotectants (Douzou, 1977), and the EAL substrate, aminoethanol, forms a mesodomain at the added concentrations of 100 mM (0.6% v/v) used to prepare the cob(II)alamin-substrate radical pair for EPR samples. Freezing of aqueous samples in the presence of cryosolvents generally maintains biological structure and function on the molecular to organellar levels (Douzou, 1977; Franks, 1998) It has long been recognized that cosolvents suppress aggregation in samples prepared for EPR spectroscopy (Leigh Jr. & Reed, 1971).

4.2 Trapping protocols

The cob(II)alamin-substrate radical pair samples are prepared by first accumulating the intermediate during steady-state turnover at 277 K, and then rapidly lowering the temperature to trap the radical pair (Wang et al., 2015; Zhu & Warncke, 2008). The standard procedure for manual cryotrapping of steady-state intermediates in EAL (Warncke et al., 1999; Zhu & Warncke, 2008) starts with manual mixing, on ice, of the prepared EAL-AdoCbl holoenzyme solution (~0.27 mL; in a 12 × 75 mm disposable culture tube, P/N 14–961-26: Fisher Scientific) with substrate (~0.03 mL of 0.1–1.0 M stock; added with an automatic pipette, under vortex mixing; elapsed time, ~5 s). The sample is then removed from the test tube with a long-stem glass transfer pipette (9-in. Pasteur Pipet, P/N 13-678-6B; Fisher Scientific), and discharged into the middle region of a 4 mm outer diameter EPR tube. The open, top end of the EPR tube is held firmly between thumb and index finger, and two high-to-low height, sweeping "whip" motions of the tube move the solution contents to the bottom, and the tube is then plunged into liquid nitrogen-chilled isopentane (T = 133 K) to trap the cob(*II*)alamin-substrate radical pair state. The total elapsed time from substrate mixing to isopentane immersion by this procedure is ≤15 s. The procedure can be performed in a cold room, or at room temperature, by keeping all components on ice. Samples are stored in liquid nitrogen, prior to use.

The cooling rate during trapping is relevant to conservation of the accumulated paramagnetic state, because, at some temperature during the trapping, diffusive exchange of substrate/product with the active site ceases. If the cooling rate is much greater than the rate of substrate radical reaction (step 3, Fig. 1), then the total accumulated population is trapped. If the cooling rate and reaction rate are comparable, a proportion of the steady-state level of cob(II)alamin-substrate radical pair will decay. In the case of EAL, the decay is to the EPR-silent, diamagnetic product state. The mean cooling rate of the EPR sample for the above standard cryotrapping procedure is 10 K/s. The loss of radical intermediate during cooling is partially mitigated by the slowing of the decay reaction, itself, with decreasing temperature. Cooling rate is also an adjustable parameter, that can be used to dissect mechanism. Methods for the variation of the cooling rate from 0.07 to 400 K/s have been described (Wang et al., 2015). For the EAL substrate radical samples, no significant effects of cooling rate variation over this range on the temperature-dependence of the decay reaction kinetics are observed.

4.3 Time-resolved, full-spectrum EPR measurements

The X-band EPR measurements were performed on a Bruker ELEXSYS E500 EPR spectrometer, by using a Bruker ER4123SHQE X-band cavity resonator, and Bruker ER4131VT system for temperature control. A nitrogen flow cooling system was used, with a 15 cm heating coil installed in a transfer Dewar (length ~1 m, P/N 1100104, Bruker), approximately 60 cm from the EPR cavity Dewar insert. The system used vent gas from a 250 L liquid nitrogen Dewar to drive N_2 gas flow (N_2 flow rate regulated by using a ball flow gauge; King Instrument Co.) through a coil immersed in a liquid nitrogen bath in a ~4 L Dewar. This system allows effectively unlimited run time, under the requirement of periodic replenishment of the liquid nitrogen in the bath. For this system, the uncertainty in the temperature value was approximately ±0.5 K (Wang et al., 2015).

EPR samples were held at a staging temperature of 160 or 180 K in the microwave cavity, under control of a ER4131VT cryostat system, and the microwave bridge was tuned. *T*-steps from 160 or 180 K to the decay measurement temperature was initiated by changing the ER4131VT temperature set-point. Once the sample temperature stabilized at the set point, the pre-set auto-tune/auto-scan mode of the spectrometer was triggered, and the sample was auto-tuned at the high-temperature set point, followed immediately by repeated, continuous spectrum acquisition. The time from initiation of the temperature step to the start of acquisition of the first spectrum was 30–60 s. The first EPR spectrum was collected upon stabilization at the set point. The acquisition time of an individual spectrum was typically 24 s for the sweep across the radical component of the EPR spectrum (record length, 1024 points; dwell time equal to sweep time divided by 1000), with a time constant equal to 10–20% of the dwell time, but could be as short as 6 s (with corresponding shorter dwell time and time constant) for more rapid decays at higher temperatures. Fig. 4 shows substrate radical decay plots of amplitude versus time at different temperatures, and a representative

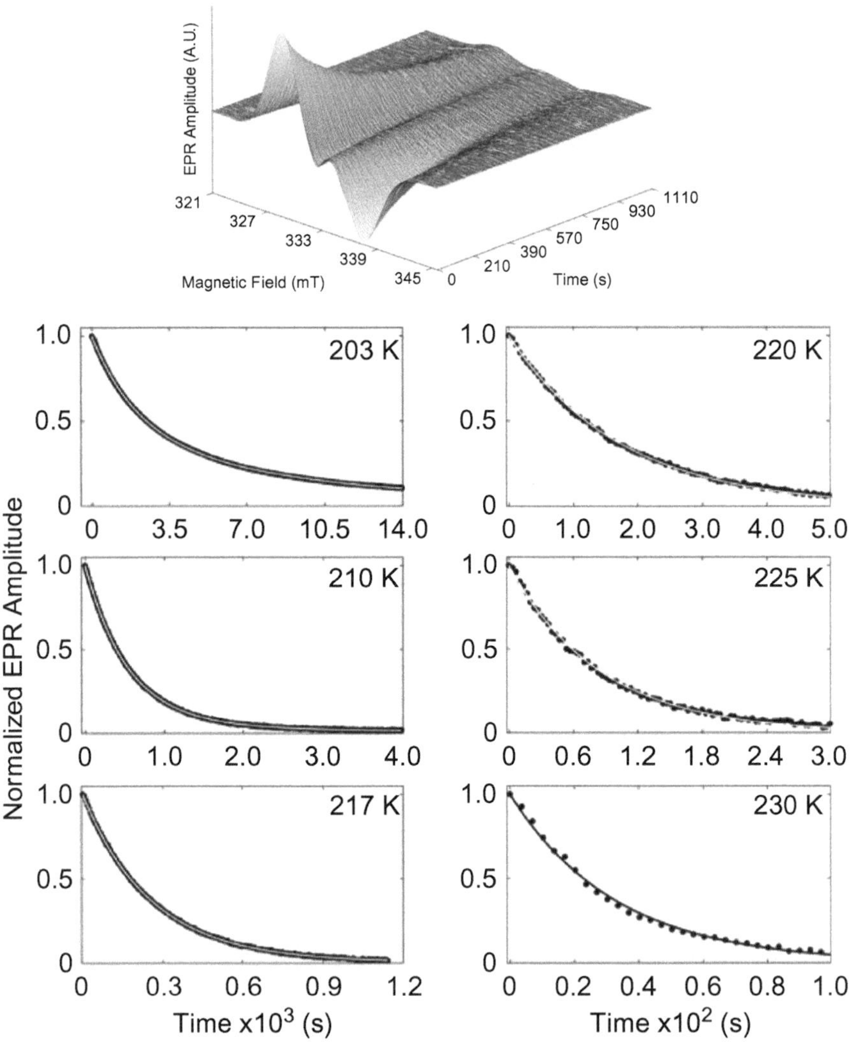

Fig. 4 Time-dependence of the ^{2}H-aminoethanol substrate radical EPR spectrum at 210 K (top), and representative dependences of EPR amplitude as function of time at different temperature values (Kohne et al., 2019). Biexponential ($T \leq 217$ K) and mono-exponential ($T \geq 220$ K) fitting curves are shown in gray overlay.

full-substrate radical EPR spectrum time series. The decay kinetics were fitted by using in-house programs run on the MATLAB (MathWorks, Natick, MA) platform.

5. Temperature-dependence of substrate radical decay reaction in ethanolamine ammonia-lyase

The temperature dependence of the observed first-order rate constants for decay of the natural isotopic abundance, ^{1}H-substrate radical displays a piecewise-continuous pattern over the temperature range of 203 to 295 K on the Arrhenius plot shown in Fig. 5 (Kohne et al., 2017, 2019).

The temperature-dependence is characterized by the following regions, with decreasing temperature: (Kohne et al., 2019) (1) Monoexponential decay (rate constant, k_{obs}) extends from 295 to 220 K. The linear relation indicates that E_a and A are temperature-independent, and thus, that the native energy surface prevails over the entire range. (2) An abrupt kinetic bifurcation leads to biexponential decay starting at 219 K, with slow and fast observed first-order rate constants, $k_{obs,s}$ and $k_{obs,f}$. The kinetic behavior in the plateau region over 219–217 K represents a transformation of the reaction coordinate by

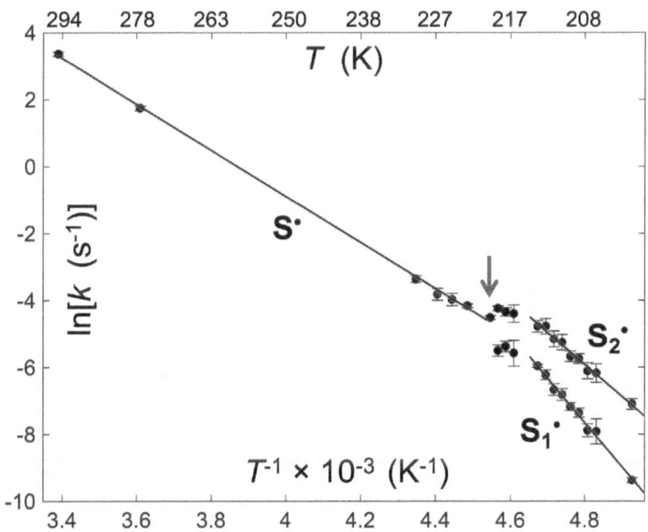

Fig. 5 Arrhenius plot of observed first-order rate constants for substrate radical decay reactions, from states $S^{\bullet}$, $S_1^{\bullet}$ and $S_2^{\bullet}$ (Kohne et al., 2017). Arrow denotes position of the kinetic bifurcation. Mean rate constants and standard deviations, from at least three separate decay measurements, are shown.

compensating changes in activation enthalpy and entropy, and will be described, elsewhere. (3) Following a concave-down kink, the biexponential decay continues along two distinct Arrhenius relations, from 214 to 203 K. The monoexponential kinetics over 295–220 K correspond to decay from a single macrostate, $\mathbf{S}^\bullet$. The biexponential kinetics below the bifurcation evidence the presence of two substrate radical substates, $\mathbf{S}_1^\bullet$ and $\mathbf{S}_2^\bullet$. Evidence for two separate decay channels from $\mathbf{S}_1^\bullet$ and $\mathbf{S}_2^\bullet$, rather than the earlier-proposed single-channel through $\mathbf{S}_2^\bullet$ (Kohne et al., 2017), came from partial $^1H/^2H$ isotope effects on the hydrogen transfer step, HT2, which uniformly shift the entire piecewise-continuous Arrhenius relation down by two-fold in rate constant (Kohne et al., 2019). The partitioning of the $\mathbf{S}^\bullet$ macrostate arises from abrupt emergence of a (free) energy barrier to interconversion of the $\mathbf{S}_1^\bullet$ and $\mathbf{S}_2^\bullet$ substates, over 220–219 K, as depicted in the state and reaction step profile in Fig. 6. These (Kohne et al., 2017) and subsequent

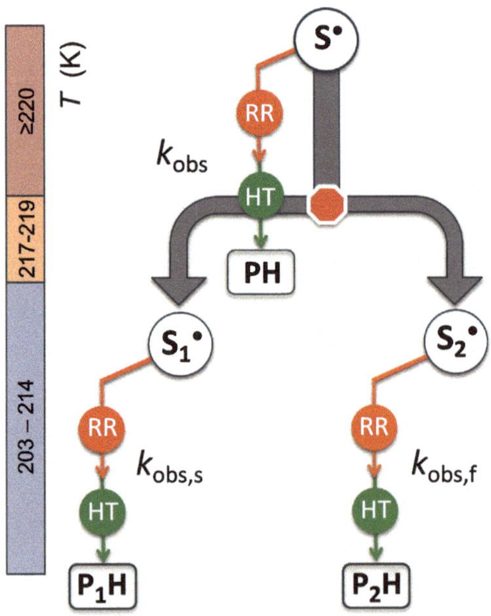

Fig. 6 Hierarchy of resolved states and pathways for natural abundance (^{1}H-) aminoethanol substrate radical decay and their emergence from manipulation of temperature in EAL (Kohne et al., 2019). Regions of temperature dependence of the kinetics are indicated. Red octagon indicates effective block of $\mathbf{S}_1^\bullet$, $\mathbf{S}_2^\bullet$ substate interconversion within the substrate radical macrostate, $\mathbf{S}^\bullet$, at the kinetic bifurcation. For radical rearrangement (RR) and hydrogen transfer (HT) steps, orange indicates contribution to rate limitation of decay, and green indicates no rate limitation. HT contributes to rate limitation under conditions of ^{2}H-labeling of substrate and C5′-methyl group of 5′-deoxyadenosine (Kohne et al., 2019).

(Kohne et al., 2019; Ucuncuoglu & Warncke, 2018) results led to proposed roles for the substates: $\mathbf{S}_1^{\bullet}$ represents a "capture" protein configuration that stabilizes the nascent substrate radical following radical pair separation (Fig. 1). $\mathbf{S}_2^{\bullet}$ represents a protein configuration that is "enabled" for the adiabatic chemical step of radical rearrangement. This "capture and enable" strategy is a paradigm for creation and control of high-reactivity (radical) intermediates in enzymes. In the context of the "ensemble perspective," (Hammes et al., 2011) the configurational microstates, that compose the progress coordinate through the $\mathbf{S}_1^{\bullet}$ and $\mathbf{S}_2^{\bullet}$ substates of the $\mathbf{S}^{\bullet}$ macrostate, are thermalized and conformationally-sampled, during reaction of the substrate radical at $T \geq 220\,\text{K}$.

6. Interpretation: Contributions of select protein configurational dynamics to EAL enzyme catalysis

6.1 Interpretation of the bifurcation: Select collective fluctuations conduct the $S_1^{\bullet}$ to $S_2^{\bullet}$, protein configurational state change

The barrier to conversion of $\mathbf{S}_1^{\bullet}$ to $\mathbf{S}_2^{\bullet}$, at $T \leq 217\,\text{K}$ on the time scale of the decay reaction, in the direction of the progress coordinate, arises from temperature-dependent contributions of the protein and coupled solvent dynamics. The structure of the cob(II)alamin-substrate radical pair in the $\mathbf{S}_1^{\bullet}$ and $\mathbf{S}_2^{\bullet}$ states is detectably homogeneous, in terms of the radical pair separation, R_{ee}, and mutual geometry, as indicated by the EPR lineshape (Kohne et al., 2017), and by ^{2}H ESEEM of the substrate radical and its active site interaction with the C5′-methyl of the 5′-deoxyadenosyl group (U. Twahir and K. Warncke, unpublished). The intrinsic chemistry associated with the rearrangement reaction is also not expected to abruptly change under the conditions. The presence of the "capture" and "enable" states, $\mathbf{S}_1^{\bullet}$ and $\mathbf{S}_2^{\bullet}$, is revealed through the difference in their decay reactivity. The $\mathbf{S}_1^{\bullet}$ and $\mathbf{S}_2^{\bullet}$ states continue to react, below the bifurcation, although they follow different Arrhenius relations at $T \leq 217\,\text{K}$, relative to $T \geq 220\,\text{K}$. Thus, configurational fluctuations are present below the bifurcation, that conduct the reactions, indicating that there has not been a catastrophic cessation of all functional dynamics, at $T < 220\,\text{K}$. This behavior is consistent with the understanding of the temperature-dependence of the different classes of fluctuations in condensed phase systems,(Angell, 1995) and crucial to the interpretation. In particular, although the non-linear Arrhenius decrease in the frequency of collective fluctuations (α-fluctuation type) and the cluster character (Johari-Goldstein β-fluctuation type) may eliminate their contribution on the measurement time scale, the local, incremental, β-fluctuations, which display linear Arrhenius dependence, persist, with relatively high frequencies, to low temperatures ($T < 200K$), to provide a possible pathway. Therefore, the barrier to, and observed quenching of $\mathbf{S}_1^{\bullet}$, $\mathbf{S}_2^{\bullet}$ interconversion, even in the presence of the subclass of the incremental, β-fluctuations, indicates that a class of *collective* protein and coupled solvent fluctuation makes a select dynamical contribution to the $\mathbf{S}_1^{\bullet}$ to $\mathbf{S}_2^{\bullet}$ transition on the progress coordinate in EAL.

6.2 Interpretation of the change in Arrhenius slope and intercept at the bifurcation: Select collective fluctuations contribute to the radical rearrangement reactions of $S_1^{\bullet}$ and $S_2^{\bullet}$

The Arrhenius relations in Fig. 5 refer to the chemical, rearrangement reaction of the substrate radical from $\mathbf{S}_1^{\bullet}$ and $\mathbf{S}_2^{\bullet}$. Under the assumption that the structure of the system does not change significantly, as evidenced by results presented in Section 6.1, the energies of the protein and coupled solvent configurations do not change. Therefore, the energy surface for the reactions is maintained from $T \geq 220$ to $T \leq 217$ K, through the bifurcation. Thus, the change in the Arrhenius relation at the bifurcation indicates that collective fluctuations, that contribute to the chemical reaction steps of $\mathbf{S}_1^{\bullet}$ and $\mathbf{S}_2^{\bullet}$ above the bifurcation ($T \geq 220$ K), are eliminated (on the time scale of the reactions) at the bifurcation. Below the bifurcation, for $T \leq 217$ K, incremental fluctuations conduct the reactions.

The change in activation parameters for the substrate radical reaction, across the bifurcation, supports this model. The slope and intercept of the linear segments of the Arrhenius relations increase with decreasing temperature, as shown in Fig. 5. This was proposed to be caused by an increase in the number of states along the reaction path, in the transition state region, relative to the minima of $\mathbf{S}^{\bullet}$ (Kohne et al., 2017). These changes are consistent with a change in the class of fluctuations and connected configurations, from collective to incremental, that changes the topography of the pathway through configuration space in the transition state region. The topology of the progress coordinate on the energy surface above the bifurcation temperature is relatively smooth, which reflects the collective fluctuations of the protein and coupled solvent among a relatively small number of configurations stabilized by multiple concerted displacements of protein and coupled solvent elements. Below the bifurcation temperature, the progress coordinate through the transition state region is populated by additional configurations, that represent the distribution of incremental microstates.

The incremental microstates are connected by the lower-order, incremental fluctuations, that sequentially achieve the result of the collective configuration changes, in physical space. In the configuration space representation, the incremental changes (incremental fluctuations) represent the components of the collective change (projection of the collective fluctuation onto the subspace of incremental fluctuations). A simplified representation of a change in conducting fluctuations from collective to incremental is presented in Fig. 7.

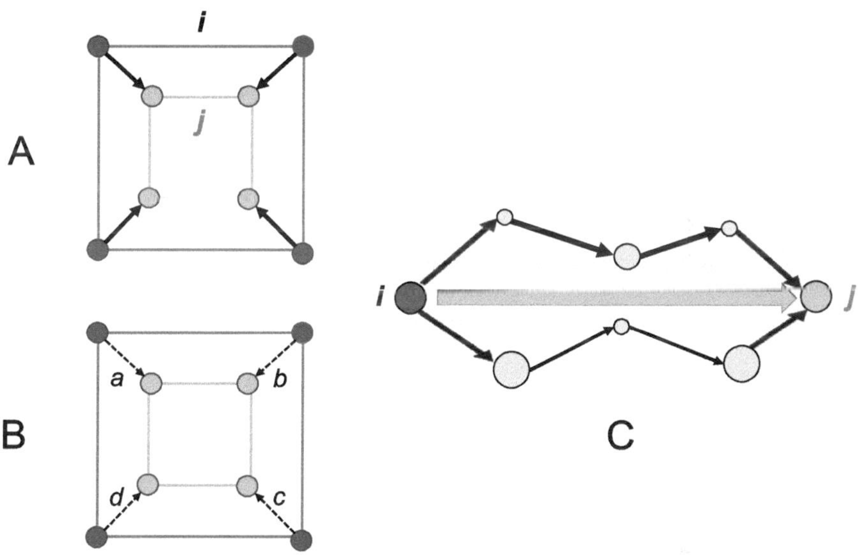

Fig. 7 Real space and corresponding configuration space depiction of a directional shift in configuration achieved by collective and constituent incremental fluctuations. The configuration change represents one segment of a multi-segment traverse through the transition state region. In the model, the collective fluctuation provides a one-step "portal" between the native configurations that conduct the collective movements of structure in real space, and that create the select pathway through configuration space. Fluctuations are reversible, but are depicted here as unidirectional, for clarity. (A) Real space: Collective fluctuation (*dark arrows*) changes positions of groups from configuration state, *i* (position of groups, dark circles) to configuration state, *j* (position of groups, *gray circles*) (B) Real space: Incremental fluctuations *a, b, c,* and *d* (dashed arrows) change the positions of groups from state *i* to *j*, one-by-one. (C) Configuration space: Collective fluctuation (*bold gray arrow*) provides a one-step "portal" through configuration space. Incremental fluctuations (*black arrows*) and intermediate configurations (*light gray* points) represents projections of the collective change onto the incremental change subspace. Two of the twenty-four possible incremental paths (with respect to initial step and subsequent order of execution) are depicted.

The energy along the pathway followed through the transition state region is raised, because the incremental configurations manifest only partially the stabilization achieved by the collective changes of structure and associated electrostatic charges. Thus, the E_a is raised, as observed. The larger number of configurational microstates below the bifurcation leads to a higher configurational entropy for the transition state region, and therefore, a larger entropy of activation, $\Delta S^{\ddagger}$. In accord with transition state theory, a finite $\Delta S^{\ddagger}$ term resides formally in the prefactor, A, of the Arrhenius rate expression (Cornish-Bowden, 1981). Thus, remodeling of the pathway through the transition state region by the bifurcation is consistent with the observed increase in A, for both the $k_{obs,s}$ ($\mathbf{S}_1^{\bullet}$ decay) and $k_{obs,f}$ ($\mathbf{S}_2^{\bullet}$ decay) relations, relative to the $\mathbf{S}^{\bullet}$ relation. Overall, the results indicate that select collective fluctuations contribute to the native chemical reaction step from the $\mathbf{S}_2^{\bullet}$ state, at $T \geq 220$ K. Incremental fluctuations conduct the chemical reaction steps at $T \leq 217$ K.

7. Role of solvent coupling in select protein configurational dynamics

A change in the motional properties of hydrated proteins has been observed over a temperature range of approximately 220 ± 20 K (Doster, 2008; Lewandowski et al., 2015; Ringe & Petsko, 2003). A concave-down (E_a inversely proportional to T) to linear (E_a independent of T) Arrhenius "crossover" behavior in protein hydration water is associated with this event (Chen et al., 2006). The EAL protein in the frozen aqueous system is surrounded by the protein-associated domain (PAD; akin to the protein hydration layer) and aqueous-cosolvent mesodomain, as shown in Fig. 3. Our EPR measurements reveal a change in sample dielectric permittivity,ε'', of the solvent around EAL at the bifurcation temperature, that is detected as a change in magnetic susceptibility (Dalal, Eaton, & Eaton, 1981; McRae et al., 2020; Poole, 1983) of the protein-interior paramagnets, cob(II)alamin and substrate radical, bound at the active site (Ionescu et al., 2021). A microscopic origin of the change in ε'' is suggested by comparison with the frequency and temperature dependences of the wide frequency range of dielectric relaxations in protein powder samples at water contents ($h \geq 0.3$ g H_2O/g protein), that lead to surface hydration equivalent to approximately one to two water molecule thickness. At a frequency of 9.5 GHz, and temperature range of 200–250 K, a class of dielectric relaxation in these systems is assigned to coupling of water with protein surface groups

(Nakanishi & Sokolov, 2015; Panagopoulou et al., 2012; Shinyashiki et al., 2009), which occurs on length scales of ≤ 3 Å (Hong, Smolin, Lindner, Sokolov, & Smith, 2011; Nickels, Sakai, & Sokolov, 2013). The high-frequency (relative to the α- and β-like fluctuations) and temperature-dependence of this class are characteristic of a fluctuation, observed in a variety of hydrated materials, that is assigned to water-polar surface group fluctuations (Cerveny, Alegria, & Colmenero, 2008; Ngai, Capaccioli, Ancherbak, & Shinyashiki, 2011), with characteristics of the Johari-Goldstein-β class of local cluster fluctuations (Capaccioli et al., 2007). Overall, these results suggest that the fluctuations responsible for the change in ε'' and temperature-dependence of the EPR amplitude (magnetic susceptibility) at the microwave frequency of 9.5 GHz originate from coupled solvent-protein, cluster motions around EAL. The crossover from collective to incremental fluctuations of this class is a microscopic mechanistic basis of the kinetic bifurcation in EAL.

The effect of the coupled protein and hydration water motions is communicated through the protein interior to the active site. When the fluctuations impact the enzyme active site, the collective fluctuations provide the lowest-barrier routes to configurations that selectively accomplish the interconversion of $\mathbf{S}_1^{\bullet}$ and $\mathbf{S}_2^{\bullet}$, and contribute selectively to achieving the lowest-barrier chemical reactions of $\mathbf{S}_1^{\bullet}$ and $\mathbf{S}_2^{\bullet}$, on the time scale of the decay reactions. The site-to-site movement of CO through myoglobin in response to the hydration water-coupled "β_h" fluctuations (Frauenfelder et al., 2009) represents detection of the percolation of these configuration changes, throughout a protein. A large number of experimental and computational studies that show contributions of protein surface and internal motions to events in enzymes have recently been reviewed and annotated (Agarwal, 2019).

8. Conclusions

8.1 Perspectives on contributions of protein dynamics to coenzyme B_{12}-dependent mechanistic enzymology and enzyme catalysis in general

The "ensemble view" of enzyme catalysis incorporates protein dynamics as the unresolved fluctuations among configurational states at different hierarchical levels, generic in the sense of degree of collectiveness, to achieve the thermalized, equilibrium distribution of protein configurations among canonical intermediate states, and bridging substates and microstates, that guide the system through the catalytic cycle (Hammes et al., 2011).

However, selectivity in dynamical contributions to enzyme function is detected in the statistical ensembles by correlations with the temperature-dependent properties of collective fluctuations. Ensemble kinetic studies of probe molecule (CO) interactions with myoglobin, distinguish contributions of different classes of fluctuations of solvent to "functions" of CO entry and exit, and protein-internal migration, according to their degree of collectiveness. In addition to the small-molecule binding and cofactor manipulation functions of myoglobin, the representative enzyme, EAL, conducts a complex choreography of central adiabatic reaction steps (Fig. 1) that convert substrate to product, through a radical mechanism, with mechanistic themes in common with coenzyme B_{12}-dependent (Banerjee, 1999, 2003; Brown, 2005; Frey, 2010; Marsh & Melendez, 2012; Toraya, 2003) and other (Frey, Hegeman, & Ruzicka, 2008; Sayler et al., 2019; Vey & Drennan, 2011) radical enzymes. The experimental, analytical and interpretive approaches described in this chapter advance our understanding of coenzyme B_{12}-dependent enzyme catalysis (and enzyme catalysis more generally) by identifying particular, or "select," contributions of dynamics to catalysis. These contributions favor specific higher-order collective movements, relative to lower-order or incremental fluctuations, to events along the reaction progress coordinate.

8.2 Contributions of protein and coupled solvent dynamics to the molecular mechanism of ethanolamine ammonia-lyase

The application of the methodology to the coenzyme B_{12}-dependent EAL enzyme leads to the following principal conclusions:

(1) Two sequential substrate radical intermediate substates, $\mathbf{S}^{\bullet}_1$ and $\mathbf{S}^{\bullet}_2$, are revealed within the canonical intermediate macrostate, $\mathbf{S}^{\bullet}$, by a bifurcation transition in the first-order decay kinetics, over 220–219 K. $\mathbf{S}^{\bullet}_1$ represents the protein and coupled solvent configurational disposition for "capture" or stabilization of the nascent radical formed by the radical pair separation process. $\mathbf{S}^{\bullet}_2$ represents the configuration enabled for the chemical step of radical rearrangement. The $\mathbf{S}^{\bullet}_1$ and $\mathbf{S}^{\bullet}_2$ substates abide by the concepts of resemblance of protein configurations to pre- and succeeding macrostates, and the preconditioning of principal steps in the catalytic cycle by protein "conformational sampling." In general, the low-temperature approach has promise to reveal intermediate substates and select configurational interconversion dynamics, that are undetected at room temperature.

(2) The native path-forward protein configurational change from $\mathbf{S}_1^{\bullet}$ to $\mathbf{S}_2^{\bullet}$ is conducted by select collective protein fluctuations, that are coupled, through the protein, to fluctuations of protein surface groups and hydration water (protein-associated solvent domain). This select collective coupling is obligatory, on the time scale of the radical decay reactions. The $\mathbf{S}_1^{\bullet}$ and $\mathbf{S}_2^{\bullet}$ substates are not distinguished at elevated, natural temperatures, because the rates of the select collective fluctuations are much greater than the rates of substate reaction.

(3) The select collective fluctuations make a partial contribution to barrier lowering on the native substrate radical reaction progress coordinate, which is represented by the decay reaction of $\mathbf{S}_2^{\bullet}$. Decay from the capture substate, $\mathbf{S}_1^{\bullet}$, is an informative, non-native process. The continuity, and increases in both slope ($\propto E_a$) and ordinate-intercept ($\propto \Delta S^{\ddagger}$), of the Arrhenius relation, through the bifurcation with descending temperature, suggest the following: (a) Above the bifurcation temperature, the select collective fluctuations provide a portal through the transition state region, characterized by relatively few configuration states, that are connected by the select collective cluster fluctuations. (b) Below the bifurcation temperature, incremental fluctuations that additively compose the select collective fluctuations conduct the system through the transition state region, which leads to a transition state region topography that is punctuated by multiple, higher-energy incremental configurational microstates.

8.3 Future directions

Each canonical state in the enzyme catalytic cycle in Fig. 1 corresponds to the same real-space constellation of active site residues, surrounding protein, and interacting solvent. In the coenzyme B_{12}-dependent and other enzymes, the protein must therefore orchestrate multiple molecular events that comprise the core chemical reactions by using the same basic architecture. Our hypothesis is that the requisite diversity of function can be achieved by selectivity in the contributions of protein dynamics and coupled solvent dynamics. The identification of the broad class of select collective fluctuation and configurational contributions to events in EAL catalysis, presented here, is relatively low in resolution. The future direction is to enhance the resolution of select fluctuation types, in terms of their cluster size and frequency band. This can be accomplished by applying the experimental methods described here; by correlating function with the dynamics of different probe molecules in the surrounding solvent and labeled at protein surface sites (for example, EPR spin probes and spin labels); by variation of solvent to examine the chemical versus fundamental physical aspects of mechanisms of dynamical coupling; and by high-resolution structural studies to determine the physical junctures of the coupling among protein, solvent and reactants. Understanding the roles of select dynamics in enzyme catalytic processes will facilitate the design and implementation of protein devices to benefit human health and drive molecular technological advances.

Acknowledgments

Research reported in this publication was supported by the National Institutes of Diabetes and Digestive and Kidney Diseases (R01 DK054514) and National Institute of General Medical Sciences (R01 GM142113) of the National Institutes of Health. The purchase of the Bruker E500 EPR spectrometer was funded by the National Center for Research Resources of the National Institutes of Health (RR 17767), and by Emory University.

References

Adam, G., & Gibbs, J. H. (1965). On the temperature dependence of cooperative relaxation properties in glass-forming liquids. *The Journal of Chemical Physics*, *43*(1), 139.

Agarwal, P. K. (2019). A biophysical perspective on enzyme catalysis. *Biochemistry*, *58*, 438–449.

Agmon, N. (2004). Coupling of protein relaxation to ligand binding and migration in myoglobin. *Biophysical Journal*, *87*, 1537–1543.

Anderson, C. J., & Kendall, M. M. (2017). Salmonella enterica Serovar typhimurium strategies for host adaptation. *Frontiers in Microbiology*, *8*, 1983.

Angell, C. A. (1995). Formation of glasses from liquids and biopolymers. *Science*, *267*, 1924–1935.

Arakawa, T., Kita, Y., & Timasheff, S. N. (2007). Protein precipitation and denaturation by dimethyl sulfoxide. *Biophysical Chemistry*, *131*, 62–70.

Austin, R. H., Beeson, K. W., Eisenstein, L., Frauenfelder, H., & Gunsalus, I. C. (1975). Dynamics of ligand binding to myoglobin. *Biochemistry*, *14*, 5355–5373.

Babior, B. M., Moss, T. H., Orme-Johnson, W. H., & Beinert, H. (1974). Mechanism of action of ethanolamine Ammonia-Lyase, a B12-dependent enzyme - participation of paramagnetic species in catalytic deamination of 2-Aminopropanol. *The Journal of Biological Chemistry*, *249*(14), 4537–4544.

Bandarian, V., & Reed, G. H. (1999). Ethanolamine Ammonia-Lyase. In R. Banerjee (Ed.), *Chemistry and biochemistry of B12* (pp. 811–833). New York: John Wiley and Sons.

Bandarian, V., & Reed, G. H. (2002). Analysis of the Electron paramagnetic resonance Spectrum of a radical intermediate in the Coenyzme B12-dependent ethanolamine Ammonia-Lyase catalyzed reaction of S-2-Aminopropanol. *Biochemistry*, *41*, 8580–8588.

Banerjee, R. (1999). *Chemistry and biochemistry of B12*. New York: Wiley.

Banerjee, R. (2003). Radical carbon skeleton rearrangements: Catalysis by coenzyme B12-dependent mutases. *Chemical Reviews*, *103*, 2083–2094.

Banerjee, R. (2017). Introduction to the thematic Minireview series: Host-microbiome metabolic interplay. *The Journal of Biological Chemistry*, *293*, 8544–8545.

Bender, G., Poyner, R. R., & Reed, G. H. (2008). Identification of the substrate radical intermediate derived from ethanolamine during catalysis by ethanolamine Ammonia-Lyase. *Biochemistry*, *47*, 11360–11366.

Berman, H. M., Westbrook, J., Feng, Z., Gilliland, G., Bhat, T. N., Weissig, H., et al. (2000). The protein data bank. *Nucleic Acids Research*, *28*, 235–242.

Blumenfeld, L. A., Davydov, R. M., Magonov, S. N., & Vilu, R. O. (1974). Studies on the conformational changes of metalloproteins induced by electrons in water-ethylene glycol solutions at low temperatures. Haemoglobin. *FEBS Letters*, *49*, 246–249.

Boehr, D. D., Dyson, H. J., & Wright, P. E. (2006). An NMR perspective on enzyme dynamics. *Chemical Reviews*, *106*, 3055–3079.

Boehr, D. D., McElheny, D., Dyson, H. J., & Wright, P. E. (2006). The dynamic energy landscape of dihydrofolate reductase catalysis. *Science*, *313*, 1638–1642.

Bovell, A. M., & Warncke, K. (2013). The structural model of Salmonella typhimurium ethanolamine Ammonia-Lyase directs a rational approach to the assembly of the functional $[(EutB\text{–}EutC)_2]_3$ oligomer from isolated subunits. *Biochemistry*, *52*, 1419–1428.

Brown, K. L. (2005). Chemistry and enzymology of vitamin B12. *Chemical Reviews*, *105*, 2075–2149.

Bruice, T. C. (2002). A view at the millenium: The efficiency of enzyme catalysis. *Accounts of Chemical Research*, *35*, 139–148.

Canfield, J. M., & Warncke, K. (2002). Geometry of reactant centers in the Co^{II}-substrate radical pair state of coenzyme B12-dependent ethanolamine deaminase determined by using orientation-selection-ESEEM spectroscopy. *The Journal of Physical Chemistry. B*, *106*, 8831–8841.

Capaccioli, S., Ngai, K. L., & Shinyashiki, N. (2007). The Johari-Goldstein β-relaxation of water. *The Journal of Physical Chemistry. B*, *111*, 8197–8209.

Cerveny, S., Alegria, A., & Colmenero, J. (2008). Universal features of water dynamics in solutions of hydrophilic polymers, biopolymers, and small glass forming materials. *Physical Review E*, *77*. 031803/031801-031803/031805.

Chen, S.-H., Liu, L., Fratini, E., Baglioni, P., Faraone, A., & Mamontov, E. (2006). Observation of fragile-to-strong dynamic crossover in protein hydration water. *Proceedings of the National Academy of Sciences*, *103*(24), 9012–9016.

Cornish-Bowden, A. (1981). *Fundamentals of enzyme kinetics*. Boston: Butterworths and Company, Ltd.

Curtis, J. E., Nanda, H., Khodadadi, S., Cicerone, M., Lee, H. J., McAuley, A., et al. (2012). Small-angle neutron scattering study of protein crowding in liquid and solid phases: Lysozyme in aqueous solution, frozen solution, and carbohydrate powders. *The Journal of Physical Chemistry. B*, *116*, 9653–9667.

Dalal, D. P., Eaton, S. S., & Eaton, G. R. (1981). The effects of lossy solvents on quantitative EPR studies. *Journal of Magnetic Resonance*, *44*, 415–428.

Davydov, R., & Hoffman, B. M. (2011). Active intermediates in heme mono reactions as revealed by cryoreduction/annealing, EPR/ENDOR studies. *Archives of Biochemistry and Biophysics*, *507*, 36–43.

Davydov, R., Kofman, V., Fujii, H., Yoshida, T., Ikedo-Saito, M., & Hoffman, B. (2002). Catalytic mechanism of heme oxidase through EPR and ENDOR of cryoreduced oxy-heme oxygenase and its Asp 140 mutants. *Journal of the American Chemical Society*, *124*, 1798–1808.

Doster, W. (2008). The dynamical transition of proteins, concepts and misconceptions. *European Biophysics Journal*, *37*(5), 597–602.

Douzou, P. (1977). *Cryobiochemistry: An introduction*. New York: Academic Press.

Dyre, J. C., Hechsher, T., & Niss, K. (2009). A brief critique of the Adam-Gibbs entropy model. *Journal of Non-Crystalline Solids*, *355*, 624–627.

Faust, L. P., & Babior, B. M. (1992). Overexpression, purification, and some properties of the adocbl-dependent ethanolamine ammonia-lyase from Salmonella-typhimurium. *Archives of Biochemistry and Biophysics*, *294*(1), 50–54.

Fenimore, P. W., Frauenfelder, H., McMahon, B. H., & Young, R. D. (2004). Bulk-solvent and hydration-shell fluctuations, similar to α- and β-fluctuations in glasses, control protein motions and functions. *Proceedings of the National Academy of Sciences*, *101*, 14408–14413.

Franks, F. (1998). Freeze-drying of bioproducts: Putting principles into practice. *European Journal of Pharmaceutics and Biopharmaceutics*, *45*, 221–229.

Franks, F. (2003). Scientific and technological aspects of aqueous glasses. *Biophysical Chemistry*, *106*, 251–261.

Frauenfelder, H., Chen, G., Berendzen, J., Fenimore, P. W., Jansson, H., McMahon, B. H., et al. (2009). A unified model of protein dynamics. *Proceedings of the National Academy of Sciences*, *106*, 5129–5134.

Frey, P. A. (2010). Cobalamin coenzymes in enzymology. In L. Mander, & H.-W. Lui (Eds.), *Vol. 7. Comprehensive natural products II chemistry and biology* (pp. 501–546). Oxford UK: Elsevier.

Frey, P. A., Hegeman, A. D., & Ruzicka, F. J. (2008). The radical SAM superfamily. *Critical Reviews in Biochemistry and Molecular Biology*, *43*(1), 63–88.

Gerfen, G. J. (1999). EPR spectroscopy of B12-dependent enzymes. In R. Banerjee (Ed.), *Chemistry and biochemistry of B12* (pp. 165–195). New York: John Wiley and Sons.

Glowacki, D. R., Harvey, J. N., & Mulholland, A. J. (2012). Taking Ockham's razor to enzyme dynamics and catalysis. *Nature Chemistry*, *4*, 169–176.

Goldstein, M. (2011). The past, present, and future of the Johari-Goldstein relaxation. *Journal of Non-Crystalline Solids*, *357*, 249–250.

Hammes, G. G., Benkovic, S. J., & Hammes-Schiffer, S. (2011). Flexibility, diversity, and cooperativity: Pillars of enzyme catalysis. *Biochemistry*, *50*.

Henzler-Wildman, K. A., & Kern, D. (2007). Dynamic personalities of proteins. *Nature*, *450*, 964–972.

Herschlag, D., & Natarajan, A. (2013). Fundamental challenges in mechanistic enzymology: Progress toward understanding the rate enhancements of enzymes. *Biochemistry*, *52*, 2050–2067.

Hong, L., Smolin, N., Lindner, B., Sokolov, A. P., & Smith, J. C. (2011). Three classes of motion in the dynamic neutron-scattering susceptibility of a globular protein. *Physical Review Letters*, *107*. 148102/148101-148102/148104.

Hubei, A., Bischof, J. C., & Aksan, A. (2009). Freezing-induced phase separation and spatial microheterogeneity in protein solutions. *The Journal of Physical Chemistry. B*, *113*, 10081–10087.

Ionescu, A., Li, W., Nforneh, B., & Warncke, K. (2021). Coupling of ethanolamine ammonia-lyase protein and solvent dynamics characterized by the temperature dependence of EPR spin probe mobility and dielectric permittivity. *The Journal of Chemical Physics*, *154*, 175101.

Jencks, W. P. (1969). *Catalysis in chemistry and enzymology*. New York: McGraw-Hill Book Co., Inc.

Johari, G. P., & Goldstein, M. (1970). Viscous liquids and the glass transition. II. Secondary relaxations in glasses of rigid molecules. *The Journal of Chemical Physics*, *53*(6), 2372–2388.

Kamerlin, S. C. L., & Warshel, A. (2010). At the dawn of the 21st century: Is dynamics the missing link for understanding enzyme catalysis? *Proteins*, *78*, 1339–1375.

Ke, S. C. (2003). Spin-spin interaction in ethanolamine deaminase. *Biochimica et Biophysica Acta*, *1620*, 267–272.

Khoroshun, D. V., Warncke, K., Ke, S.-C., Musaev, D. G., & Morokuma, K. (2003). Internal degrees of freedom, structural motifs and conformational energetics of 5'-deoxyadenosyl radical: Implications for function in adenosylcobalamin-dependent enzymes. A computational study. *Journal of the American Chemical Society*, *125*, 570–579.

Kohne, M., Li, W., Zhu, C., & Warncke, K. (2019). Deuterium kinetic isotope effects resolve low-temperature substrate radical reaction pathways and steps in B12-dependent ethanolamine ammonia-lyase. *Biochemistry, 58*, 3683–3690.

Kohne, M., Zhu, C., & Warncke, K. (2017). Two dynamical regimes of the substrate radical rearrangement reaction in B12-dependent ethanolamine ammonia-lyase resolve contributions of native protein configurations and collective configurational fluctuations to catalysis. *Biochemistry, 56*, 3257–3264.

Körber, C. (1988). Phenomena at the advancing ice-liquid interface: Solutes, particles and biological cells. *Quarterly Reviews of Biophysics, 21*(2), 229–298.

Kraut, D. A., Carrol, K. S., & Herschlag, D. (2003). Challenges in enzyme mechanism and energetics. *Annual Review of Biochemistry, 75*, 517–571.

Leigh, J. S., Jr., & Reed, G. H. (1971). Electron paramagnetic resonance studies in frozen aqueous solutions. Elimination of freezing artifacts. *The Journal of Physical Chemistry, 75*(9), 1202–1204.

Lewandowski, J. R., Halse, M. E., Blackledge, M., & Emsley, L. (2015). Direct observation of hierarchical protein dynamics. *Science, 348*, 578–581.

Liu, Y. H., Fujita, T., Aji, D. P. B., Matsuura, M., & Chen, M. W. (2014). Structural origins of Johari-Goldstein relaxation in a metallic glass. *Nature Communications, 5*, 1–7.

LoBrutto, R., Bandarian, V., Magnusson, O. T., Chen, X., Schramm, V. L., & Reed, G. H. (2001). 5'-deoxyadenosine contacts the substrate radical intermediate in the active site of ethanolamine ammonia-lyase: ^{2}H and ^{13}C electron nuclear double resonance studies. *Biochemistry, 40*, 9–14.

Lukoyanov, D., Barney, B. M., Dean, D. R., Seefeldt, L. C., & Hoffman, B. M. (2007). Connecting nitrogenase intermediates with the kinetic scheme for N_2 reduction by a relaxation protocol and identification of the N_2 binding state. *Proceedings of the National Academy of Sciences, 104*, 1451–1455.

Ma, B., & Nussinov, R. (2010). Enzyme dynamics point to stepwise conformational selection in catalysis. *Current Opinion in Chemical Biology, 14*, 652–659.

Mallamace, F., Broccio, M., Corsaro, C., Faraone, A., Wanderlingh, U., Liu, L., et al. (2006). The fragile-to-strong dynamic crossover transition in confined water: Nuclear magnetic resonance results. *The Journal of Chemical Physics, 124*, 161102.

Marsh, E. N. G., & Melendez, G. D. R. (2012). Adenosylcobalamin enzymes: Theory and experiment begin to converge. *Biochimica et Biophysica Acta, 1824*, 1154–1164.

McRae, C. R. H., Wang, H., Gao, J., Vissers, M. R., Brecht, T., Dunsworth, A., et al. (2020). Materials loss measurements using superconducting microwave resonators. *The Review of Scientific Instruments, 91*, 091101.

Nagel, Z. D., & Klinman, J. P. (2009). A 21st century revisionist's view at a turning point in enzymology. *Nature Chemical Biology, 5*, 543–550.

Nakanishi, M., & Sokolov, A. P. (2015). Protein dynamics in a broad frequency range: Dielectric spectroscopy studies. *Journal of Non-Crystalline Solids, 407*, 478–485.

Nforneh, B., & Warncke, K. (2017). Mesodomain and protein-associated solvent phases with temperature-tunable (200-265 K) dynamics surround ethanolamine ammonia-lyase in globally polycrystalline aqueous solution containing dimethyl sulfoxide. *The Journal of Physical Chemistry. B, 121*, 11109–11118.

Nforneh, B., & Warncke, K. (2019). Control of solvent dynamics around the B12-dependent ethanolamine ammonia-lyase enzyme in frozen aqueous solution by using dimethyl sulfoxide modulation of mesodomain volume. *The Journal of Physical Chemistry. B, 123*, 5395–5404.

Ngai, K. L., Capaccioli, S., Ancherbak, S., & Shinyashiki, N. (2011). Resolving the ambiguity of the dynamics of water and clarifying its role in hydrated proteins. *Philosophical Magazine, 91*, 1809–1835.

Nickels, J. D., Sakai, V. G., & Sokolov, A. P. (2013). Dynamics in protein powders on the nanosecond-picosecond time scale are dominated by localized motions. *The Journal of Physical Chemistry. B, 117*, 11548–11555.

Panagopoulou, A., Kyritsis, A., Shinyashiki, N., & Pissis, P. (2012). Protein and water dynamics in bovine serum albumin-water mixtures over wide ranges of compositions. *The Journal of Physical Chemistry. B, 116*, 4593–4602.

Poole, C. P. (1983). *Electron spin resonance* (2nd Ed.). Mineola, NY: Dover Publications, Inc.

Richard, J. P. (2013). Enzymatic rate enhancements: A review and perspective. *Biochemistry, 52*, 2009–2011.

Ringe, D., & Petsko, G. A. (2003). The "glass transition" in protein dynamics: What it is, why it occurs, and how to exploit it. *Biophysical Chemistry, 105*(2–3), 667–680.

Roof, D. M., & Roth, J. R. (1988). Ethanolamine Utilization in Salmonella-Typhimurium. *Journal of Bacteriology, 170*(9), 3855–3863.

Sayler, R. I., Stich, T. A., Joshi, S., Cooper, N., Shaw, J. T., Begley, T. P., et al. (2019). Trapping and electron paramagnetic resonance characterization of the 5′ dAdo• radical in a radical S-adenosyl methionine enzyme reaction with a non-native substrate. *ACS Central Science, 5*, 1777–1785.

Semialjac, M., & Schwartz, H. (2003). Computational study on mechanistic details of the aminoethanol rearrangement catalyzed by the vitamin B12-dependent ethanolamine ammonia lyase: His and asp/Glu acting simultaneously as catalytic auxilliaries. *The Journal of Organic Chemistry, 68*, 6967–6983.

Sharma, P. K., Chu, Z. T., Olsson, M. H. M., & Warshel, A. (2007). A new paradigm for electrostatic catalysis of radical reactions in vitamin B12 enzymes. *Proceedings of the National Academy of Sciences, 104*(23), 9661–9666.

Shibata, N., Sueyoshi, Y., Higuchi, Y., & Toraya, T. (2018). Direct participation of a peripheral side chain of a corrin ring in coenzyme B12 catalysis. *Angewandte Chemie, 130*(26), 7956–7961.

Shibata, N., Tamagaki, H., Hieda, N., Akita, K., Komori, H., Shomura, Y., et al. (2010). Crystal structures of ethanolamine Ammonia-lyase complexed with coenzyme B-12 analogs and substrates. *The Journal of Biological Chemistry, 285*, 26484–26493.

Shinyashiki, N., Yamamoto, W., Yokoyama, A., Yoshinari, T., Yagihara, S., Kita, R., et al. (2009). Glass transitions in aqueous solutions of protein (bovine serum albumin). *The Journal of Physical Chemistry. B, 113*, 14448–14456.

Staib, L., & Fuchs, T. M. (2014). From food to cell: Nutrient exploitation strategies of enterpathogens. *Microbiology, 160*, 1020–1039.

Stillinger, F. H. (1995). A topographic view of supercooled liquids and glass formation. *Science, 267*, 1935–1939.

Sun, L., Groover, O. A., Canfield, J. M., & Warncke, K. (2008). Critical role of arginine 160 of the EutB protein subunit for active site structure and radical catalysis in coenzyme B12-dependent ethanolamine ammonia-lyase. *Biochemistry, 47*(20), 5523–5535.

Thiennimitr, P., Winter, S. E., Winter, M. G., Xavier, M. N., Tolstikov, V., Huseby, D. L., et al. (2011). Intestinal inflammation allows Salmonella to use ethanolamine to compete with the microbiota. *Proceedings of the National Academy of Sciences, 108*(42), 17480–17485.

Timasheff, S. N. (2002). Protein-solvent preferential interactions, protein hydration, and the modulation of biochemical reactions by solvent compounds. *Proceedings of the National Academy of Sciences, 99*, 9721–9726.

Toraya, T. (2003). Radical catalysis in coenzyme B12-dependent isomerization (eliminating) reactions. *Chemical Reviews, 103*, 2095–2127.

Ucuncuoglu, N., & Warncke, K. (2018). Protein configurational states guide radical rearrangement catalysis in ethanolamine ammonia-lyase. *Biophysical Journal, 114*, 2775–2786.

Vey, J. L., & Drennan, C. L. (2011). Structural insights into radical generation by the radical SAM superfamily. *Chemical Reviews, 111*, 2487–2506.

Vos, M. H., Lambry, J.-C., Robles, S. J., Youvan, D. C., Breton, J., & Martin, J.-L. (1991). Direct observation of vibrational coherence in bacterial reaction centers using femtosecond absorption spectroscopy. *Proceedings of the National Academy of Sciences*, *88*, 8885–8889.

Wang, M., & Warncke, K. (2008). Kinetic and thermodynamic characterization of Co^{II}-substrate radical pair formation in coenzyme B12-dependent ethanolamine ammonia-lyase in a cryosolvent system by using time-resolved, full-spectrum continuous-wave Electron paramagnetic resonance spectroscopy. *Journal of the American Chemical Society*, *130*, 4846–4858.

Wang, M., & Warncke, K. (2013). Entropic origin of cobalt-carbon bond cleavage catalysis in adenosylcobalamin-dependent ethanolamine ammonia-lyase. *Journal of the American Chemical Society*, *135*, 15077–15084.

Wang, M., Zhu, C., Kohne, M., & Warncke, K. (2015). Resolution and characterization of chemical steps in enzyme catalytic sequences by using low-temperature and time-resolved, full-spectrum EPR spectroscopy in fluid cryosolvent and frozen solution systems. *Methods in Enzymology*, *563*(Part A), 59–94.

Warncke, K., Schmidt, J. C., & Ke, S.-C. (1999). Identification of a rearranged-substrate, product radical intermediate and the contribution of a product radical trap in vitamin B12 coenzyme-dependent ethanolamine deaminase catalysis. *Journal of the American Chemical Society*, *121*, 10522–10528 (erratum: (2008) ibid., 130, 6055).

Wetmore, S. D., Smith, D. M., Bennet, J. T., & Radom, L. (2002). Understanding the mechanism of action of B12-dependent ethanolamine ammonia-lyase: Synergistic interactions at play. *Journal of the American Chemical Society*, *124*, 14054–14065.

Yabukarski, F., Beil, J. T., Pinney, M. M., Doukov, T., Powers, A. S., Fraser, J. S., et al. (2020). Assessment of enzyme active site positioning and tests of catalytic mechanisms through X-ray-derived conformational assemblies. *Proceedings of the National Academy of Sciences*, *117*(52), 33204–33215.

Yang, H., McDaniel, E. C., Impano, S., Byer, A. S., Jodts, R. J., Yokoyama, K., et al. (2019). The elusive 5′-deoxyadenosyl radical: Captured and characterized by electron paramagnetic resonance and electron nuclear double resonance spectroscopies. *Journal of the American Chemical Society*, *141*(30), 12139–12146.

Zhu, C., & Warncke, K. (2008). Reaction of the Co^{II}-substrate radical pair catalytic intermediate in coenzyme B_{12}-dependent ethanolamine ammonia-lyase in frozen aqueous solution from 190 to 217 K. *Biophysical Journal*, *95*, 5890–5900.

CHAPTER ELEVEN

Investigating radical pair reaction dynamics of B_{12} coenzymes 1: Transient absorption spectroscopy and magnetic field effects

Joanna A. Hughes[a], Samantha J.O. Hardman[b], Valentina Lukinović[c], Jonathan R. Woodward[d], and Alex R. Jones[e,*]

[a]Laboratory of Ultrafast Spectroscopy, ISIC, and Lausanne Centre for Ultrafast Science (LACUS), École Polytechnique Fédérale de Lausanne, Lausanne, Switzerland
[b]Manchester Institute of Biotechnology, The University of Manchester, Manchester, United Kingdom
[c]Institute of Biochemistry, Carleton University, Ottawa, ON, Canada
[d]Graduate School of Arts and Sciences, The University of Tokyo, Tokyo, Japan
[e]Biometrology, Department of Chemical and Biological Sciences, National Physical Laboratory, Middlesex, United Kingdom
*Corresponding author: e-mail address: alex.jones@npl.co.uk

Contents

Methods in Enzymology, Volume 669
ISSN 0076-6879
https://doi.org/10.1016/bs.mie.2021.12.019

Abstract

B_{12} coenzymes are vital to healthy biological function across nature. They undergo radical chemistry in a variety of contexts, where spin-correlated radical pairs can be generated both thermally and photochemically. Owing to the unusual magnetic properties of B_{12} radical pairs, however, most of the reaction and spin dynamics occur on a timescale (picoseconds–nanoseconds) that cannot be resolved by most measurement techniques. Here, we describe a method that combines femtosecond transient absorption spectroscopy with magnetic field exposure, which enables the direct scrutiny of such rapid processes. This approach should provide a means by which to investigate the apparently profound effect protein environments have on the generation and reactivity of B_{12} radical pairs.

1. Introduction

Be it as signaling molecules, reaction intermediates or a source of oxidative stress, the importance of radicals to both healthy biological function and disease states is becoming increasingly apparent (Halliwell & Gutteridge, 2007). In their role as cofactors to a range of metabolic enzymes, B_{12} coenzymes have a strong association with radical biochemistry (Golding & Buckel, 2012). When substrates bind to 5′-deoxyadenosylcobalamin (AdoCbl)-dependent enzymes, for example, a transient cob(*II*)alamin/Ado$^{\bullet}$ radical pair (RP) is generated. The Ado$^{\bullet}$ then facilitates enzyme catalysis by abstracting a H-atom from the substrate, thus triggering radical rearrangement reactions (Marsh, Patterson, & Li, 2010). Computational data (Bucher, Sandala, Durbeej, Radom, & Smith, 2012) are consistent with a discrete Ado$^{\bullet}$ and useful inferences have been made from stopped flow (e.g., Bandarian & Reed, 2000; Jones, Hay, Woodward, & Scrutton, 2007; Marsh & Ballou, 1998; Padmakuma, Padmakuma, & Banerjee, 1997) and electron paramagnetic resonance (EPR) data (e.g., Bender, Poyner, & Reed, 2008; Zhu & Warncke, 2010). Owing to the transience of the Ado$^{\bullet}$ radical, however, its direct measurement during turnover has proven elusive. It thus remains a significant challenge to investigate directly the apparently profound impact of the protein environment on the creation and reaction dynamics of the cob(*II*)alamin/Ado$^{\bullet}$ RP under ambient conditions.

One strategy is to exploit the light-sensitivity of B_{12} coenzymes (Jones, 2017). Using laser pulses to initiate a reaction can give access to an experimental time resolution on the order of femtoseconds, which is not available to time-resolved methods for initiating thermally-driven reactions like those of enzymes (e.g., rapid mixing, ~milliseconds). Following photoexcitation in solution and when bound to dependent enzymes, two of the biologically-active B_{12} derivatives, AdoCbl and methylcobalamin (MeCbl), both produce cob(*II*)alamin/alkyl RPs equivalent to those described above. In the case of AdoCbl in aqueous solution (Fig. 1), the cob(*II*)alamin/Ado$^•$ geminate RP is generated photolytically on a picosecond timescale, with a proportion recombining within a few nanoseconds and the remainder diffusing apart to give solvent-separated RPs (Walker, Shiang, Anderson, Pullen, & Sension, 1998). The probability that any given RP will recombine to the diamagnetic precursors or separate to produce free radicals is to a significant extent dictated by its spin dynamics, which in turn are strongly influenced by the RP environment (Woodward, 2002). Thus, we have a route into scrutinizing the impact of the protein environment on the reaction dynamics of B_{12} or similar RPs through their photochemical spin dynamics (Jones, 2016). In recent years, a photobiological role has been discovered for B_{12} photochemistry in bacterial transcriptional regulation (Jones, 2017; Jost et al., 2015; Kutta et al., 2015; Ortiz-Guerrero, Polanco, Murillo, Padmanabhan, & Elías-Arnanz, 2011; Padmanabhan, Jost, Drennan, & Elías-Arnanz, 2017).

This chapter describes a method that combines ultrafast transient absorption (TA) spectroscopy with exposure of the sample to tunable magnetic fields (MF). This combination enables one to probe the influence of spin on the early reaction dynamics (<10 ns) of the geminate RP following AdoCbl photolysis. Chapter "Investigating radical pair reaction dynamics of B12 coenzymes 2: Time-resolved electron paramagnetic resonance

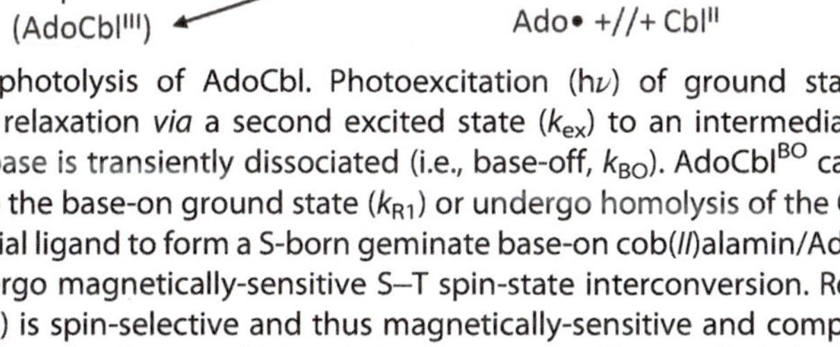

Fig. 1 The photolysis of AdoCbl. Photoexcitation (hν) of ground state AdoCbl is followed by relaxation *via* a second excited state (k_{ex}) to an intermediate where the lower axial base is transiently dissociated (i.e., base-off, k_{BO}). AdoCbl^{BO} can then either relax back to the base-on ground state (k_{R1}) or undergo homolysis of the Co–C bond to the upper axial ligand to form a S-born geminate base-on cob(*II*)alamin/Ado RP (k_H). This RP can undergo magnetically-sensitive S–T spin-state interconversion. Recombination of the RP (k_R) is spin-selective and thus magnetically-sensitive and competes with diffusive separation (or "escape," k_E) and the formation of free radicals (indicated by //).

spectroscopy" by Lukinović et al. is related and describes the use of time-resolved EPR spectroscopy to probe longer-lived (tens–hundreds of nanoseconds) RPs following photolysis of MeCbl, which includes both cob(*II*)alamin/methyl geminate pairs and pairs of methyl free radicals (so-called *f*-pairs) that form following diffusive encounters (Lukinović et al., 2019). Together, the two chapters provide a foundational experimental approach for investigating the photochemical RP reactions of B_{12} coenzymes across timescales that capture a variety of spin dynamic mechanisms (Hughes et al., 2019; Lukinović et al., 2019). The intention is that this foundation can be built upon and methods adapted to help probe the influence of the protein environment on B_{12} radical reactions (e.g., Jones, Levy, Hay, & Scrutton, 2013).

To properly understand the methods we describe, one must appreciate how MF exposure influences the course of RP reactions (Steiner & Ulrich, 1989). When a pair of radicals is generated from non-radical precursors (e.g., by bond homolysis, electron transfer or H-atom abstraction), the spins of the resulting unpaired electrons are initially correlated. RPs are thus either singlet-born (S, antiparallel spins, ↑↓) or triplet-born (T, parallel spins, ↑↑) and this polarized spin-state is conserved from the multiplicity of the RP precursors. Before the system relaxes to equilibrium (i.e., a Boltzmann distribution of S and T) the RP spin-state can evolve with time such that the spin-polarized S and T states interconvert (e.g., for AdoCbl see Fig. 1). This process can be driven by various magnetic interactions such as hyperfine couplings between the unpaired electron spins and magnetic nuclei or by the difference in molecular *g*-value (a dimensionless parameter analogous to the chemical shift in NMR) of each radical of the pair. While the spins remain correlated, S–T interconversion is a coherent, oscillatory process that is quantum mechanical in origin. If, during S–T interconversion, the radicals of the pair (or at least the unpaired spins) separate in space and then reencounter, the probability of RP recombination is spin-selective. Assuming a S diamagnetic ground state (by far the most common case), Pauli exclusion states that S RPs can recombine, whereas T RPs cannot. The application of MF to a spin-correlated RP system can influence the extent and or rate of S–T interconversion and thus the probability of RP recombination (e.g., for AdoCbl see Fig. 1). Under the correct conditions (Kim et al., 2021), RP reactions can therefore be magnetically-sensitive.

The photochemical reaction dynamics of B_{12} coenzymes are duly magnetically-sensitive through a RP mechanism (Fig. 1) (Chagovetz & Grissom, 1993; Jones, Woodward, & Scrutton, 2009). Relative to most common, organic RPs, however, those generated from B_{12} have quite exotic magnetic properties. Both the hyperfine coupling and molecular *g*-values of the cob(*II*)alamin radical are very large, which, unusually, result in S–T interconversion for the cob(*II*)alamin/alkyl RPs that occurs on the ultrafast timescale (<10 ns) (Hughes et al., 2019; Jones, Hardman, Hay, & Scrutton, 2011). This opens up a wider range of experimental approaches than are usually employed to probe RP chemistry, with magnetic field effects (MFEs) on B_{12} photochemistry having been measured on the transient dynamics of the earliest geminate pairs following photoexcitation by femtosecond laser pulses and the steady-state dynamics of RPs under continuous illumination over minutes (Chagovetz & Grissom, 1993; Hughes et al., 2019; Jones, 2017; Jones et al., 2007, 2009, 2011; Woodward, Foster, Jones, Salaoru, & Scrutton, 2009). Such approaches have informed on the impact of rapid radical quenching by the substrate in typical B_{12} enzymes (where MFEs are removed) (Jones et al., 2007) and the influence of local protein motions on the transient reaction dynamics of Ado$^{\bullet}$ (Jones et al., 2011, 2013).

MF exposure presents additional experimental challenges to those usually encountered with ultrafast TA spectroscopy and specialist approaches are employed for data analysis and simulation to account for the underlying spin dynamics. We therefore present a detailed protocol for measuring and analyzing MFEs on the ultrafast RP reaction following the photolysis AdoCbl in solution to enable others less familiar with spin chemistry to exploit spin dynamic effects for the investigation of B_{12} reaction mechanism.

2. Materials, equipment and reagents

Sample preparation

- 5′-deoxyadenosylcobalamin (Coenzyme B_{12}, Merck)
- Buffer ingredients: HEPES, deionized water, glycerol
- Nitrogen for degassing
- Dark room
- Anaerobic glove box for preparing samples
- UV–visible spectrophotometer (Cary 60, Agilent)

MF generation and measurement

- DZ0X8-N52, axially magnetized neodymium disc magnets (maximum flux = 1.48 T, K&J Magnetics)
- Custom built magnet holder (Fig. 2C)
- STD18-0404 transverse Hall probe and an 5180 Gaussmeter (FW Bell)

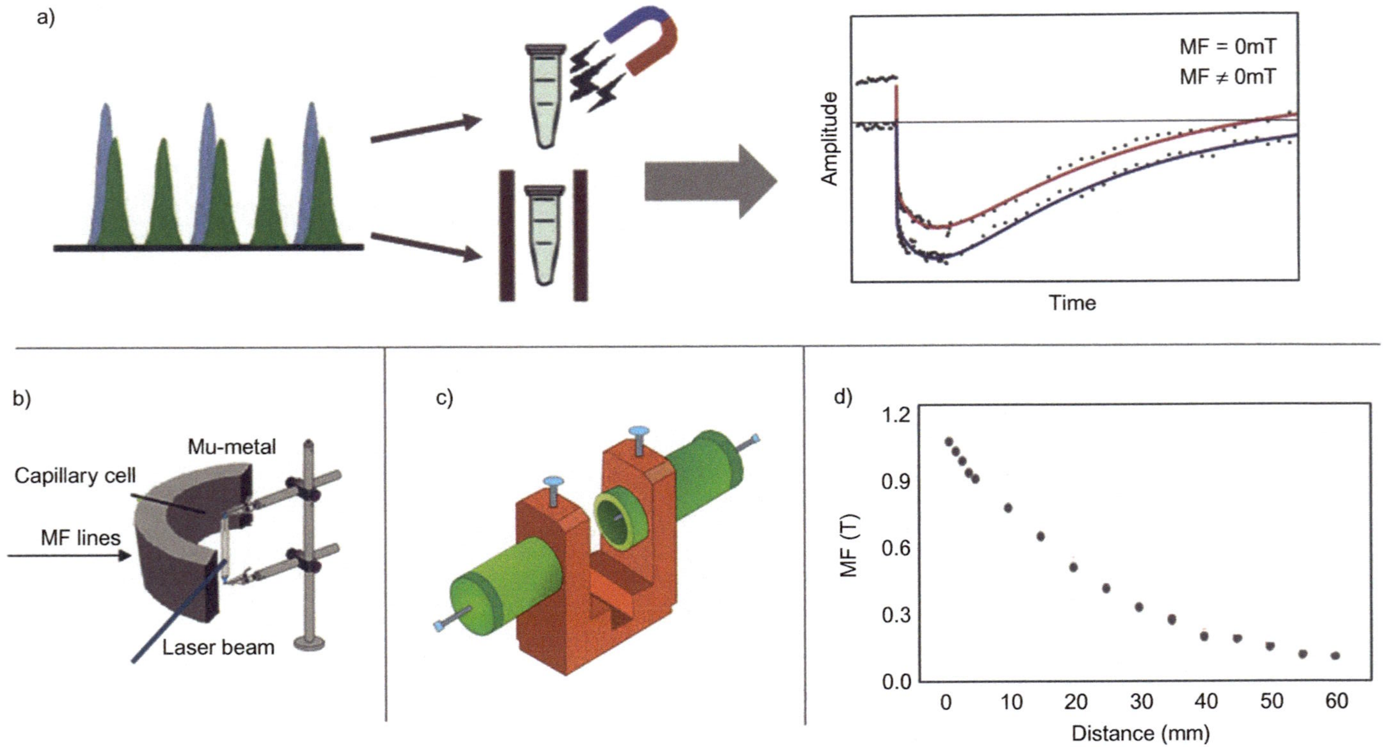

Fig. 2 (A) Laser pulse sequence used in the experiments (blue pump, green probe); the electronically controlled shutter allows for alternation between the "MF" and "shield" samples (see main text). Examples of data acquired for AdoCbl in both the "MF" and "shield" optical paths. (B) Capillary flow cell and semi-cylindrical mu-metal shielding in the "shield" optical path. (C) Holder for mounting the permanent magnets and varying their distance from the sample cuvette. The red section illustrates the chassis into which hollow brass barrels (green) containing the magnets are screwed. The extent to which the barrels are screwed into the chassis control the distance of the magnets from the cuvette and hence the MF magnitude at the sample position. (D) Measurements of the MF generated at the sample position between pairs of neodymium disc magnets as a function of the distance between the two magnets.

Sample flowing system

- 2 mm pathlength square profile quartz capillary
- PTFE Tubing (VWR internal diameter 0.8 mm)

Transient absorption instrument and laser excitation

- Ti:Sapphire amplifier system (Solstice Ace, Newport Spectra Physics) producing 6 mJ, 100 fs pulses at 1 kHz centered at 800 nm
- Optical Parametric Amplifiers with wavelength extension units (TOPAS Prime OPAs and NIR-UVis units, Light Conversion)
- Femtosecond TA Spectrometer (Helios, Ultrafast Systems)
- Lock-in amplifier (SR830, SRS)
- Two Biased Si Photodetectors (DET100A, Thorlabs)
- LabVIEW computer software written in-house for instrument control and data collection

3. Method details: Instrumental setup and sample preparation

3.1 Modified transient absorption spectrometer

The magnitude of MFEs on the kinetics of (bio)chemical RP reactions measured at the molecular level tend to be quite small. For example, anaerobic continuous wave photolysis of AdoCbl under MF exposure resulted in a maximum of ~18% decrease in the rate of accumulation of the cob(*II*) alamin radical when compared to the rate in the absence of an applied field (Jones et al., 2009). Because of the non-linear response to field dose, when collecting a full MF-dependence most percentage MFEs measured are significantly smaller than this. Moreover, it was necessary to acquire these data in a highly viscous solvent (67% glycerol w/w) to increase the lifetime of the geminate RP and thus enhance the measured MFE. Under less amenable conditions the magnetically-induced change can be smaller still. It was therefore necessary to adapt standard TA instrumentation to enable the efficient acquisition of high quality data to confidently discern potentially subtle MFEs on the ultrafast, photochemical reaction and spin dynamics of B_{12} cofactors such as AdoCbl (Hughes et al., 2019).

In standard ultrafast TA experiments two repetitively pulsed beam paths are used; a pump beam is used to photoexcite the sample (from an OPA or harmonic of a laser), and a probe beam, which can be either a single color (from an OPA or harmonic of a laser) or white light continuum (WLC, typically generated in sapphire or CaF_2), is used to interrogate the absorbance of the sample. The pump beam passes through an optical chopper, which removes alternate pulses from the train, and detection electronics monitor changes in signal from the probe beam that occur at the frequency of the optical chopper (i.e., 500 Hz in this case). The detectors for this method do not provide the time resolution of the experiment (often averaging of 1–2 s is used) rather the probe beam path passes via an optical delay line, which is used to delay the arrival of the beams relative to one another, with sub-fs resolution. Most pump/probe TA instrumentation includes two probe beam paths to account for the pulse-to-pulse variations, which are a common feature of ultrafast laser systems. In path 1 ("sample path") the pump beam excites the sample and the probe beam then passes through it at a given time delay to reach the first detector. In path 2 ("reference path") part of the probe beam that has been split off prior to reaching the sample is monitored by a second detector to provide a reference signal. TA detection may either use single point detectors and a lock-in amplifier if a single-color probe is used, or array (CCD) detectors with more complex detection electronics if a white light continuum is used as the probe. The latter benefits from collection of the complete spectrum, which can contain useful chemical information. We use the former here, however, which tends to benefit from much improved signal/noise ratios, an important feature when attempting to measure potentially small MFEs. The peak absorption of the cob(*III*)alamin in ground state AdoCbl (~530 nm) is known to bleach significantly during homolysis and conversion to the cob(*II*)alamin radical (Chen & Chance, 1990); absorbance changes can therefore be monitored around this wavelength without the need to probe the complete spectrum.

To enable measurement of MFEs on ultrafast timescales, we used the optical routing, probe delay line, and pump optical chopper from a commercially produced femtosecond TA spectrometer (Helios, Ultrafast Systems). One OPA output was used as the pump beam, and another for the probe beam. The pump and probe beams were diverted before the original TA sample position via a periscope to a secondary optical set-up. Here, the pump beam is split equally to excite samples on two "sample paths," and the probe beam is split twice, into three beams: two "sample paths" and a single reference path. In the first sample path, the pump and probe beams pass through a sample exposed to tunable MF ("MF path"); in the second, the pump and probe beams pass through a sample shielded from the magnets in the neighboring path by mu-metal ("shield path"). Mu-metal is a ferromagnetic alloy with high permeability such that it serves as a low reluctance path for the field lines. A representation of the pump and probe beams and the two samples paths in shown in Fig. 2A. The sample in both the "MF" and "shield" paths flows in parallel through quartz capillary cells from a common reservoir (the shielded path and cell are shown in Fig. 2B). The sample in each path is therefore identical apart from the MF exposure condition and sample flow mitigates the effects of photobleaching. Despite having three probe beam paths, the instrumentation has only two detectors, one to detect the reference signal and one "sample detector." A shutter is placed before the sample positions such that either the "MF" or "shield" sample probe beam path is blocked. Thus, detection of signal from "MF" and "shield" samples can be easily alternated with no other disturbance to the experimental set-up. This may sound like a minor advantage but when monitoring very small changes any fractional adjustment to the sample position or overlap of pump and probe beams in the sample can completely obscure a MFE. This arrangement enables efficient and reproducible acquisition of time-resolved datasets with and without an applied-MF. Data were acquired using a Labview program written in-house.

3.2 Magnetic field-generating apparatus

As mentioned in Section 1, B_{12} coenzymes have magnetic properties that deviate significantly from the range usually encountered for organic radicals. The Co ion results in substantial values both for the average hyperfine coupling (7.33 mT) and molecular *g*-tensor (2.105) (Hughes et al., 2019). There are thus large differences in these parameters between the cob(*II*)alamin radical and alkyl radicals. One consequence of these large differences is that they drive spin-state mixing on sub-nanosecond timescales (Hughes et al., 2019; Jones et al., 2011), hence the manifestation of MFEs on ultrafast kinetic parameters. Another consequence is that AdoCbl, at the time of writing, remains the only published example where a MFE owing to the Δg-mechanism (i.e., when spin-state interconversion is dominated by the difference in molecular *g*-value) has been observed at applied MFs < 1 T (Hughes et al., 2019). Δg is usually very small and very large MFs are usually required for the corresponding MFEs to be non-negligible. The very large Δg for the cob(*II*)alamin/Ado$^{\bullet}$ RP led to this apparently unprecedented observation and brought the manifestation of Δg-mechanism MFEs down to an applied MF value that meant a full MF-dependence could be acquired using a single piece of MF-generating apparatus (Hughes et al., 2019).

A holder was designed and constructed that enables pairs of neodymium disc magnets to be mounted at variable distances from the sample (Fig. 2C). Each magnet (or combination of magnets) is inserted into one of a pair of hollow, threaded brass barrels, which are screwed into a U-shaped aluminum chassis such that they are positioned on either side of the capillary flow cell. The distance between each magnet and the cell can be reduced by screwing the barrels further into the chassis and the distance between the magnets corresponds to the MF magnitude at the sample position (Fig. 2D). The maximum MF achieved at the sample position using this arrangement is 950 mT, which corresponds to a ~2 mm separation between the magnets. The chassis needs to be physically strong enough not to distort when the magnets are this close together as the attraction between them becomes considerable. Due to space limitations, the lowest MF we achieved using the 1.5″ disc magnets was 5 mT, which corresponds a ~9 cm separation between the magnets. Lower MF could of course be achieved using smaller magnets.

Although electromagnets produce a homogenous field over a larger volume, there are various reasons why we chose to generate MF using pairs of permanent magnets. First, electromagnets can produce heat and vibrations, which introduce confounding differences between the "MF" and "shield" samples. Although coils can be water-cooled, for example, these effects are difficult to remove entirely, especially at higher MF exposures. This is not an issue when using permanent magnets. Second, the smaller homogeneous MF volume produced by permanent magnets is not a significant concern for the TA measurements described here. The measurement volume is defined by the overlap of the pump and probe beams, which is typically in the range of 100–200 μm. Third, there are various health and safety reasons for avoiding the use of electromagnets where possible. High voltages and currents are required to generate MF of up to 1 T. Moreover, if water cooling is required, which would almost certainly be the case at these MF values, the combination of water and high voltages represents a significant hazard.

3.3 Sample preparation and data acquisition conditions

Samples were prepared in 20 mM HEPES/50% glycerol (v/v), pH = 7.5. The nature of the lower axial ligand in B_{12} coenzymes has been demonstrated to have a significant influence on the quantum yield of the photolysis. Around neutral pH, photoexcitation of AdoCbl leads to homolysis at the upper axial position with near unity quantum yield (Walker et al., 1998). At acidic pH, the nitrogen that coordinates the lower axial 5,6-dimethylimidazole ligand to the Co ion becomes protonated and is replaced at the lower axial position by water molecule. This opens a new channel for the rapid non-radiative decay to the ground state (i.e., internal conversion) (Lodowski, Jaworska, Garabato, & Kozlowski, 2015), which competes with the upper axial bond homolysis, substantially reducing the yield of RPs (~0.12 at pH=2) (Peng et al., 2010). To maximize the observed MFE it is therefore necessary to acquire data on samples at or slightly above neutral pH.

The addition of a significant quantity of glycerol in the buffer solution enhances the magnitude of the MFE further. When a geminate RP is first generated in isotropic solution, it must diffuse apart to some degree to reduce the energy gap between the S and T spin states of the RP and allow them to coherently interconvert (Steiner & Ulrich, 1989). Without this interconversion MFEs will not be observed. If the members of the RP diffuse apart too rapidly, however, the likelihood of the geminate pair recombining is reduced. Recombination is the spin-selective step and again, without it MFEs are not observed. The introduction of a viscogen like glycerol allows diffusion, and hence spin-state interconversion, but prolongs the lifetime of the geminate RP sufficiently to increase the recombination probability and enhance the MFE. This so-called RP "cage effect" can also be achieved by a protein environment, as previously reported for MFEs on the photolysis of AdoCbl in one of its dependent enzymes, ethanolamine ammonia lyase (Jones et al., 2009).

The photochemistry of B_{12} coenzymes is highly oxygen sensitive, where both the cob(*II*)alamin and alkyl radicals can be rapidly quenched unless maintained under anaerobic conditions (Schwartz & Frey, 2007). We therefore assembled the optical setup described above in a discrete lidded enclosure, the internal environment of which is constantly purged with $N_{2(g)}$. The enclosure also keeps the samples from being exposed to the ambient light and thus mitigates spurious photolysis. The fact that the "MF" and "shield" samples are flowed in parallel and data collection from each alternated using a remotely controlled shutter allows for all measurements to be performed without the need to access the chamber and disturb the ambient conditions.

The number, distribution and averaging of time delays in the TA experiment will depend on the signal quality, kinetics and photostability of the sample and to some degree must be determined empirically. Some general principles apply, however: (1) some data should be collected pre $t=0$ to identify the rise of the earliest time signal; (2) sufficient data should be collected during signal rise and decay to enable reliable kinetic fitting and following decay to ensure the entire signal is captured; (3) if the decay kinetics contain multiple phases on different orders of temporal magnitude, it is worth considering arranging the time delays on a logarithmic as opposed to a linear timebase to ensure sufficient data at early times; (4) to mitigate the influence of photobleaching or other sample degradation artefacts in the kinetic analysis, data at different delay times should be collected in a randomized order as opposed to sequentially with increasing time.

4. Step-by-step protocol: Measuring MFEs on the ultrafast photochemical dynamics of AdoCbl

4.1 Day 1

4.1.1 Sample preparation

(a) Prepare 20 mM HEPES buffer/50% glycerol solutions using deionized water and adjust to pH = 7.5 with $NaOH_{(aq)}$. Filter the buffers (0.2 μm) and degas them for at least 40 min by bubbling $N_{2(g)}$ then store in the anaerobic glove box at least 24 h prior to sample preparation with the lid of the buffer flask ajar.

4.2 Day 2

(b) In the anaerobic glovebox and under filtered ambient light (wavelengths >600 nm), dissolve the desired amount of AdoCbl in an appropriate volume of buffer solution and confirm the concentration spectrophotometrically ($\varepsilon_{525nm} = 8 \times 10^3\ M^{-1}\ cm^{-1}$).

(c) Dilute the sample to achieve an optical density of ~0.5 at the pump wavelength in the 2 mm path length quartz capillary cuvette. Use a black, lidded tube as the sample reservoir to protect the photolabile AdoCbl from ambient light. Produce at least 2 aliquots of 1.5 mL of sample.

(d) When mounting the sample within the optical arrangement, pierce two small holes in the lid of the sample reservoir such that the outlet and inlet flow tubing is held securely in place and seal with parafilm.

4.2.1 Beam alignment, signal optimization and setting of MF

Before beginning, the laser should be switched on and the OPAs set to produce 375 nm (pump beam) and 532 nm (probe beam). The lengths of the two beam paths must be identical at the two sample positions, which should be set when the experiment is first constructed.

(a) Align the probe beam on the delay stage so that the alignment of the onward beam path is invariant for all positions of the retroreflector (the moveable mirror, the position of which sets the various delay times).

(b) Remove the capillary in the "MF" path. Adjust the MF by changing the distance between the barrels and confirm its magnitude at the position of beam overlap using a Hall probe and Gaussmeter.

(c) Insert the empty quartz capillaries in the "MF" and "shield" lines but do not yet flow the sample. Ensure the reference, "MF" and "shield" probe beams reach their respective detectors and that their intensities are matched using a neutral density filter.

(d) Direct the pump beams towards the "MF" and "shield" paths, spatially overlap with the probe beams at the sample positions and attenuate the pump power to 40 μW at each sample position.

(e) Start flowing one aliquot of the sample through the capillaries, this will be a sacrificial alignment sample; once bubble-free flow is achieved, unblock the beams and refine the spatial overlap of pump and probe beam in both paths to optimize the signal magnitude (with the delay stage at a position you are certain is where probe beam arrives after pump beam).

(f) To determine $t = 0$, gradually adjust the position of the retroreflector from the beginning of the delay stage until a large signal change is observed. Make note of this value for use when determining delay times during data acquisition.

(g) Block the probe and pump beams, flow the now depleted alignment sample from the flow system. Place a fresh sample reservoir in position and purge the enclosure with $N_{2(g)}$ for around half an hour. Start flowing the sample and begin data acquisition.

(h) It is best practice to check the signal magnitude every hour or two during the experimental day to check for misalignment and sample degradation. Because the sample is continuously cycled, a UV–visible spectrum of the working sample should be acquired periodically and replaced with fresh sample if degradation is evident (typically every 1–2 h).

4.2.2 Data acquisition

(a) Set the desired delay times in the data acquisition software following the principles described above.

(b) Unblock the pump and probe beams and start data acquisition for a given MF exposure condition.

(c) Repeat data collection for the same MF magnitude at least three times (preferably on the same day) to allow calculation of experimental error. When necessary, adjust the MF magnitude (2b) and exchange the sample (2 g–h).

4.3 Subsequent days

(a) Repeat steps 2 and 3 for each MF magnitude until a full MF-dependence data set has been acquired. In a similar way to how the delay times are set, data at different MF magnitudes should be acquired in a randomized order as opposed to sequentially for increasing MF. This will mitigate systematic errors in the analysis of the MF-dependence related to changes in instrument performance with time.

5. Data analysis

5.1 Kinetic analysis

In general, the magnitude of a MFE on a particular parameter is determined as the difference between the parameter from data acquired under a given MF exposure condition and the same parameter from a corresponding dataset where no MF is applied (e.g., Jones, 2016). For a given MF exposure condition, the data acquired using the protocol described above result in a pair of kinetic traces (i.e., one from the "MF" path and one the "shield" path, Fig. 2A) both of which correspond to the generation and spin-selective recombination of cob(*II*)alamin/Ado$^{\bullet}$ RPs following the photolysis of AdoCbl. Each kinetic trace can be fit to the sum of three exponentials using the first order rate equation, yielding species lifetimes that correspond to the decays of an excited state (τ_1) and base-off intermediate (τ_2) and the "close" RP (i.e., competition between spin-selective, geminate recombination and cage escape) (τ_3) (c.f., Fig. 1) (Hughes et al., 2019).

The fractional MFE on each lifetime at a given MF can be calculated by dividing the parameter from the "MF" data (τ_{on}) by the parameter from the "shield" data (τ_{off}). A result, within error, of greater than 1 means a magnetically-induced increase in that parameter while a result of less than 1 means a magnetically-induced reduction. The fractional MFE (i.e., τ_{on}/τ_{off}) can then be plot as a function of increasing MF (Figs. 3A). The lifetimes of species that precede RP formation (τ_1 and τ_2) show no MF-dependence, as expected. By contrast, the lifetime that corresponds to the RP reaction dynamics (τ_3) yields a clear, non-linear response to increasing MF, as predicted by the RPM (Fig. 3B) (Woodward, 2002). The nature of this non-linearity contains mechanistic detail about the RP reaction and spin dynamics and its shape can be simulated to elucidate this detail. An overview of such simulations is given in the following section.

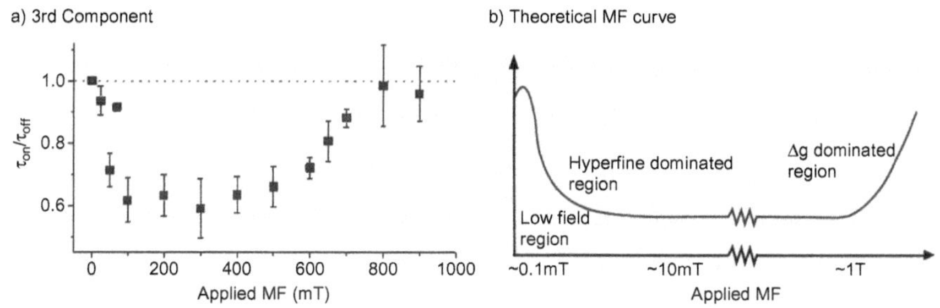

Fig. 3 (A) Ratio of lifetimes of the S-born, geminate cob(*II*)alamin/Ado$^{\bullet}$ RP in the presence (τ_{on}) and absence (τ_{off}) of an applied MF as a function of increasing MF magnitude. (B) Theoretically predicted effect of increasing MF on a component lifetime for a S-born RP. *Panel (B) is adapted from Woodward, J. R. (2002). Radical pairs in solution.* Progress in Reaction Kinetics and Mechanism, *27, 165–207.*

5.2 Analysis of the magnetic field response

At the time of writing, it remains the case that the analysis MFEs on RP reactions remains in the domain of specialist spin chemists and spin physicists. While it is possible and not uncommon for a non-specialist to use open-source or commercial software to simulate many kinds of experimental spectra—for example, vibrational spectra using Density Functional Theory (e.g., Gaussian (Frisch et al., 2016) or ORCA (Neese, 2012)) or magnetic resonance spectra using packages like Easyspin (Stoll & Schweiger, 2006)—it is more challenging to simulate the MF-dependence of RP reactions (so-called MARY (Magnetically Affected Reaction Yield) spectra). The introduction of the open-source SPINACH package (Hogben, Krzystyniak, Charnock, Hore, & Kuprov, 2011) for spin dynamic simulations improved the situation somewhat, removing the need to write custom code. There currently exist no detailed tutorials or explanations for using the codebase for RP spin dynamic simulations, however, despite the inclusion of example code in the SPINACH distribution. Therefore, collaboration with someone experienced in spin dynamic simulations of RPs to undertake the analysis is highly recommended. Regardless, we provide some details on how our simulations were performed in Hughes et al. (2019) using the SPINACH codebase, which will allow mechanical reproduction and may serve as a useful starting point to explore this kind of simulation.

Spin dynamic simulations in SPINACH are performed in Liouville Space (Bain, 2006) but can, in principle, be performed simply with a knowledge of the necessary magnetic and kinetic parameters, without a need to understand, in detail, the underlying quantum mechanics. Simulation of

MARY curves requires calculation of the time-evolution of the RP density matrix under the operation of a quantum mechanical superoperator called the Liouvillian according to the Liouville von Neumann equation. The time-evolution allows the calculation of product yields and the procedure is then repeated for each MF value, allowing the MARY curve to be determined. For basic RP spin dynamic simulations, this superoperator is composed of three terms. The first is a Hamiltonian superoperator responsible for coherent S-T state mixing, which includes the magnetic parameters of each component radical of the pair. Second, is a chemical reaction operator, which describes the spin-selective chemical reactions through the S and T reaction channels. The third is a relaxation operator that describes incoherent spin relaxation of the RPs. This third operator is optional, while the first two operators are a minimum requirement to determine the MARY spectrum using this approach.

Input parameters for the Hamiltonian operator are the molecular *g*-values and hyperfine couplings of each of the radicals in the RP. For the RP generated by photolysis of AdoCbl, the values for the Ado$^{\bullet}$ radical are well-established (Bussandri, Kiarie, & Van Willigen, 2002) and were fixed in the simulation. For the cob(*II*)alamin radical, a literature value for the hyperfine coupling constant of ^{59}Co is known (Baumgarten, Lubitz, & Winscom, 1987) and was fixed, while the molecular *g*-value was kept as a free parameter for fitting the simulation to the experimental data. Indeed, this method can be regarded as a way of measuring the *g*-value of the cob(*II*)alamin radical at ambient temperature (Hughes et al., 2019). For the reaction model, the very simplest model (the "exponential model" (Brocklehurst, 1976)) was employed. It is simple in SPINACH to introduce more complicated reaction models; primarily the "Haberkorn" model (Haberkorn, 1976), where different reaction rates for the S and T reaction can be specified. The exponential model ultimately gave a sufficiently good match with the experimental data and allowed a reduction of the number of free parameters. A single exponential RP recombination rate constant was used, consistent with published kinetic analysis (Walker et al., 1998), and varied for best fit with the experimental data.

Proper treatment and analysis of incoherent electron spin relaxation is a research field in its own right (e.g., Fay, Lindoy, & Manolopoulos, 2021) and so again a simple approach aimed at reducing the number of free parameters was employed. The use of a Lindblad operator (Havel, 2003) allows generalized effects of spin-lattice and spin-spin relaxation to be included, based on single rate coefficients—i.e., there is no underlying physical mechanism determined in this approach. As our primary interest here is in the determination of the *g*-value, this approach is the most suitable as it again minimizes free parameters while providing relatively realistic changes to the density matrix. Thus, our simulations were undertaken with this approach and a total of 4 undetermined parameters—the cob(*II*)alamin *g*-value, the RP recombination rate coefficient and rate coefficients for spin-lattice and spin-spin relaxation. These values were constrained to remain physically reasonable and as each of them has quite different effects on the shape of the MF-dependence, a unique set of parameters could be established that gave the best fit to the experimental data (Fig. 3C). Ultimately, a *g*-value at ambient temperature of 2.105 was determined for the cob(*II*)alamin radical (Hughes et al., 2019).

6. Advantages and limitations

The main advantage of the method we describe here is that it enables observation of early time B_{12} RP reaction and spin dynamics that are otherwise inaccessible at ambient temperatures. Reactions involving the fleeting Ado$^{\bullet}$ radical can therefore be captured, as can the ultrafast effects of the Co(II) ion on the spin-state of the RP. The optical setup described also enables such MFE data to be collected efficiently with minimal disturbance to the experimental conditions. Furthermore, the flow cell system used allows the sample volume to be kept to a minimum. These features are particularly pertinent when studying biological systems like B_{12} cofactors, which are sensitive to various ambient factors (light, oxygen, heat, vibrations etc.) and whose dependent proteins can be expensive (in terms of money and time) to produce in significant quantities.

The straightforward alternation between collecting data from the "MF" and "shield" paths helps mitigate against variations in both the laser performance and sample condition with time. Despite being a simpler approach to detection than using a WLC, probing absorption changes at a single wavelength enables both highly sensitive lock-in detection and the use of higher powers at the probe wavelength by comparison. Detection using a WLC is also known to be more susceptible to laser instabilities.

The use of permanent magnets has the advantage that it removes the confounding effects of heating and vibration that are a significant problem when using an electromagnet to produce larger MF. This is particularly important when studying biological systems. Varying and calibrating the MF using the instrumentation described here is slightly more complex than for a typical electromagnet, which can be achieved by computer control. This is only a minor inconvenience, however, and certainly worth tolerating because of the lack of confounding variables. Moreover, the issue of the smaller volume of homogeneous field generated by permanent magnets is negated by the small volume being probed in such TA experiments.

7. Summary

We describe a method that combines pump/probe TA measurement and exposure of samples to MF of <1 T to investigate magnetically-induced changes to ultrafast (<10 ns) kinetic parameters following the photolysis of AdoCbl. Emphasis has been placed on protocols that enable data to be collected both in the presence and absence of an applied MF with as little disturbance to the experimental setup as possible. This not only results in time-efficient data collection and processing, but also MF exposure conditions where confounding variables—light, vibrations, heat, oxygen, laser fluctuations—are kept to a minimum. In doing so, we can increase confidence that any changes are a result of the MF as opposed to anything else. This method, in combination with that described in chapter "Investigating radical pair reaction dynamics of B12 coenzymes 2: Time-resolved electron paramagnetic resonance spectroscopy" by Lukinović et al., can be exploited to investigate the sub-μs reaction and spin dynamics that are intrinsic to the RPs of B_{12} coenzymes. It should be possible for the interaction between these rapid processes and the RP environment—protein or otherwise—to be probed with relatively minor adaptations to the protocol.

Acknowledgments

We would like to thank all of those who have been involved in related work over the years, including Nigel Scrutton, Darren Graham, Sam Hay, Derren Heyes, Henry Russell (all The University of Manchester), Greg Greetham and Mike Towrie (Central Laser Facility). We apologize if we have missed anyone out. A.R.J. thanks Matt Hall and Max Ryadnov (both NPL) for valuable feedback on the manuscript. The instrumentation described was developed at the Ultrafast Biophysics Facility, Manchester Institute of Biotechnology, funded by BBSRC Alert14 Award No. BB/M011658/1. J.A.H. and V.L. acknowledge Early Stage Researcher funding from the European Union's Seventh Framework Programme FP7-PEOPLE-2013-ITN through the "MAGnetic Innovation in Catalysis" (MAGIC) Initial Training Network (Grant Agreement No. 606831). A.R.J. acknowledges the National Measurement System of the Department for Business, Energy and Industrial Strategy for funding.

References

Bain, A. D. (2006). Operator formalisms: An overview. *Concepts in Magnetic Resonance: Part A, 28A*(6), 369–383. https://doi.org/10.1002/cmr.a.20067.

Bandarian, V., & Reed, G. H. (2000). Isotope effects in the transient phases of the reaction catalysed by ethanolamine ammonia lyase: Determination of the number of exchangeable hydrogens in the enzyme-cofactor complex. *Biochemistry, 39*, 12069–12075.

Baumgarten, M., Lubitz, W., & Winscom, C. J. (1987). EPR and ENDOR studies of cobaloxime(II). *Chemical Physics Letters, 133*(2), 102–108.

Bender, G., Poyner, R. R., & Reed, G. H. (2008). Identification of the substrate radical intermediate derived from ethanolamine during catalysis by ethanolamine ammonia-lyase. *Biochemistry, 47*(43), 11360–11366.

Brocklehurst, B. (1976). Spin correlation in the geminate recombination of radical ions in hydrocarbons. *Journal of the Chemical Society, Faraday Transactions 2, 72*, 1869.

Bucher, D., Sandala, G. M., Durbeej, B., Radom, L., & Smith, D. M. (2012). The elusive 5′-deoxyadenosyl radical in coenzyme-B_{12}-mediated reactions. *Journal of the American Chemical Society, 134*(3), 1591–1599.

Bussandri, A. P., Kiarie, C. W., & Van Willigen, H. (2002). Photoinduced bond homolysis of B_{12} coenzymes, an FT-EPR study. *Research on Chemical Intermediates, 28*(7-9), 697–710.

Chagovetz, A. M., & Grissom, C. B. (1993). Magnetic field effects in adenosylcob(III)alamin photolysis: Relevance to B_{12} enzymes. *Journal of the American Chemical Society, 115*, 12152–12157.

Chen, E., & Chance, M. R. (1990). Nanosecond transient absorption spectroscopy of coenzyme B_{12}. Quantum yields and spectral dynamics. *The Journal of Biological Chemistry, 265*(22), 12987–12994.

Fay, T. P., Lindoy, L. P., & Manolopoulos, D. E. (2021). Spin relaxation in radical pairs from the stochastic Schrödinger equation. *The Journal of Chemical Physics, 154*(8), 084121. https://doi.org/10.1063/5.0040519.

Frisch, M. J., Trucks, G. W., Schlegel, H. B., Scuseria, G. E., Robb, M. A., Cheeseman, J. R., et al. (2016). *Gaussian 16 Rev. C.01*. Wallingford, CT: Gaussian, Inc.

Golding, B. T., & Buckel, W. (2012). Radical enzymes. In C. Chatgilialoglu, & A. Studer (Eds.), *Chemical biology: Vol 3. Encyclopedia of radicals in chemistry, biology and materials* (pp. 1501–1546). Chichester, UK: John Wiley and Sons, Inc.

Haberkorn, R. (1976). Density matrix description of spin-selective radical pair reactions. *Molecular Physics, 32*(5), 1491–1493. https://doi.org/10.1080/00268977600102851.

Halliwell, B., & Gutteridge, J. M. C. (2007). *Free radicals in biology and medicine* (4th ed.). Oxford, UK: University Press.

Havel, T. F. (2003). Metric matrix embedding in protein structure calculations, NMR spectra analysis, and relaxation theory. *Magnetic Resonance in Chemistry, 41*(S1), S37–S50.

Hogben, H. J., Krzystyniak, M., Charnock, G. T. P., Hore, P. J., & Kuprov, I. (2011). Spinach—A software library for simulation of spin dynamics in large spin systems. *Journal of Magnetic Resonance, 208*(2), 179–194. https://doi.org/10.1016/j.jmr.2010.11.008.

Hughes, J. A., Hardman, S. J. O., Scrutton, N. S., Graham, D. M., Woodward, J. R., & Jones, A. R. (2019). Observation of the Δg mechanism resulting from the ultrafast spin dynamics that follow the photolysis of coenzyme B_{12}. *The Journal of Chemical Physics, 151*(20), 201102. https://doi.org/10.1063/1.5127258.

Jones, A. R. (2016). Magnetic field effects in proteins. *Molecular Physics, 114*(11), 1691–1702. https://doi.org/10.1080/00268976.2016.1149631.

Jones, A. R. (2017). The photochemistry and photobiology of vitamin B_{12}. *Photochemical & Photobiological Sciences, 16*, 820–834.

Jones, A. R., Hardman, S. J. O., Hay, S., & Scrutton, N. S. (2011). Is there a dynamic protein contribution to the substrate trigger in coenzyme B_{12}-dependent ethanolamine ammonia lyase? *Angewandte Chemie, International Edition, 50*(46), 10843–10846.

Jones, A. R., Hay, S., Woodward, J. R., & Scrutton, N. S. (2007). Magnetic field effect studies indicate reduced geminate recombination of the radical pair in substrate-bound adenosylcobalamin-dependent ethanolamine ammonia lyase. *Journal of the American Chemical Society*, *129*(50), 15718–15727.

Jones, A. R., Levy, C., Hay, S., & Scrutton, N. S. (2013). Relating localized protein motions to the reaction coordinate in coenzyme B_{12}-dependent enzymes. *The FEBS Journal*, *280*(13), 2997–3008.

Jones, A. R., Woodward, J. R., & Scrutton, N. S. (2009). Continuous wave photolysis magnetic field effect investigations with free and protein-bound alkylcobalamins. *Journal of the American Chemical Society*, *131*, 17246–17253.

Jost, M., Fernández-Zapata, J., Polanco, M. C., Ortiz-Guerrero, J. M., Chen, P. Y.-T., Kang, G., et al. (2015). Structural basis for gene regulation by a B_{12}-dependent photoreceptor. *Nature*, *526*, 536–541.

Kim, Y., Bertagna, F., D'Souza, E. M., Heyes, D. J., Johannissen, L. O., Nery, E. T., et al. (2021). Quantum biology: An update and perspective. *Quantum Reports*, *3*(1), 80–126.

Kutta, R. J., Hardman, S. J. O., Johannissen, L. O., Bellina, B., Messiha, H. L., Ortiz-Guerrero, J. M., et al. (2015). The photochemical mechanism of a B_{12}-dependent photoreceptor protein. *Nature Communications*, *6*, 7907. https://doi.org/10.1038/ncomms8907.

Lodowski, P., Jaworska, M., Garabato, B. D., & Kozlowski, P. M. (2015). Mechanism of Co–C bond photolysis in methylcobalamin: Influence of axial base. *The Journal of Physical Chemistry. A*, *119*(17), 3913–3928. https://doi.org/10.1021/jp5120674.

Lukinović, V., Woodward, J. R., Marrafa, T. C., Shanmugam, M., Heyes, D. J., Hardman, S. J. O., et al. (2019). Photochemical spin dynamics of the vitamin B_{12} derivative, methylcobalamin. *The Journal of Physical Chemistry. B*, *123*(22), 4663–4672. https://doi.org/10.1021/acs.jpcb.9b01969.

Marsh, E. N. G., & Ballou, D. P. (1998). Coupling of cobalt-carbon homolysis and hydrogen atom abstraction in adenosylcobalamine-dependent glutamate mutase. *Biochemistry*, *37*, 11864–11872.

Marsh, E. N. G., Patterson, D. P., & Li, L. (2010). Adenosyl radical: Reagent and catalyst in enzyme reactions. *Chembiochem*, *11*(5), 604–621.

Neese, F. (2012). The ORCA program system. *WIREs Computational Molecular Science*, *2*(1), 73–78. https://doi.org/10.1002/wcms.81.

Ortiz-Guerrero, J. M., Polanco, M. C., Murillo, F. J., Padmanabhan, S., & Elías-Arnanz, M. (2011). Light-dependent gene regulation by a coenzyme B_{12}-based photoreceptor. *Proceedings of the National Academy of Sciences of the United States of America*, *108*(18), 7565–7570.

Padmakuma, R., Padmakuma, R., & Banerjee, R. (1997). Evidence that cobalt-caron bond homolysis is coupled to hydrogen atom abstraction from substrate in methylmalonyl-CoA mutase. *Biochemistry*, *36*, 3713–3718.

Padmanabhan, S., Jost, M., Drennan, C. L., & Elías-Arnanz, M. (2017). A new facet of vitamin B_{12}: Gene regulation by cobalamin-based photoreceptors. *Annual Review of Biochemistry*, *86*(1), 485–514. https://doi.org/10.1146/annurev-biochem-061516-044500.

Peng, J., Tang, K.-C., McLoughlin, K., Yang, Y., Forgach, D., & Sension, R. J. (2010). Ultrafast excited-state dynamics and photolysis in base-off B_{12} coenzymes and analogues: Absence of the trans-nitrogenous ligand opens a channel for rapid nonradiative decay. *The Journal of Physical Chemistry. B*, *114*(38), 12398–12405. https://doi.org/10.1021/jp104641u.

Schwartz, P. A., & Frey, P. A. (2007). 5'-Peroxyadenosine and 5'-peroxyadenosylcobalamin as intermediates in the aerobic photolysis of adenosylcobalamin. *Biochemistry*, *46*(24), 7284.

Steiner, U. E., & Ulrich, T. (1989). Magnetic field effects in chemical kinetics and related phenomena. *Chemical Reviews*, *89*, 51–147.

Stoll, S., & Schweiger, A. (2006). EasySpin, a comprehensive software package for spectral simulation and analysis in EPR. *The Journal of Magnetic Resonance*, *178*(1), 42–55. https://doi.org/10.1016/j.jmr.2005.08.013.

Walker, L. A., Shiang, J. J., Anderson, N. A., Pullen, S. H., & Sension, R. J. (1998). Time-resolved spectroscopic studies of B_{12} coenzymes: The photolysis and geminate recombination of adenosylcobalamin. *Journal of the American Chemical Society*, *120*(29), 7286–7292.

Woodward, J. R. (2002). Radical pairs in solution. *Progress in Reaction Kinetics and Mechanism*, *27*, 165–207.

Woodward, J. R., Foster, T. J., Jones, A. R., Salaoru, A. T., & Scrutton, N. S. (2009). Time-resolved studies of radical pairs. *Biochemical Society Transactions*, *37*(2), 358–362.

Zhu, C., & Warncke, K. (2010). Kinetic isolation and characterization of the radical rearrangement step in coenzyme B12-dependent ethanolamine ammonia-lyase. *Journal of the American Chemical Society*, *132*(28), 9610–9615. https://doi.org/10.1021/ja907769g.

CHAPTER TWELVE

Investigating radical pair reaction dynamics of B_{12} coenzymes 2: Time-resolved electron paramagnetic resonance spectroscopy

Valentina Lukinović[a], Joanna A. Hughes[b], Jonathan R. Woodward[c], and Alex R. Jones[d,*]

[a]Institute of Biochemistry, Carleton University, Ottawa, Ontario, Canada
[b]Laboratory of Ultrafast Spectroscopy, ISIC, and Lausanne Centre for Ultrafast Science (LACUS), École Polytechnique Fédérale de Lausanne, Lausanne, Switzerland
[c]Graduate School of Arts and Sciences, The University of Tokyo, Tokyo, Japan
[d]Biometrology, Department of Chemical and Biological Sciences, National Physical Laboratory, Middlesex, United Kingdom
*Corresponding author: e-mail address: alex.jones@npl.co.uk

Contents

Methods in Enzymology, Volume 669
ISSN 0076-6879
https://doi.org/10.1016/bs.mie.2021.12.020

Abstract

The chemistry of B_{12} coenzymes is highly sensitive to the nature of their upper axial ligand and can be further tuned by their environment. Methylcobalamin, for example, generates RPs photochemically but undergoes non-radical biochemistry when bound to its dependent enzymes. Owing to the transient nature of the reaction intermediates, it remains a challenge to investigate how their environment controls reactivity. Here, we describe how to use time-resolved electron paramagnetic spectroscopy to directly monitor the generation and evolution of transient radicals that result from the photolysis of a B_{12} coenzyme. This method produces evolving, spin-polarized spectra that are rich in mechanistic detail.

1. Introduction

As detailed in chapter "Investigating radical pair reaction dynamics of B_{12} coenzymes 1: Transient absorption spectroscopy and magnetic field effects" by Hughes et al., reactions catalyzed by B_{12}-dependent enzymes serve as an excellent example of the vital role transient radicals can play in biological function. Electron paramagnetic resonance (EPR) spectroscopy is one of the most powerful techniques at our disposal to investigate such radicals because it directly probes the unpaired electron spins of paramagnetic species and can yield rich detail about the environment and reactivity of those spins (Hagen, 2009; Roessler & Salvadori, 2018). Indeed, EPR has provided valuable mechanistic insight about a variety of 5′-deoxyadenosylcobalamin (AdoCbl)-dependent enzymes, including the identification of various radical intermediates (Bandarian & Reed, 2002; Bender, Poyner, & Reed, 2008; Licht, Gerfen, & Stubbe, 1996; Padmakumar & Banerjee, 1995; Warncke, 2005; Zhu & Warncke, 2010). Owing to the transient nature of many radicals, however, various experimental strategies must be employed to enable their observation by EPR. Often radicals are cryo-trapped by arresting reactions mid-cycle through rapid freezing [e.g., (Bender et al., 2008)] followed by data acquisition at cryogenic temperatures. Spin-traps are also employed, where transient radicals are captured by forming covalent adducts with species with more stable and thus longer-lived paramagnetic centers, which are then detected by EPR in what is essentially a proxy measurement (Davies, 2016). Although strategies such as these elucidate valuable chemical (e.g., identifying radical intermediates) and structural (e.g., inter-radical distances) information, dynamic detail is often lost and, in the case of spin-traps, inferences must be made indirectly.

Direct detection of transient radicals by EPR under ambient conditions is possible, however, and has been realized for B_{12} coenzymes through their photochemistry (Bussandri, Kiarie, & van Willigen, 2002; Lukinović et al., 2019; Sakaguchi, Hayashi, & I'Haya, 1990). As discussed in chapter "Investigating radical pair reaction dynamics of B_{12} coenzymes 1: Transient absorption spectroscopy and magnetic field effects" by Hughes et al., two of the active derivatives of B_{12}, AdoCbl and methylcobalamin (MeCbl), form cob(*II*)alamin/alkyl radical pairs (RPs) following photolysis of the upper axial ligand (Jones, 2017) (for MeCbl, see Fig. 1). Such RPs are analogous to those generated upon substrate binding to AdoCbl-dependent enzymes (Reed, 2004). MeCbl-dependent enzymes, on the other hand, do not proceed via radical intermediates. Interestingly, the photophysical events that precede photolysis of the Co-Me bond (Walker et al., 1998) are also different from those for photolysis of the Co-Ado bond (Walker, Shiang, Anderson, Pullen, & Sension, 1998). These contrasting responses to light absorption imply differences in electronic structure that are likely to be tuned by a protein environment to favor different chemistry.

Flash photolysis, time-resolved EPR (TREPR) is a powerful technique that can be used to follow the formation, dynamics, and lifetime of

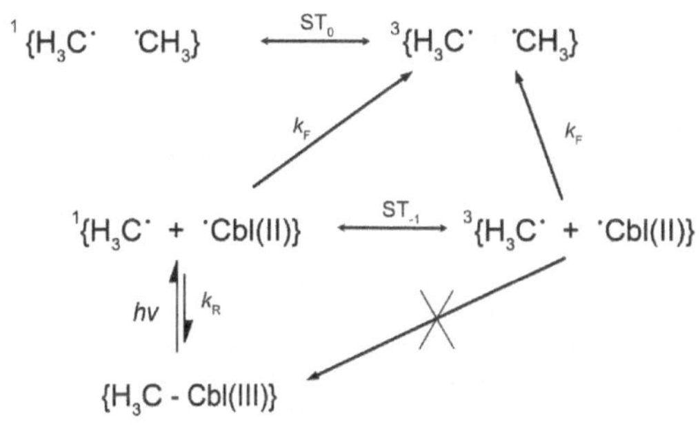

Fig. 1 Schematic presentation of the photochemical reaction and spin dynamics of the cob(*II*)alamin/Me RP $\{H_3C^\bullet + {}^\bullet Cbl(II)\}$ following the photolysis of MeCbl. $h\nu$—absorption of a photon resulting in homolysis; k_R—rate of RP recombination; $^1\{\}$ and $^3\{\}$—singlet and triplet spin states of the RP, respectively; ST_{-1} and ST_0 describe the origin of spin polarization; k_F—rate of methyl radical diffusion to form methyl – methyl *f*-pairs. These *f*-pairs are by definition triplet-born (T-born), as methyl radicals that encounter in the S spin-state will react to form ethane. Various reactions are possible for the methyl radical; however, these are not illustrated here for the sake of simplicity. *Adapted from Lukinović, V., Woodward, J. R., Marrafa, T. C., Shanmugam, M., Heyes, D. J., Hardman, S. J. O., Scrutton, N. S., Hay, S., Fielding, A. J., & Jones, A. R. (2019). Photochemical spin dynamics of the vitamin B 12 derivative, Methylcobalamin.* The Journal of Physical Chemistry B, *123(22), 4663–4672. https://doi.org/10.1021/acs.jpcb.9b01969.*

short-lived, photochemically-generated radical species with nanosecond resolution (Bittl & Weber, 2005). As with transient absorption spectroscopy of a similar time resolution, the laser has two main roles: to send a light pulse to activate the photochemistry and to send a corresponding electrical (trigger) pulse to the spectrometer to define $t=0$ for data acquisition. Not only does TREPR enable the direct detection of transient radicals and their environment, but it can also provide rich and varied mechanistic detail. This includes how the radical(s) are formed, the preceding photophysics, radical diffusion, transfer of spins down electron transfer chains and radical reactions. This is particularly true when a pair of spin-correlated radicals are formed simultaneously (Woodward, 2002). Initially, the unpaired electron spins of such RPs are 100% polarized either in the singlet state (S, antiparallel spins, ↑↓) or in a triplet state ($T_0/T_{-1}/T_{+1}$, parallel spins, ↑↑). Further polarization can develop and evolve as a result of the subsequent spin dynamics. This spin polarization represents a key benefit over conventional EPR, which relies on absorption of photons by populations of spins at equilibrium (i.e., a Boltzmann distribution). Spin polarized signals are usually enhanced in amplitude by comparison, often by several orders of magnitude, substantially improving the sensitivity of the technique. In addition, the polarization can have both net (all hyperfine lines) and multiplet (i.e., per hyperfine line) contributions, which may be in absorption or emission, leading to complex line intensity patterns that contain significant mechanistic detail, which can be extracted through simulations.

The potential of spin-polarized spectra acquired by TREPR to elucidate a range of mechanistic detail is nicely illustrated by data from the photolysis of MeCbl (Fig. 1) published in Lukinović et al. (2019). Owing to fast relaxation of the cob(*II*)alamin radical at room temperature, TREPR data can only be acquired directly from the methyl radical. The cob(*II*)alamin does make its presence felt, however, because the spin dynamics of the geminate RP contribute to the spin polarized signals observed for the methyl radical. The time-resolved spectra of the methyl radical thus deviate from the 1:3:3:1 pattern expected at equilibrium. Initially (80 ns) the four lines are all in absorption and are distorted toward low field, a feature that becomes more marked as a function of time until the lowest field line disappears almost entirely at 360 ns. Following simulation, these data confirm that the geminate cob(*II*)alamin/methyl RP is predominantly S-born (something previously not clear (Bussandri et al., 2002; Jaworska, Lodowski, Andruniow, & Kozlowski, 2007)) and initially generates polarization through coherent

ST_{-1} spin-state mixing. As this initial polarization relaxes (with a T_1 of 75 ± 19 ns) the methyl radicals rapidly diffuse to become free radicals that can then encounter one another, generating methyl-methyl *f*-pairs, which are by definition T-born (Fig. 1). These *f*-pairs generate further (and different) polarization through coherent ST_0 spin-state mixing, which begins to dominate the signal at later times (hundreds of ns). In this chapter, we will describe in detail the protocol that enables such signals to be captured and that can be adapted to explore the impact of different environments—including that of a protein—on the B_{12} RP dynamics.

This chapter is related to chapter "Investigating radical pair reaction dynamics of B_{12} coenzymes 1: Transient absorption spectroscopy and magnetic field effects" by Hughes et al., which describes a method for using a combination of transient absorption spectroscopy and magnetic field effects to probe the ultrafast (<10 ns) RP and spin dynamics following the photolysis of B_{12} coenzymes (Hughes et al., 2019). Together, they provide a combination of methods for probing environmental effects of B_{12} chemistry over a range of timescales and reactive conditions (Hughes et al., 2019; Lukinović et al., 2019). In practice, several experimental challenges present themselves when attempting to acquire good quality TREPR spectra. Moreover, the polarized signals from different spin dynamic mechanisms (e.g., from the RP precursors, geminate pairs and *f*-pairs) are summed in the overall spectra, which requires non-trivial analysis to deconvolute the different contributions. We therefore provide as much detail as possible to make this powerful approach more accessible to those not expert in EPR spectroscopy.

2. Equipment, materials and reagents

Continuous flow apparatus:

- single piston HPLC pump (Gilson 305)
- quartz tubing (O.D. 3 mm, I.D. 1 mm)

Time-resolved EPR setup with laser excitation:

- EPR spectrometer (Bruker E580X with a Bruker ER4118X-MD5 dielectric resonator)
- delay generator (Stanford DG535)
- oscilloscope (Tektronix MDO4014B-3 Mixed Domain Oscilloscope)
- laser (Brilliant blue laser q-switched Nd:Yag)
- Si detector (Thorlabs DETa)

Sample preparation:

- methylcobalamin (Sigma)
- buffer components (HEPES, Formedium Ltd.; glycerol—Merck)
- sample vessel with Suba-Seal cap
- anaerobic glove box (Belle Technologies)
- dark room with safe light
- argon or nitrogen for degassing the buffers and sample during the experiment
- UV–visible spectrophotometer (Cary 60, Agilent)

3. Method details: Instrumental setup and sample preparation

3.1 Instrument setup: Continuous flow TREPR

Several approaches can be taken to enable the acquisition of EPR data with high spectral and time resolution. Here, we describe the use of Fourier transform (FT)-EPR, where one can measure transient paramagnetic signals in the time domain as a free induction decay (FID). To enhance signal quality, one must continuously flow fresh sample through the resonator to enable signal accumulation and averaging. Commercially available products for continuous flow are not compatible with resonators used for time-resolved EPR studies; we therefore designed and assembled a bespoke flow-cell system where the sample solution is pumped through quartz tubing in the resonator to waste. Important factors that must be considered during the flow cell assembly are pump type, quartz tubing diameter, and the flow rate through the resonator. Continuous, pulseless flow of the sample through the resonator can be achieved using a HPLC pump, which also allows one to minimize the sample volume, an important feature when studying biomolecules.

The dimension of the quartz tubing that can be used is dependent on the dielectric constant of the solvent. Solvents with high dielectric constants can result in attenuation of the microwaves in the resonator, which can be mitigated by using narrower quartz tubing to reduce the volume of the solvent inside the resonator. After testing, we found that quartz tubing with an outer diameter of 3 mm and an internal diameter of 1 mm was optimal for time-resolved experiments in aqueous solutions when using a 9 GHz EPR spectrometer. The internal diameter of the quartz tubing to some degree dictates the flow rate that can be used, alongside the area of the laser beam and repetition rate of the laser (Eqs. 1 and 2). For example, the volume,

V, of sample in a 1 mm internal diameter tube (i.e., of radius, *r*, of 0.5 mm) exposed to a laser flash with beam size 3 mm:

$$V = r^2 * \pi * 3\ \text{mm} = (0.5\ \text{mm})^2 * \pi * 3\ \text{mm} = 2.35\ \text{mm}^3 \tag{1}$$

The minimum flow rate, Q, for this tubing that will ensure a volume of fresh sample is excited with each laser pulse when using a repetition rate of 10 Hz is then:

$$Q = 2.35\ \text{mm}^3 \times 10\ \text{Hz} = 23.56\text{mm}^3/\text{s} = 1.41\ \text{mL}/\min \tag{2}$$

3.2 Trigger sequences and delay after flash timing

To keep the experimental deadtime to a minimum, the radicals must be detected as soon as possible after their creation. The laser flash initiates photolysis of MeCbl, while the microwave pulse initiates magnetization of the resulting radicals and generation of the FID. Each microwave pulse is therefore synchronized with each laser pulse using a delay generator. Briefly, and with reference to Fig. 2, the trigger for the laser flashlamp is held at a constant value. The position of the pulse triggering the Q-switch (blue line) can vary slightly depending on the laser power, but typically it is around 331 μs after

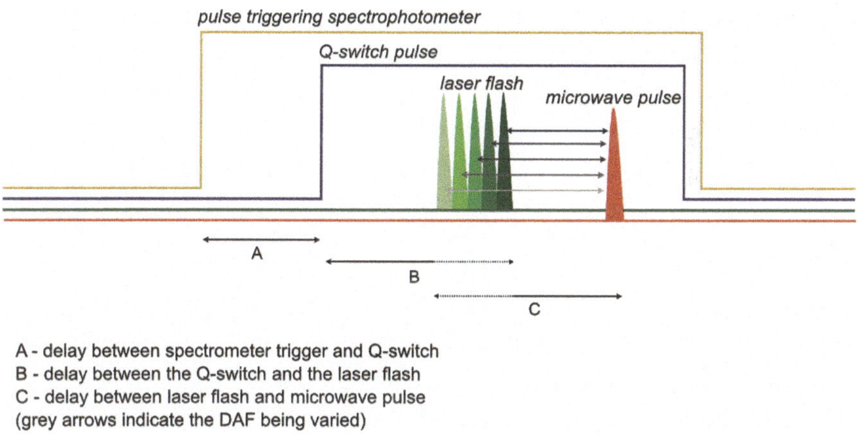

Fig. 2 Positions of the triggers and pulses relevant to defining the DAF between the laser pulse and microwave pulse. The sequences illustrate pulses observed on an oscilloscope: pulse triggering the spectrometer (yellow); Q-switch trigger (blue); laser flash (green); the microwave pulse in the Signal Channel after quadrature detection (red). A delay between red and green pulses—the DAF—can be varied (grey arrows). Pulses sent from the delay generator were followed using an oscilloscope and the laser flash was detected using a Si detector.

the flash lamp for a 20 mJ pulse energy. The precise timing of the resulting laser flash (green pulse) is monitored using a photodiode. The position of the pulse triggering the spectrometer (yellow line) is then varied with respect to the laser flash, which determines the effective "delay after flash" (DAF; i.e., of the microwave pulse in the resonator, red pulse, after the laser pulse). Invariably, the spectrometer must be triggered ahead of the Q-switch to achieve shorter DAFs (in our case less than ~320 ns) because of the time it takes for the electronics of EPR spectrometer to generate the microwave pulse.

For optimization, all the pulse timings, including the first time point where the signal can be detected, are recorded on an oscilloscope. Using this approach, an effective deadtime of as little as 40 ns can be achieved, which corresponds to the delay between the signal on the Transmitter Monitor (the spectrometer channel detecting a pulse before the resonator) and the Signal Channel (signal after detection that contains the FID). All experiments that monitor the FID with different DAFs must account for this instrumental deadtime, which originates from internal delays related to amplification and detection.

Varying the DAF (Fig. 2) enables one to build up a picture of the reaction and spin dynamics of the RP following photolysis with nanosecond time resolution. Data acquisition follows a pump-probe principle where each spectrum at each DAF is acquired from fresh sample and averaged for a defined number of pump pulses before changing to the next DAF; thus, time-course data are gradually accumulated. The pulse positions can be varied from: applying the microwave (probe) pulse before the laser (pump) flash (i.e., red before green, Fig. 2) to acquire the baseline signal; overlapping the two pulses (i.e., $t = 0$); to increasing the delay of the microwave pulse following the laser flash (i.e., red after green, Fig. 2). The optimum pulse sequence for early radical detection should be determined at a single resonant frequency. The intensity of the signal and relaxation of the resonance lines are monitored in the time domain (i.e., FIDs, Fig. 3A), the nature of which varies with DAF. After the initial appearance of the methyl radical, for example, the signal magnitude evolves with increasing DAF, peaking at 80 ns (Fig. 3B). The time domain signal at each DAF is then converted into the frequency domain by Fourier transform, where the peak intensities correspond to those in the time domain and the frequencies to the FID. It is worth noting that each "DAF" also accounts for the 40 ns instrument deadtime discussion above; e.g., a DAF of 80 ns corresponds to the relative delay between the laser flash and microwave pulse detected in the signal channel (Fig. 2) of 120 ns.

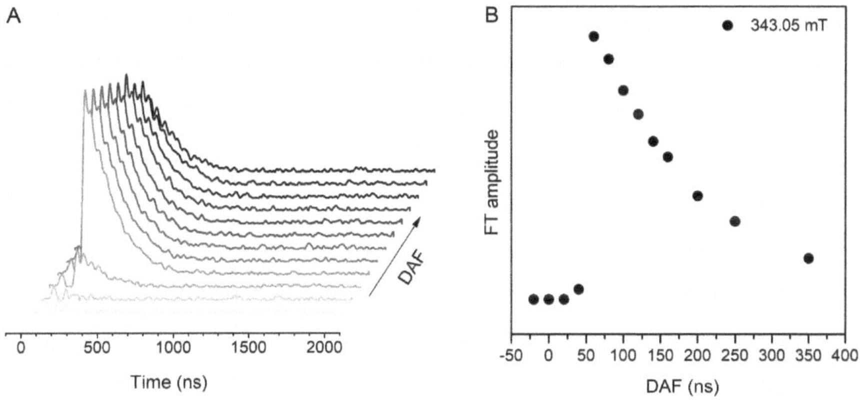

Fig. 3 (A) Time domain signals (i.e., FIDs) of the −31 MHz peak (which corresponds to 343.05 mT) from the methyl radical at varying DAF. (B) The peak FT amplitude as a function of DAF. DAF = 0 when the laser flash and microwave pulse overlap in time. Data were acquired following photoexcitation at 532 nm, 20 mJ pulse energy, of a flowing solution of 0.5 mM MeCbl in HEPES, pH 7.5, 50% glycerol at room temperature.

3.3 Sample preparation and data acquisition conditions

We will now describe the sample conditions that enable the acquisition of high quality TREPR spectral data following the photolysis of MeCbl in aqueous solution at room temperature. MeCbl solutions are prepared under filtered ambient light (wavelengths >600 nm) to avoid spurious photolysis. The buffer used was an anaerobic solution of 20 mM HEPES, pH 7.5, with either 0% or 50% glycerol. Complete removal of oxygen is required because it can quench both the methyl and cob(*II*)alamin radicals, which will interfere with the RP dynamics and attenuate the FT-EPR signal. Comparison between data from solutions ± glycerol enable investigation into the impact of solvent viscosity on the RP dynamics. Following the photolysis of MeCbl, for example, viscosity both impacts the evolution of the polarization pattern with time and results in an enhancement of the resonance line intensities, presumably owing to the RP cage effect (i.e., the higher viscosity increases the lifetime of the geminate RP before diffusive separation) (Lukinović et al., 2019).

During data acquisition, a three-neck bottle served as a sample reservoir: one neck can be used to deliver the sample to the resonator, with the second and third used to continuously purge the sample with argon gas during the experiment (10 min before and then throughout data acquisition). Each neck was sealed with Suba-seals and parafilm to ensure that the sample is kept under anaerobic conditions. The bottle was also wrapped in aluminum foil to limit exposure to ambient light.

4. Step-by-step protocol: Measuring MeCbl RP dynamics using TREPR

Day 1

1. Sample preparation

a) Prepare 20 mM Hepes buffer solutions (with or without 50% glycerol as required) using deionized water and adjust to pH 7.5. Filter the buffers (0.2 μm) and degas them for at least 40 min by bubbling $N_{2(g)}$ with stirring, then store in the anaerobic glove box for at least 24 h prior to sample preparation with the lid of the buffer flask ajar

Day 2

b) Weigh out the required amount of MeCbl (use as purchased without further purification) and transfer to the sample reservoir (three necked flask)

c) Under safe light (wavelengths >600 nm) and in the anaerobic glove box, dissolve the sample in an appropriate volume of the degassed buffer (step a) and measure the concentration using a spectrophotometer ($\varepsilon_{520\ \mathrm{nm}} = 7.7 \times 10^3\ \mathrm{M^{-1}\ cm^{-1}}$)

d) Seal the flask necks using Suba-seals and parafilm to ensure anaerobic conditions are retained outside the glove box. Wrap the vessel with aluminum foil to protect the photolabile MeCbl from ambient light. The sample can now be taken outside the anaerobic glove box to the EPR spectrometer

e) Once the sample reservoir is mounted near the resonator, purge the sample with argon or nitrogen gas by inserting a needle attached to the gas line into one neck of the sample reservoir and a free needle into another as a gas outlet. This will ensure retention of anaerobic conditions during measurement

2. Continuous flow instrument setup

a) Insert the quartz tubing attached to the silicone tubing into the resonator

b) Prime the HPLC pump with buffer before inserting the pump tubing into the sample vessel. The other end (i.e., after passing through the resonator) should lead to an empty vessel for waste

3. TREPR measurements

a) Connect the laser and the EPR spectrometer to a delay generator, which is used to trigger each instrument and to control the DAF timing

b) Connect both the laser and spectrometer to the oscilloscope to monitor the positions of the microwave and laser pulses, help optimize the pulse sequences, and inform the DAF timings

c) Use the second (532 nm) or third (355 nm) harmonic of Q-switched Nd: Yag laser with a repetition rate of 10 Hz for photoexcitation. (The photochemistry of MeCbl is known to be wavelength-dependent (Shiang, Walker, Anderson, Cole, & Sension, 1999)). We used a 20 mJ pulse energy. Ensure that the laser beam is focused on the flow cell within the optical window of the resonator

d) Use the Bruker Xepr software to input the microwave pulse and detection parameters. We used a $\pi/2$ pulse of 16 ns to generate the FID and 8 dB microwave attenuation for detection

e) Start the laser firing and the pump to flow the sample through the resonator. Because of the bandwidth of the resonator at room temperature, it was not possible to collect data for all the resonance lines of the methyl radical in parallel. Data for each one of the four resonance lines was therefore collected at their respective field values. It is best to start with the highest intensity line as it is the easiest one to detect. Once the resonance field for that position is identified, the remaining field positions can be calculated based on the known hyperfine parameters (for the methyl radical ~2.2 mT (Bussandri et al., 2002; van Slageren et al., 2000))

f) Collect the FIDs for each resonance line and each DAF using four-step phase cycling in the Xepr PulseSPEL program, which cancels out offset artefacts. For the methyl radical, data were acquired and averaged for 100 FIDs for each field position and DAF. The FIDs were collected with 1 ns time resolution

g) Collect the off-resonance time domain signal, which is later used in signal processing. This is completed by recording the time domain spectrum at a field position far from the resonance field position of methyl radical (for example, 20 mT before the first or last field position of methyl radical)

h) Fourier transform the collected FIDs for each resonance line and DAF into the frequency domain spectrum using Bruker Xepr software as described below. Complete spectra of the methyl radical can then be assembled using software such as Matlab or OriginPro (code presented below)

Critical: When the exact resonance field for each line is found, the FID will have the form of single exponential decay. If the field position is not in resonance, but close to it, the FID will instead appear as an oscillation. This is important for later processing, as an oscillating FID is likely to result in an erroneous intensity in the frequency domain. Obtaining correct intensities is vital when identifying polarization patterns through simulations and hence to interpreting physically meaningful mechanisms.

5. Assembly and analysis of the TREPR spectra

Once the data are acquired, a full spectrum must be assembled at each DAF and the evolution of the polarization pattern with time can then be analyzed using simulations. Different combinations of line intensities and phases (i.e., absorptive or emissive) result from different polarization mechanisms (and combinations thereof), which in turn inform on the mechanisms that generate the RPs and their reactive dynamics.

5.1 Construction of FT-EPR spectra

As mentioned above, the EPR spectrum of the methyl radical comprises a quartet of lines: the resonance from the unpaired electron split by hyperfine coupling to the three equivalent ^{1}H nuclei (Hagen, 2009). Owing to the limited resonator bandwidth at room temperature, however, each line must be collected separately, and the spectrum assembled post-acquisition. Processing of the signal from the time to the frequency domain in Xepr involves: (1) subtracting the off-resonance signal; (2) baseline correction using a 0th order polynomial (a constant); and (3) phasing the traces to obtain pure absorption or emission (real) and dispersion (imaginary) signals. Afterward, a "left-shift" of the time domain spectrum is completed by shifting the data points until the first point of the FID is at the beginning of the left-hand side of the viewport in Xepr. This step is completed before FT, in order to remove any potential ring-down signal, which corresponds to the time needed for the instrument to dissipate the high energy originating from the microwave pulse (Weber, 2008). Finally, FT of the data is carried out after zero-filling, which is used to increase the number of points, by increasing the length of the time domain spectra. This results in smoother spectra and an improved signal-to-noise ratio (Bowman, Krzyaniak, Cruce, & Weber, 2013).

5.2 Code for plotting the data

Once the FIDs are converted into the frequency domain, complete spectra can be assembled, which are typically plotted as signal amplitudes as a function of magnetic field (or sometimes frequency). Each signal is separated by the magnitude of the hyperfine coupling between the unpaired electron and any magnetic nuclei present. Detailed code written and executed in MATLAB for assembling complete FT-EPR spectra of the methyl radical is copied below and contains functions for: (1) importing the data and converting from frequency to field values; (2) building the X-axis in the field domain and defining junctions for each of the four resonance lines; and (3) assembling each of the peaks to the corresponding junctions/field values. An example for one FT-EPR spectrum of the methyl radical is presented below:

(1)
```
[exp_freq1,exp_intensity1]=eprload('FT2_0p3mM_d400_3414p6.DSC';
exp_field=1000*exp_freq1/2.8025;
field_position1=3414.6;
[exp_freq2,exp_intensity2] = eprload('FT_0p3mM_d400_3437p2.
DSC');
field_position2=3437.2;
[exp_freq3,exp_intensity3] = eprload('FT_0p3mM_d400_3459p8.
DSC');
field_position3=3459.8;
[exp_freq4,exp_intensity4] = eprload('FT_0p3mM_d400_3482p55.
DSC');
field_position4=3482.55;
```

(2)
```
f_step=exp_field(2)-exp_field(1);
L=length(exp_field);
X_axis=3300:f_step:3600;
step_position1=round((field_position1-3300)/f_step);
step_position2=round((field_position2-3300)/f_step);
step_position3=round((field_position3-3300)/f_step);
step_position4=round((field_position4-3300)/f_step);
junction1=round((field_position1-3300)/f_step +(field_position2-
field_position1)/(f_step*2));
junction2=round((field_position3-field_position2)/f_step +junction1);
junction3=round((field_position4-field_position3)/f_step +junction2);
```

(3)
```
Y_axis=zeros(1,length(X_axis));
Y_axis(1:junction1)=exp_intensity1(floor(L/2 +(step_position2-ste-
p_position1)/2)-junction1:floor(L/2 +(step_position2-step_position1)/2)-1);
Y_axis(junction1:junction2)=exp_intensity2(L/2-(junction2-junction1)/2:
L/2 +(junction2-junction1)/2);
Y_axis(junction2:junction3)=exp_intensity3(L/2-(junction3-junction2)/2:
L/2 +(junction3-junction2)/2);
Y_axis(junction3:length(Y_axis))=exp_intensity4(floor(L/2-(step_position4-
step_position3)/2):L/2 +length(Y_axis)-junction3-floor((step_position4-step
_position3)/2)-1);
plot(X_axis,Y_axis)
```

5.3 Simulation of stick spectra

Interpretation of the data by simulation involves computing the stick EPR spectra for different possible electron spin polarization mechanisms and then determining which combinations can potentially reproduce the experimentally observed polarization patterns. For example, the different mechanisms that were considered for the photolysis of MeCbl were (Lukinović et al., 2019): Triplet Mechanism (TM) polarization (absorptive and emissive); radical pair mechanism (RPM) ST_0 polarization for the geminate methyl/cob(*II*)alamin RPs and for the methyl/methyl f-pairs; and RPM ST_{-1} polarization for the methyl/cob(*II*)alamin geminate RPs. Owing to the data acquisition conditions and what is known about the photolysis of MeCbl (Jones, 2017), this set comprises the mechanisms that are realistic possibilities for this reaction.

The first step is to calculate the equilibrium EPR stick spectrum to identify the line frequencies and the equilibrium polarization pattern. While the original simulations performed used custom, home-made simulation software, the same result is readily achieved using the literature data for molecular *g*-values and hyperfine couplings for the methyl radical and performing the simulation using the open source software package EasySpin (Stoll & Schweiger, 2006). The resulting 1:3:3:1 spectrum has the same line intensities as those produced by absorptive TM polarization and can be simply multiplied by −1 to generate the pattern for emissive polarization (not shown). Once one has the resonant line positions, the polarization patterns for other mechanisms can be determined (Fig. 4) using the calculation processes described in the literature. In the case of RPM ST_0 polarization, line intensities are calculated by assigning the two radicals as methyl and cob(*II*)alamin and using the method described in Adrian (1971). This is perhaps most simply achieved using a small Python program. The same code can then be used to compute the intensities for the methyl/methyl *f*-pairs, assuming a T-born RP (Fig. 1). Finally, the ST_{-1} polarization pattern can be computed using the approach described in Buckley and McLauchlan (1987) for the methyl/cob(*II*)alamin geminate RP.

5.4 Simulation and interpretation of the experimentally observed spectra

Initially, the spin-state of RPs is highly polarized, and that polarization evolves during the early time spin dynamics (typically <1 μs). The result for the methyl radical is a spectrum that is distorted away from the equilibrium 1:3:3:1 pattern (Lukinović et al., 2019). To elucidate the origin of this distortion, the stick spectra simulated in the previous section can be combined in an attempt to simulate the experimentally observed patterns. The strategy for assigning polarization mechanisms to the observed data is to initially assume only contributions from geminate RPs. No combination of the stick spectra from TM, ST_0 and ST_{-1} geminate pairs, however, can be used to reproduce

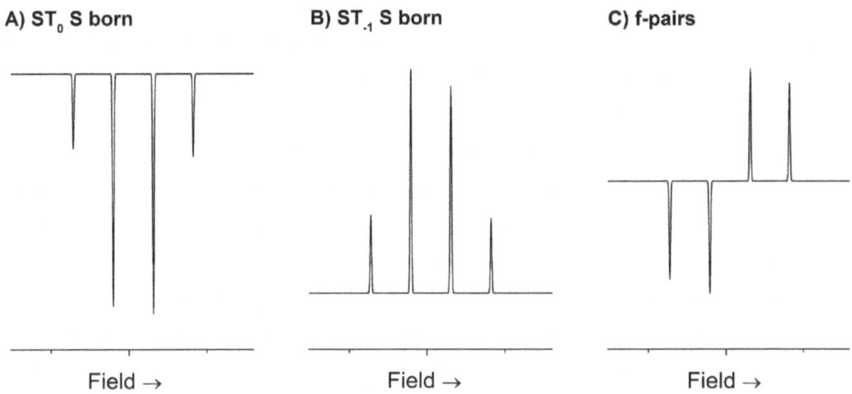

Fig. 4 Theoretical spectra for spin polarization mechanisms of the methyl radical, originating from mixing of S and T_0 (A) or S and T_{-1} (B) states from S-born (S) pairs and T-born *f*-pairs (C). The negative hyperfine value for methyl radical is taken into account.

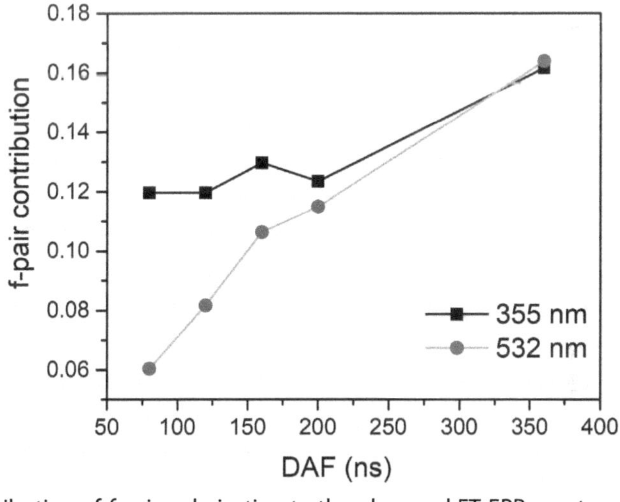

Fig. 5 Contribution of *f*-pair polarization to the observed FT-EPR spectra of the methyl radical as a function of time at different excitation wavelengths. This was completed in combination with S-born ST_{-1} RPM polarization from geminate RPs. Data were acquired for 1 mM MeCbl, in HEPES buffer pH 7.5, 0% glycerol after photoexcitation at 532 or 355 nm, 20 mJ.

the observed spectra. Therefore, the next step is to consider the possible contribution from *f*-pairs, which was found to enable good fits to the experimental data in combination with polarization from ST_{-1} geminate pairs. As a result, in Lukinović et al. (2019) we were able to distinguish between different relative contributions from *f*-pairs as a function of time when MeCbl photolysis is triggered at different excitation wavelengths (Fig. 5).

6. Advantages and limitations

The major advantage of TREPR is that it facilitates early detection of radical species at room temperature. Not only does this enable observation of very short-lived species under ambient conditions, but it also gives access to spin-polarized spectra and hence enhanced signal amplitudes and rich mechanistic detail. The main limitation is the narrow resonator bandwidth at room temperature. As illustrated by MeCbl, this can result in the need to collect data for each resonance line separately, which gives greater opportunity for experimental error and increases the risk of misinterpretation owing to spurious differences between line intensities. This risk can be mitigated, however, by controlling the signal phase by calibration using a standard sample (*e.g.*, benzophenone, see below) and by taking care to find the appropriate resonance field position.

7. Optimization and troubleshooting

An important aspect is to determine the phase of the observed FID signal. The phase of each signal (i.e., absorption (A) or emission (E)) is key to the interpretation of spin-polarized spectra, which can vary from low to high field and between early and later DAFs. To determine that the measured phase of the methyl radical FIDs (and therefore its corresponding FT peak in FT-EPR spectra) are truly in A or E, we compared it to the FID of a standard sample of known polarization patterns. We started by phasing the microwave pulse (followed on an oscilloscope) in an upward position and used that as the internal reference. As a standard, we used the photochemistry of benzophenone in isopropanol, which at concentrations ≤ 10 mM is dominated by a low to high field E/A pattern from the benzophenone ketyl/propan-2-oyl RP (Jones & Woodward, 2006). We observed that when the microwave pulse is phased in an upward position as described above, the observed FIDs showed the lower field lines in A and higher in E, which is opposite in phase from those expected for the benzophenone ketyl/propan-2-oyl RP. Each FID was therefore multiplied by -1 to obtain the FT in the correct phase. Another area where careful optimization is key, is the synchronization of the several pieces of equipment to minimize and precisely define the instrument deadtime and ensure accurate timing of the various pulses. Unless aerobic photochemistry is of interest, like for the B_{12}-dependent photoreceptor, CarH (Jost et al., 2015; Kutta et al., 2015; Ortiz-Guerrero, Polanco, Murillo, Padmanabhan, & Elias-Arnanz, 2011), another area of the experimental setup that must be optimized is the necessity for working under anaerobic conditions.

8. Summary

We have described a method for using TREPR to measure the spin-polarized spectra of transient methyl radicals following the photolysis of MeCbl in solution. Careful consideration has been given to synchronizing the laser flash and microwave pulses to enable detection of the earliest signal possible and to accurately determine the instrument deadtime. The protocol has also detailed how to ensure the correct phase and amplitude of each resonant line and therefore give confidence in the resultant data analysis and interpretation. For the analysis itself, we have described how to build the polarized spectra from the data collected and simulate it using combinations of physically reasonable polarization mechanisms. In combination with the method described in chapter "Investigating radical pair reaction dynamics of B_{12} coenzymes 1: Transient absorption spectroscopy and magnetic field effects" by Hughes et al., this protocol enables one to build a detailed account of the spin and reaction dynamics from the earliest geminate RPs to *f*-pairs following diffusive reencounter. Adapting these methods should provide new insights into the influence of a protein environment on these different dynamics.

Acknowledgments

We would like to thank all of those who have been involved in related work over the years, including Nigel Scrutton, Sam Hay, Muralidharan Shanmugam, Steven Rigby, Derren Heyes, Samantha Hardman, Teresa Marrafa (all The University of Manchester) and Alistair Fielding (Liverpool John Moores University). We apologize if we have missed anyone out. A.R.J. thanks Natalie Belsey and Mike Shaw (both NPL) for valuable feedback on the manuscript. V.L. and J.A.H. acknowledge Early Stage Researcher funding from the European Union's Seventh Framework Program FP7-PEOPLE-2013-ITN through the "MAGnetic Innovation in Catalysis" (MAGIC) Initial Training Network (Grant Agreement No. 606831). A.R.J. acknowledges the National Measurement System of the Department for Business, Energy and Industrial Strategy for funding.

References

Adrian, F. J. (1971). Theory of anomalous electron spin resonance spectra of free radicals in solution. Role of diffusion-controlled separation and reencounter of radical pairs. *The Journal of Chemical Physics*, *54*(9), 3918–3923.

Bandarian, V., & Reed, G. H. (2002). Analysis of the electron paramagnetic resonance Spectrum of a radical intermediate in the coenzyme B12-dependent ethanolamine ammonia-lyase catalyzed reaction of S-2-aminopropanol. *Biochemistry*, *41*(27), 8580–8588.

Bender, G., Poyner, R. R., & Reed, G. H. (2008). Identification of the substrate radical intermediate derived from ethanolamine during catalysis by ethanolamine ammonia-lyase. *Biochemistry*, *47*(43), 11360–11366.

Bittl, R., & Weber, S. (2005). Transient radical pairs studied by time-resolved EPR. *Biochimica et Biophysica Acta (BBA) - Bioenergetics*, *1707*(1), 117–126.

Bowman, M. K., Krzyaniak, M. D., Cruce, A. A., & Weber, R. T. (2013). Skew projection of echo-detected EPR spectra for increased sensitivity and resolution. *Journal of Magnetic Resonance*, *231*, 117–125. https://doi.org/10.1016/j.jmr.2013.03.011.

Buckley, C. D., & McLauchlan, K. A. (1987). The influence of ST − 1 mixing in spectra which exhibit electron spin polarization (CIDEP) from the radical pair mechanism. *Chemical Physics Letters*, *137*(1), 86–90.

Bussandri, A. P., Kiarie, C. W., & van Willigen, H. (2002). Photoinduced bond homolysis of B-12 coenzymes. An FT-EPR study. *Research on Chemical Intermediates*, *28*(7–9), 697–710.

Davies, M. J. (2016). Detection and characterisation of radicals using electron paramagnetic resonance (EPR) spin trapping and related methods. *Methods*, *109*, 21–30.

Hagen, W. R. (2009). *Biomolecular EPR spectroscopy*. CRC Press, Taylor & Francis Group.

Hughes, J. A., Hardman, S. J. O., Scrutton, N. S., Graham, D. M., Woodward, J. R., & Jones, A. R. (2019). Observation of the **Δ** *g* mechanism resulting from the ultrafast spin dynamics that follow the photolysis of coenzyme B $_{12}$. *The Journal of Chemical Physics*, *151*(20), 201102. https://doi.org/10.1063/1.5127258.

Jaworska, M., Lodowski, P., Andruniow, T., & Kozlowski, P. M. (2007). Photolysis of Methylcobalamin: Identification of the relevant excited states involved in co − C bond scission. *The Journal of Physical Chemistry. B*, *111*(10), 2419–2422.

Jones, A. R. (2017). The photochemistry and photobiology of vitamin B12. *Photochemimcal & Photobiological Sciences*, *16*(6), 820–834.

Jones, A. R., & Woodward, J. R. (2006). Alternative source of emissive CIDEP in the TREPR spectra of benzophenone in alcohols. *Molecular Physics*, *104*(10 − 11), 1551–1557.

Jost, M., Fernández-Zapata, J., Polanco, M. C., Ortiz-Guerrero, J. M., Chen, P. Y.-T., Kang, G., et al. (2015). Structural basis for gene regulation by a B12-dependent photoreceptor. *Nature*, *526*(7574), 536–541. https://doi.org/10.1038/nature14950.

Kutta, R. J., Hardman, S. J. O., Johannissen, L. O., Bellina, B., Messiha, H. L., Ortiz-Guerrero, J. M., et al. (2015). The photochemical mechanism of a B12-dependent photoreceptor protein. *Nature Communications*, *6*(1), 7907. https://doi.org/10.1038/ncomms8907.

Licht, S., Gerfen, G. J., & Stubbe, J. (1996). Thiyl radicals in ribonucleotide reductases. *Science*, *271*(5248), 477–481.

Lukinović, V., Woodward, J. R., Marrafa, T. C., Shanmugam, M., Heyes, D. J., Hardman, S. J. O., et al. (2019). Photochemical spin dynamics of the vitamin B $_{12}$ derivative, Methylcobalamin. *The Journal of Physical Chemistry B*, *123*(22), 4663–4672. https://doi.org/10.1021/acs.jpcb.9b01969.

Ortiz-Guerrero, J. M., Polanco, M. C., Murillo, F. J., Padmanabhan, S., & Elias-Arnanz, M. (2011). Light-dependent gene regulation by a coenzyme B12-based photoreceptor. *Proceedings of the National Academy of Sciences*, *108*(18), 7565–7570. https://doi.org/10.1073/pnas.1018972108.

Padmakumar, R., & Banerjee, R. (1995). Evidence from electron paramagnetic resonance spectroscopy of the participation of radical intermediates in the reaction catalyzed by Methylmalonyl-coenzyme a mutase. *Journal of Biological Chemistry*, *270*(16), 9295–9300.

Reed, G. H. (2004). Radical mechanisms in adenosylcobalamin-dependent enzymes. *Current Opinion in Chemical Biology*, *8*(5), 477–483.

Roessler, M. M., & Salvadori, E. (2018). Principles and applications of EPR spectroscopy in the chemical sciences. *Chemical Society Reviews*, *47*(8), 2534–2553. https://doi.org/10.1039/C6CS00565A.

Sakaguchi, Y., Hayashi, H., & I'Haya, Y. J. (1990). Fast formation of methyl radical from methylaquocobaloxime as studied by time-resolved optical and ESR techniques. *The Journal of Physical Chemistry*, *94*(1), 291–293.

Shiang, J. J., Walker, L. A., Anderson, N. A., Cole, A. G., & Sension, R. J. (1999). Time-resolved spectroscopic studies of B $_{12}$ coenzymes: The photolysis of Methylcobalamin is wavelength dependent. *The Journal of Physical Chemistry B*, *103*(47), 10532–10539. https://doi.org/10.1021/jp992358r.

Stoll, S., & Schweiger, A. (2006). EasySpin, a comprehensive software package for spectral simulation and analysis in EPR. *Journal of Magnetic Resonance*, *178*(1), 42–55.

van Slageren, J., Martino, D. M., Kleverlaan, C. J., Bussandri, A. P., van Willigen, H., & Stufkens, D. J. (2000). FT-EPR study of methyl radicals photogenerated from [Ru(me)(SnPh3)(CO)2(iPr-DAB)] and [Pt(me)4(iPr-DAB)]: An example of a strong excitation wavelength dependent CIDEP effect. *The Journal of Physical Chemistry. A*, *104*, 5969–5973.

Walker, L. A., Jarrett, J. T., Anderson, N. A., Pullen, S. H., Matthews, R. G., & Sension, R. J. (1998). Time-resolved spectroscopic studies of B12 coenzymes: The identification of a metastable cob(III)alamin photoproduct in the photolysis of Methylcobalamin. *Journal of the American Chemical Society*, *120*(15), 3597–3603.

Walker, L. A., Shiang, J. J., Anderson, N. A., Pullen, S. H., & Sension, R. J. (1998). Time-resolved spectroscopic studies of B12 coenzymes: The photolysis and geminate recombination of Adenosylcobalamin. *Journal of the American Chemical Society*, *120*(29), 7286–7292.

Warncke, K. (2005). Characterization of the product radical structure in the co(II)-product radical pair state of coenzyme B12-dependent ethanolamine deaminase by using three-pulse 2H ESEEM spectroscopy. *Biochemistry*, *44*(9), 3184–3193.

Weber, R. (2008). *E580 User's manual*. Bruker BioSpin Corporation.

Woodward, J. R. (2002). Radical pairs in solution. *Progress in Reaction Kinetics and Mechanism*, *27*, 165–207.

Zhu, C., & Warncke, K. (2010). Kinetic isolation and characterization of the radical rearrangement step in coenzyme B12-dependent ethanolamine ammonia-lyase. *Journal of the American Chemical Society*, *132*(28), 9610–9615.

CHAPTER THIRTEEN

Time-resolved spectroscopy: Advances in understanding the electronic structure and dynamics of cobalamins

Roseanne J. Sension[a], Taewon Chung[a], Paul Dewan Jr.[b,†], Taylor P. McClain[b], Ryan M. Lamb[a], and James E. Penner-Hahn[a,b,*]

[a]Department of Chemistry, University of Michigan, Ann Arbor, MI, United States
[b]Biophysics Program, University of Michigan, Ann Arbor, MI, United States
*Corresponding author: e-mail address: jeph@umich.edu

Contents

Abstract

Time resolved spectroscopy provides unique insight into the structure and function of cobalamins. In these experiments, the cobalamin is initially excited by a short "pump" pulse in the UV–visible region and then characterized at some later time using a short "probe" pulse. The emphasis in this chapter is on both UV–visible and X-ray probe pulses, with a particular focus on the unique information provided by the latter.

† Permanent address: University of Buffalo, Buffalo, NY, United States.

Methods in Enzymology, Volume 669
ISSN 0076-6879
https://doi.org/10.1016/bs.mie.2022.01.010

The principles of time-resolved spectroscopy are reviewed, with an emphasis on ultrafast measurements (time scales less than ~10 ps) to characterize short-lived cobalamin excited states. Several practical considerations are discussed, with a focus on the technical details that are necessary to obtain high quality, interpretable data. These include sample delivery, polarization, and excitation power. Some of the theoretical approaches to interpreting data are discussed.

1. Introduction

Many cobalamins undergo light driven photolysis. In particular, photoexcitation of alkylcobalamins (Fig. 1) in the UV–visible region of the spectrum results in bond dissociation with high quantum yields (Cole et al., 2002; Taylor, Smucker, Hanna, & Gill, 1973). The photolysis quantum yield depends both on the axial ligands to the cobalamin and on the local environment around the cobalamin (Sension et al., 2004; Sension, Harris, Stickrath, et al., 2005; Sension, Harris, & Cole, 2005; Stickrath et al., 2009). The axial ligands affect the electronic structure of the excited state, while the environment (i.e., the solvent or the protein environment) can affect both the electronic structure and the rate of diffusive separation of a geminate radical pair.

The photochemistry of cobalamins was long thought to be at best biologically irrelevant or at worst a danger to be controlled. However, recent work has shown that some organisms rely on cobalamin photochemistry to control biological functions. For example, the DNA regulatory protein

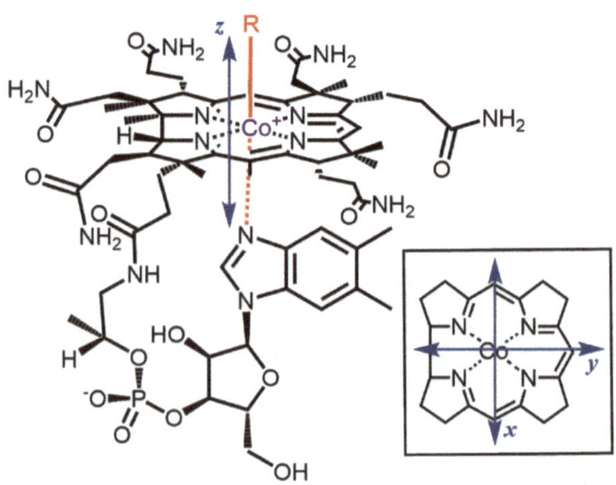

Fig. 1 Schematic of cobalamins indicating the coordinate system used in this chapter.

CarH utilizes cobalamin photochemistry to sense light levels and thus to control carotenoid biosynthesis (Padmanabhan, Jost, Drennan, & Elias-Arnanz, 2017; Padmanabhan, Perez-Castano, & Elias-Arnanz, 2019). This and related systems have brought increased attention to the factors that control the photochemistry of cobalamins and the factors that can be used to develop new light-activated systems.

Time-resolved spectroscopies with probe pulses spanning the range from IR (Jones et al., 2012; Shiang et al., 2006) to X-ray energies (Miller et al., 2017, 2018; Miller, Kaneshiro, et al., 2020; Miller, Michocki, et al., 2020; Subramanian et al., 2018) provide a window into the structure and dynamics of cobalamins. The time scales of interest range from femtoseconds to microseconds and longer. Each wavelength range provides unique insights and combined they provide a more complete picture of the dynamics. Here we will concentrate on the information available from ultrafast pump-probe transient absorption measurements using UV–visible and X-ray probes. Our focus is both on summarizing the information that is available and on highlighting some the potential experimental pitfalls in the X-ray region.

UV–visible transient absorption (UV–Vis TA) measurements provide insight into excited state lifetimes, product formation, and quantum yields. The UV–visible absorption spectrum exhibits characteristic changes with ligation and oxidation state. Early measurements focused on changes in the region around 470 nm (see Fig. 2) to identify bond dissociation and

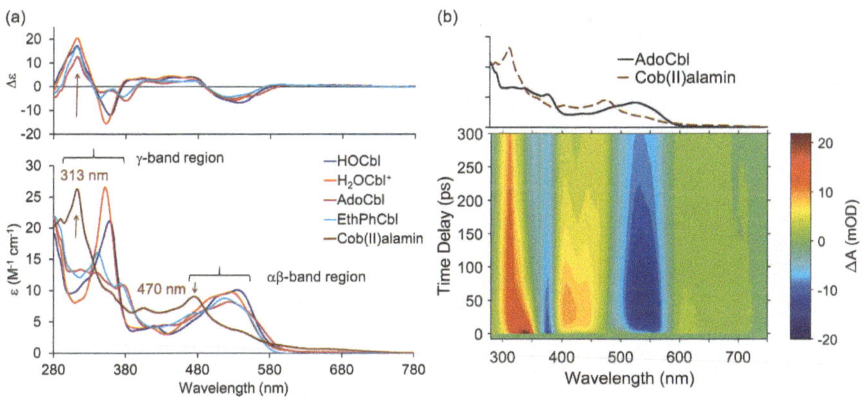

Fig. 2 (A) UV–visible spectra of select cob(III)alamins along with the spectrum of the cob(II)alamin photoproduct formed following bond dissociation. The corresponding difference spectra relative to Co(II)alamin are shown in the top panel. (B) Transient absorption spectrum following excitation of AdoCbl in water at 575 nm. The strong γ-band at 313 nm characterizes the formation of cob(II)alamin on a 100 ps time scale.

formation of cob(II)alamin (Cole et al., 2002; Lott, Chagovetz, & Grissom, 1995). However, changes in this region can be misleading. The appearance of the cob(II)alamin γ-band absorption at ca. 313 nm is a more definitive indicator of bond dissociation and radical formation. Increased absorption around 450 nm correlates with lengthening or dissociation of the axial Co—N bond (Peng et al., 2010; Salerno et al., 2020).

While UV–Vis TA is a powerful tool, as noted above the spectra are often ambiguous and changes in structure can only be inferred by comparison with known ground state compounds. In contrast, X-ray absorption (XAS) at the K-edge is sensitive to the electronic density distribution and structure of ligands around the cobalt atom (Fig. 3). Pre-edge transitions (Co 1s → 3d/valence transitions) provide insight into the electronic distribution and effective symmetry at the cobalt, the XANES (X-ray absorption near edge structure) region is sensitive to structure, and the EXAFS (extended X-ray absorption fine structure) permits precise identification of bond lengths.

In particular, time-resolved XAS (tr-XAS) in the pre-edge and XANES region provides a sensitive probe of structural evolution on the femtosecond and picosecond time scales. The position of the edge correlates with the sum of axial bond lengths and changes in the edge can be used to track structural changes with time (see Fig. 4). Changes in the pre-edge region can be used to track orbital occupancy as well as symmetry changes that influence the mixing of 3d and 4p orbitals of cobalt.

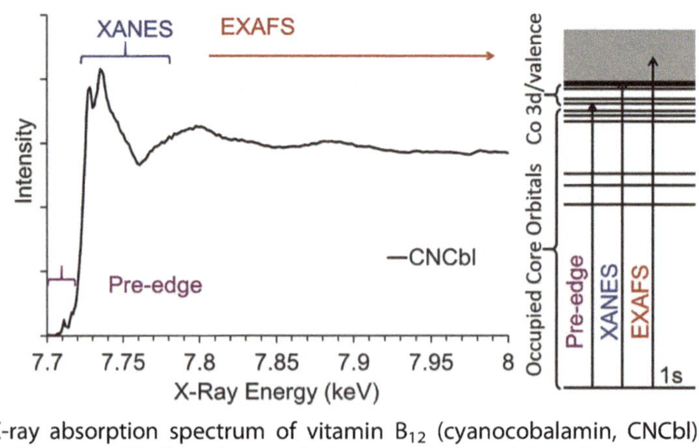

Fig. 3 X-ray absorption spectrum of vitamin B_{12} (cyanocobalamin, CNCbl) with the pre-edge (Co 1s → 3d/valence), XANES (Co 1s → ≥4p), and EXAFS regions indicated.

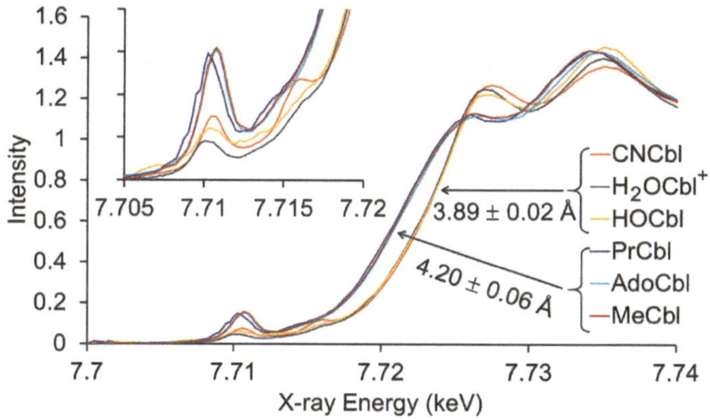

Fig. 4 Ground state XANES spectra for a series of cobalamins demonstrating the dependence of the edge on the sum of the axial bond lengths determined from X-ray crystal structures (Hannibal et al., 2009; Kratky et al., 1995; Mebs et al., 2009; Ouyang et al., 2005; Prieto et al., 2016). The inset compares the pre-edge transitions for these compounds.

In general, cobalamins are considered non-fluorescent. Emission can be observed for some compounds, however, the quantum yield is low (on the order of 10^{-5} to 10^{-6}) (Miller, Michocki, et al., 2020; Salerno et al., 2020). This is because the photoexcited molecule moves from the initially excited "bright" state, which has a significant transition dipole coupling the ground and excited state, to a "dark" state, which lacks this coupling, within 100–200 fs. Thus, complete characterization of the structural and electronic dynamics of cobalamins requires femtosecond time resolution. While femtosecond UV–visible light sources have been available for many years, it is only recently that femtosecond sources have become available in the X-ray region. The development of femtosecond X-ray free-electron lasers (XFELs), combined with the continuing improvement of picosecond transient methods at X-ray synchrotrons, now permit a wide range of sensitive measurements of cobalamin dynamics at time scales short enough to study the evolution from the initial Franck-Condon excited bright state and to follow the subsequent structural and electronic dynamics. Transient X-ray measurements are considered in more detail in the remainder of this chapter.

2. Sample delivery

Although time-resolved measurements typically rely on high peak-power illumination, the excitation pulse is typically quite short (tens of fs to perhaps 1 ps), and the repetition rates are often quite modest (hundreds of Hz to kHz). As a consequence, the average flux is often lower than in conventional, non-time-resolved spectroscopy, and acquisition of high-quality spectra requires hundreds to hundreds of thousands of pulses. Depending on the nature of the sample, this can present significant problems for sample delivery.

2.1 Photostable samples

In the simplest case, photoexcitation produces an excited state with a lifetime that is short in comparison to the repetition rate of the excitation. Since the sample is unchanged from pulse to pulse, a simple static cuvette can be used.

2.2 Photolabile samples

A more typical case would be a sample that undergoes irreversible change following excitation (or at least change that is irreversible in comparison to the repetition rate). This could be due to either direct photoconversion from the pump pulse or indirect damage from the probe pulse. The latter can be especially important in optical pump - X-ray probe experiments, where the ionizing X-ray probe beam often photo-reduces the sample.

If the sample is unstable for any of these reasons, a static sample can't be used, and it is necessary to flow the sample such that each excitation pulse "sees" fresh sample. This can be accomplished using a jet, often free-flowing in air (or more likely in He gas). For example, a syringe- or HPLC-pump together with an appropriate nozzle can produce jets with a variety of shapes, including cylinders and thin sheets. Typical conditions might use a $\sim$50 μm diameter jet. With this size, the excitation pulse transits the jet in $\sim$150 fs. Any uncertainty in the pump and probe temporal overlap is negligible in comparison with pulse length if the pulses are nearly collinear. For a 50 μm diameter jet, even a relatively modest volume flow rate of $\sim$1 mL/min gives a linear flow rate of over 8 m/s. This is convenient for spectroscopic measurements because there is effectively no movement of the sample during the measurement, but there is still complete sample replacement between measurements. For example, for measurement times as long as 1 ns the sample moves less than 0.02% relative to a $\sim$50 μm spot, while even with a 100 kHz repetition rate, the jet moves 80 μm between pulses. A schematic illustration of this is shown in Fig. 5.

If the sample eventually returns to the ground state, or if the excitation-frequency/sample-flow-rate ratio is such that only a negligible fraction of

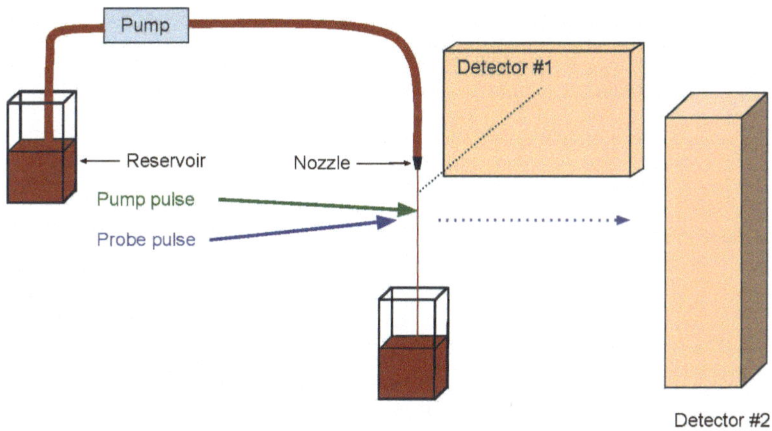

Fig. 5 Schematic illustration of a jet sample delivery. For optical-pump, optical probe, the nozzle might be configured to provide a thin sheet, detector #1 would be used to measure the transmission of the probe pulse, and detector #2 might be omitted. For optical-pump, X-ray probe, detector #1 might be used to measure X-ray fluorescence, while detector #2 would be used to measure the X-ray scattering by the jet (thus giving a normalization factor proportional to the product of the incident probe pulse intensity and the sample thickness). For samples that do not suffer substantial damage, the used sample can be collected and recycled.

the sample is illuminated, it may be possible to recycle the sample, thus minimizing the volume of solution that is required. In such a case, the best practice is to continuously monitor the UV–visible spectrum of the sample to ensure that there is no sample damage.

2.3 Minimizing sample volume

In some cases, photoconversion may be effectively irreversible or there may be only very small volumes of sample available (for example, when looking at proteins). If so, sample delivery using a jet would require unrealistic amounts of sample. One alternative is to adopt technology originally developed for ink-jet printers and deliver the sample as individual drops (Miller et al., 2019). A "drop-on-demand" system with ~50 μm diameter drops (i.e., ~65 pL/drop) operating at 1 kHz would consume only ~4 μL/min. Depending on the repetition rate, this could be several orders of magnitude more efficient than a sample jet, allowing measurements to be made on tiny amounts of sample. For example, this small volume coupled with the high peak power of the free-electron laser allowed time-resolved XANES difference spectra to be measured on tens of μL of CarH, a much smaller volume that would typically be used for static XANES measurements at a synchrotron (Miller, Kaneshiro, et al., 2020).

However, the efficiency of drop-on-demand systems comes at a cost. Because of the irregular, approximately spherical, sample geometry, it is impossible to measure transmission spectra. Rather, one is limited to fluorescence measurements (although by measuring fluorescence-detected excitation spectra, one can often approximate the absorption spectrum of the sample). A bigger limitation is the droplet-to-droplet variation in droplet volume and position (Miller et al., 2019). Even a small (μs) error in droplet arrival time can give a large variation in the fraction of the droplet that is illuminated, and even a 1% change in radius can give a signal variation as large as that expected due to the true pump-driven variability. A straightforward solution to this for visible pump/X-ray probe experiments is to use X-ray small-angle scattering from the solvent to normalize the spectra. This is equivalent to the setup shown in Fig. 5, but with discrete droplets replacing the free-flowing jet. This simultaneously corrects for variation in incident X-ray intensity, in sample position, and in droplet size.

3. Polarization dependence

The absorption spectrum of a molecule is defined in terms of a transition dipole vector that is used to calculate the probability that a photon is absorbed as a function of the orientation of the molecule relative to the electric field vector of the light. Depending on the molecular symmetry, this means that the complete absorption spectrum is not simply a scalar (absorption coefficient as a function of energy), but is in fact a tensor, with different spectra for absorption along each of three mutually orthogonal molecular axes. For molecules that can be oriented in space (i.e., for single-crystal absorption spectra) the availability of orientation-dependent absorption coefficients significantly enhances the information content of the measurements and can simplify interpretation of the spectra.

While this has been widely utilized in both X-ray and optical spectroscopy, it is not directly applicable here, since all of the measurements discussed in this chapter involve isotropic solutions. However, if a plane-polarized pump pulse is used, then only the subset of the molecules in solution that have their transition dipole aligned appropriately relative to the electric field of the excitation beam will be excited, allowing one to observe a polarized *difference* signal even for an isotropic solution. Although this orientation will relax over time, the rotational correlation time for a molecule as large as cobalamin is typically many hundreds of picoseconds or longer, so the molecules are effectively fixed in orientation over the course of a picosecond transient absorption measurement.

The observed change in absorbance (ΔA) depends on the relative orientation of the pump and the probe beam polarizations. If the pump and probe beams are oriented to have either parallel or perpendicular relative polarizations, the observed difference signal is

$$\Delta A_{\|}(t) = \frac{1}{3}\Delta A_{ISO}(t)\left[1 + \frac{4}{5}\langle P_2(\cos\theta)\rangle\right] \tag{1}$$

$$\Delta A_{\perp}(t) = \frac{1}{3}\Delta A_{ISO}(t)\left[1 - \frac{2}{5}\langle P_2(\cos\theta)\rangle\right] \tag{2}$$

where $<P_2>$ is the ensemble average of the second Legendre polynomial and θ is the angle between the pumped and probed transition dipole moments (see Miller et al., 2018, for details). For optical-pump, optical-probe transient absorption measurements, the resulting difference spectra can still be quite complex, given that there may be multiple, overlapping transitions, each having a different orientation for its transition dipole.

The situation can be quite a bit simpler for optical-pump, X-ray-probe measurements. While there are still multiple possible X-ray absorption transitions, they are well separated in energy and do not overlap. In this case, it makes sense to define the X-ray absorption transition-dipole tensor such that one axis (e.g., $\vec{a}$) is parallel to the excitation dipole moment, and thus the other two ($\vec{b}$, $\vec{c}$) are perpendicular. This sets θ to either 0° or 90°, giving

$$\Delta S_{\|} = 0.6\Delta S_a + 0.2\Delta S_b + 0.2S_c \tag{3}$$

$$\Delta S_{\perp} = 0.2\Delta S_a + 0.4\Delta S_b + 0.4S_c \tag{4}$$

which can be rearranged to give

$$\Delta S_a = 2\Delta S_{\|} - \Delta S_{\perp} \tag{5}$$

$$\Delta S_b + \Delta S_c = \Delta S_{b+c} = 3\Delta S_{\perp} - \Delta S_{\|} \tag{6}$$

where $\Delta S_{\|}$ and $\Delta S_{\perp}$ are the differences in the X-ray signal when the optical pump and X-ray probe beams are either parallel or perpendicular. For corrins, a particularly convenient choice of axes is as shown in Fig. 1, where the x and y axes are chosen to lie in the plane of the corrin ring while the z axis is in the axial direction. With these definitions, x is approximately parallel to the transition dipole for the visible $\alpha\beta$ bands, while y is approximately parallel to the transition dipole for the γ band (see Fig. 2).

For excitation between 520 nm and 550 nm, the observed difference signals ($\Delta S_{\|}$ and $\Delta S_{\perp}$) can be rearranged to give

$$\Delta S_x = 2\Delta S_{\|} - \Delta S_{\perp} \tag{7}$$

$$\Delta S_{y+z} = 3\Delta S_{\perp} - \Delta S_{\|} \tag{8}$$

while with 365 nm excitation, one gets:

$$\Delta S_y = 2\Delta S_{\|} - \Delta S_{\perp} \tag{9}$$

$$\Delta S_{x+z} = 3\Delta S_{\perp} - \Delta S_{\|} \tag{10}$$

This fortuitous arrangement of the pump transition dipole moments for cobalamins means that it is possible to determine *directly* the light-dependent change in X-ray absorption coefficient along each of the x, y, and z axes using 4 measurements (Miller et al., 2018; Miller, Michocki, et al., 2020), providing 530 and 365 nm excitation populate the same excited state. The ability to experimentally distinguish the change in absorption along the x- and y-directions is important for kinetic characterization of cobalamins. Because the "lower" benzimidazole ligand (see Fig. 1) lies approximately in the x–z plane, one can use the polarization dependence to distinguish between elongations of the "upper" and "lower" Co-ligand bonds.

4. Power

Ultrafast spectroscopic measurements are often challenged by small signals and noisy probes. While improvements in laser stability and the development of higher repetition rate systems have eliminated many of the issues for optical measurements, ultrafast X-ray measurements remain challenging. The X-ray probes are noisy and limits on available beamtime have encouraged the use of relatively high laser pump powers to increase the excited state population and thus the transient difference signal. Experiments performed near a saturation limit or reporting transient excited state populations above 50% are common (Bacellar et al., 2020; Britz et al., 2020; Katayama et al., 2019; Kinschel et al., 2020).

4.1 Consequences of high-power measurements

There are two major complications introduced by high laser powers: changes in the excited state(s) that are produced and loss of polarization information.

4.1.1 Multiphoton excitation

The most significant complication is the onset of multiphoton excitation and the formation of an ensemble of photoproducts complicating the analysis of both excited state dynamics and transient spectra. The maximum transient XANES signal for most cobalamins is observed around 7.720 keV, representing the red-shift of the XANES edge as the axial bonds elongate (Michocki et al., 2019; Miller et al., 2017, 2018; Miller, Michocki, et al., 2020; Subramanian et al., 2018). The time-dependence of a typical signal at this X-ray energy as a function of the nominal optical laser power is illustrated in Fig. 6A. The laser fluence on the sample is a function of the nominal pulse energy measured using a power meter, the size of the focus, and the presence of attenuating elements between the power meter and the sample. Because this can only be estimated, we plot the intensity versus nominal pulse energy. Two significant features are apparent in the data plotted in Fig. 6B. First, saturation of the excited state population observed between 4 and 8 ps occurs at a lower nominal pulse energy than saturation of the initial

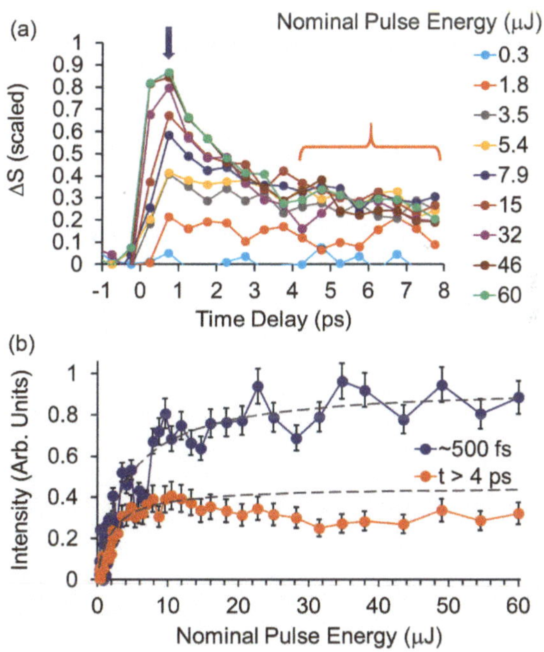

Fig. 6 Influence of laser power on transient XANES measurements. (A) Difference signal at 7.720 keV as a function of time and nominal laser power. (B) Saturation of the transient difference signal as a function of time. *Dashed lines* are fits to a saturation curve, showing the lower saturation power at long times.

excited state population. It is not sufficient to choose a single time delay to characterize saturation behavior. In addition, an increase in the nominal pulse energy above 10 μJ results in a decrease in the signal from the long-lived product(s) indicating a change in the product distribution for higher pulse energies.

Measurement of the XANES difference spectrum as a function of pulse energy is required to more fully characterize the consequences of multi-photon excitation. However, these differences can be ambiguous. There is a subtle change in the XANES region measured at 8 ps when a high energy pulse (nominal pulse energy ~40 μJ) is used compared with a much weaker pulse (nominal pulse energy 4 μJ) as illustrated in Fig. 7. However, without independent information on the appropriate scale factor, it would be hard to assign this change to structure. The difference between high and low power excitation is more significant in the pre-edge region as illustrated in the inset. The bleach is red-shifted in the high power measurement indicating the formation of photoproducts with differing electronic configurations around the cobalt as a function of the pulse energy. When performing transient XANES measurements it is important to ensure that the pulse energy falls below saturation and to resist the temptation to increase intensity to maximize signal.

4.1.2 Loss of polarization discrimination

The second drawback to increased laser power is the loss of information provided by exploiting polarization. At very high laser powers there will be a complete loss of polarization discrimination, but even at lower powers the loss

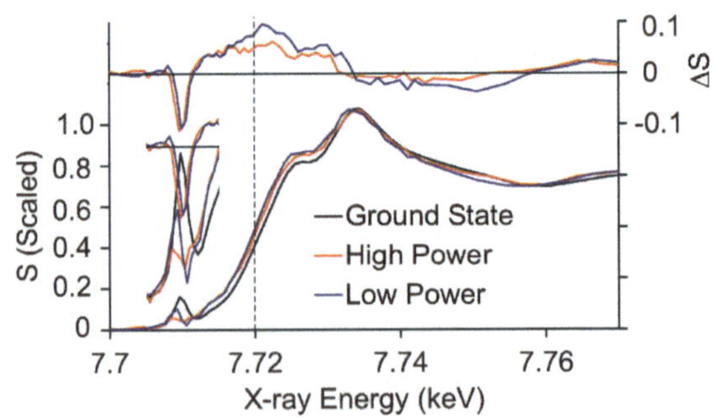

Fig. 7 The influence of laser power on the excited state spectrum. The plot compares the difference spectra and estimated excited state spectra following excitation of an alkylcobalamin with high laser power (nominal pulse energy ~40 μJ) and much lower laser power (nominal pulse energy 4 μJ). The inset highlights the pre-edge region and illustrates the difference in the bleaching of the 1 s → 3d transitions. The dashed vertical line indicates the X-ray energy for the transient measurements in Fig. 6.

of information can be significant. This is illustrated in Fig. 8. The top panel compares the x and y + z contributions at ca. 8 ps calculated using Eqs. (7) and (8) for the laser powers in Fig. 7. At very high incident laser power there is no difference in the shape of the x and y + z transients. Both represent the isotropic signal with only a difference in intensity. In contrast, there is a clear polarization dependence with a nominal pulse energy of 4 μJ (Low Power). The x contribution makes only a small contribution to both the pre-edge bleach and the edge shift monitored at ca. 7.72 keV, consistent with a bleaching of the $1s \rightarrow 3d_{z2}$ transition and elongation of the axial bonds in the z direction. The bottom panel compares the transient signal at ca. 800 fs

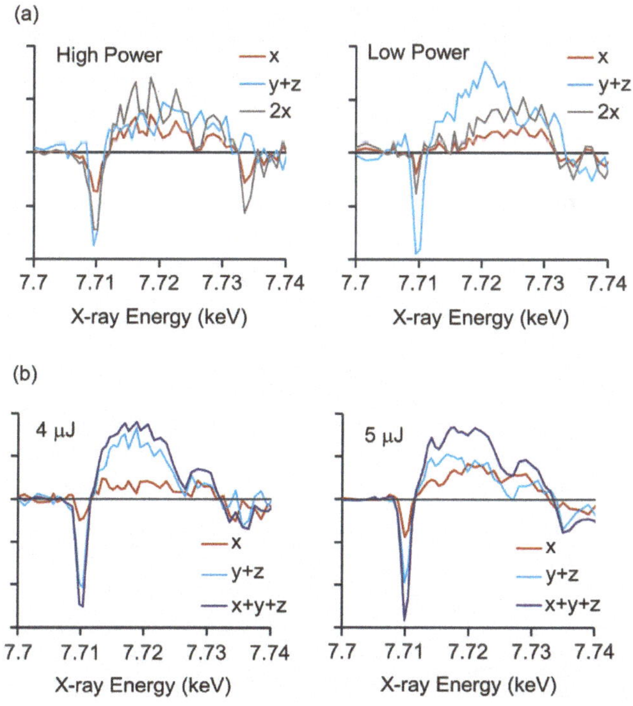

Fig. 8 The influence of laser power on the decomposition of the polarized signal. (A) At high power (nominal pulse energy ~40 μJ) there is no polarization discrimination at 8 ps and twice the "x" signal is equivalent to the "y + z" signal. At low power (nominal pulse energy 4 μJ) the pre-edge and edge regions are dominated by the y + z component, consistent with bleaching of the $1s \rightarrow 3d_{z2}$ transition and elongation of the axial bonds shifting the edge to lower energy. (B) The bottom panel illustrates the reduction in polarization discrimination even at much lower nominal pulse energies. These data at 800 fs have similar isotropic difference spectra (x + y + z), but the identification of changes in the axial direction is obscured by saturation effects. We estimate excitation fractions of 0.35 ± 0.05 and 0.44 ± 0.05 for these two measurements.

for nominal pulse energies of 4 and 5 μJ. Although both incident pulse energies fall below saturation (see Fig. 6), and the isotropic signals (x + y + z) are identical within the accuracy of the measurements, there is a significant reduction in the polarization discrimination with the 5 μJ excitation pulses. While the low power measurement plotted here allows for discrimination of x and y+z contributions, even lower pulse energies would be better. With a 4 μJ nominal pulse energy, we estimate an excitation fraction of 0.35 ± 0.05 for the data plotted in Fig. 8. In contrast, our measurements on AdoCbl, where the excitation fraction was ≈ 0.23, illustrate little or no contribution from x or y components in the pre-edge bleach (Miller, Michocki, et al., 2020).

5. Kinetics description of photoexcited cobalamins

One of the goals of ultrafast spectroscopy is to describe the time-evolution of the cobalamin following absorption of the pump pulse. A Jablonski diagram (Fig. 9, left) is a convenient way to summarize such information. This diagram, which describes the measurements for cyanocobalamin (Shiang et al., 2006), shows initial excitation with 520 nm light as a vertical excitation (i.e., forming an excited state without any change in nuclear coordinates). This initial state, sometimes referred to as the Franck-Condon state, then relaxes over ~80 fs before undergoing intersystem crossing to the S_1 excited state with a time constant of 190 fs. The S_1 excited state has a lifetime of 6.7 ps for cyanocobalamin in water. A schematic illustration of

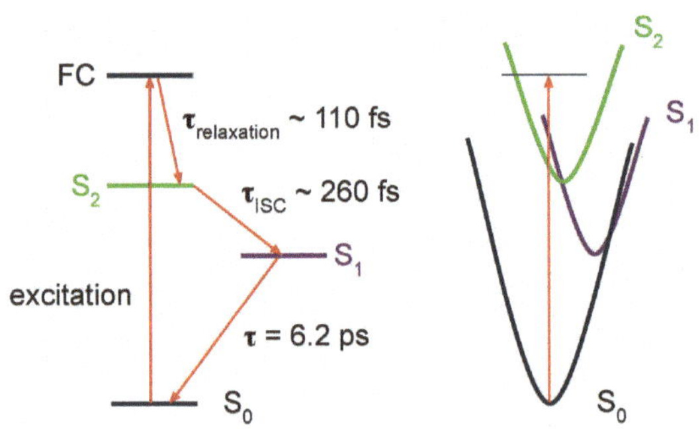

Fig. 9 Schematic illustration of the excited state behavior of cyanocobalamin. (*Left*) A Jablonski diagram, showing relative energies and lifetimes. The lifetimes are from Miller et al. (2017); these differ slightly from those in (Shiang et al. (2006). (*Right*) Schematic, one-dimensional illustration of the ground and excited state potential energy surfaces. *Adapted from Shiang, J.J., Cole, A. G., Sension, R. J., Hang, K., Weng, Y. X., Trommel, J. S., et al. (2006). Ultrafast excited-state dynamics in vitamin B-12 and related Cob(III)alamins,* Journal of the American Chemical Society, 128, *801–808. doi:10.1021/ja054374.*

the energy levels associated this decay is shown on the right, where the parabolic shapes are 1-dimensional representations of the multidimensional potential energy surfaces. The right-hand figure emphasizes the fact that the excited states will, in general, have a different nuclear arrangement than the ground state.

Implicit in the model of Fig. 9 is the assumption that the kinetic behavior can be described using conventional first-order reaction kinetics, giving rise to the half-lives indicated in the figure. For the simple unimolecular reaction $\mathbf{A} \rightarrow \mathbf{B}$ (e.g., $S_2 \rightarrow S_1$ in Fig. 9) the rate of loss of **A**, that is, $-d[A]/dt$ will depend only on the concentration of A:

$$d[A]/dt = -k[A] \tag{11}$$

Eq. (11) can be integrated to give the familiar first-order rate law

$$[A] = [A]_0 e^{-kt} \tag{12}$$

For a more complicated kinetic scheme such as that required by the data in Fig. 10, with multiple sequential first-order reactions, e.g., $\mathbf{A} \rightarrow \mathbf{B} \rightarrow \mathbf{C} \rightarrow \mathbf{S_0}$ with rate constants k_1, k_2, and k_3, it is relatively straightforward to solve this set of equations to give the number of molecules in each state as a function of time:

$$n_{s_A}(t) = e^{-k_1 t} \tag{13}$$

$$n_{s_B}(t) = \frac{k_1}{k_1 - k_2}\left(e^{-k_2 t} - e^{-k_1 t}\right) \tag{14}$$

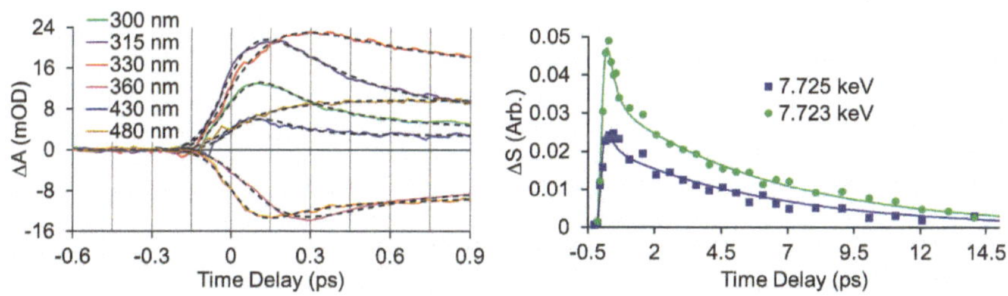

Fig. 10 Fits to the optical-pump, optical-probe transient absorption data for cyanocobalamin (*left*) and the optical-pump, X-ray probe transient absorption data at a range of different probe energies. Fits (*dashed lines*) use the kinetic model shown schematically in Fig. 9, as expressed in Eqs. (13–16). *Reproduced with permission from Miller, N. A., Deb, A., Alonso-Mori, R., Garabato, B. D., Glownia, J. M., Kiefer, L. M., et al. (2017). Polarized XANES monitors femtosecond structural evolution of photoexcited vitamin B-12.* Journal of the American Chemical Society, 139, *1894–1899. doi:10.1021/jacs.6b11295.*

$$n_{s_C}(t) = k_1 k_2 \left(\frac{e^{-k_1 t}}{(k_1 - k_2)(k_1 - k_3)} + \frac{e^{-k_2 t}}{(k_2 - k_1)(k_2 - k_3)} + \frac{e^{-k_3 t}}{(k_3 - k_1)(k_3 - k_2)} \right) \tag{15}$$

$$n_{s_0}(t) = 1 - \left(\frac{k_2 k_3 e^{-k_1 t}}{(k_1 - k_2)(k_1 - k_3)} + \frac{k_1 k_3 e^{-k_2 t}}{(k_2 - k_1)(k_2 - k_3)} + \frac{k_1 k_2 e^{-k_3 t}}{(k_3 - k_1)(k_3 - k_2)} \right) \tag{16}$$

where for convenience the scaling factor for the number of initially excited molecules has been omitted.

Descriptions such as this are widely used in describing ultrafast transient absorption measurements. We found that this model could be used to fit both the optical-pump, visible-probe and optical-pump X-ray probe data for cyanocobalamin, as shown in Fig. 10.

From consideration of Eqs. (13–16) and Fig. 10, it is clear that the experimental data cannot be fit using only rate constants (k_i); it is necessary also to refine the change in absorption coefficient at each probe energy. This can be a problem when fitting optical-pump/optical-probe data since it is extremely difficult to calculate accurate absorption coefficients for electronic excited states. Effectively, the excited state absorption coefficients as a function of wavelength become unconstrained parameters.

The ability to fit the observed data with this model does not guarantee that the model is correct. Indeed, this was the case for cyanocobalamin. Both the optical and the X-ray probe data were well described by a kinetic model with three sequential exponential lifetimes. However, when the fitted kinetic parameters were used to reconstruct the X-ray absorption spectra for each of the different species (Fig. 11, left), the predicted spectrum for intermediate **B** was unlike any known Co XANES spectrum. The calculated spectrum for **A**, which is expected to have a structure very similar to the ground state, is quite similar to that of **G**. The calculated spectrum for **C**, which is predicted to have elongations of ~10% in the Co-benzimidazole distance and ~20% in the Co-CN distance, shows a red-shift of ~2 eV and is well modeled using the predicted structural parameters (Miller et al., 2018). In comparison with the spectrum for **C**, the calculated spectrum for **B** shows nearly twice as large a shift in energy and a significant increase in peak intensity. The difference between the calculated spectra for **B** and **C** is larger than that between **G** and **C**, suggesting that

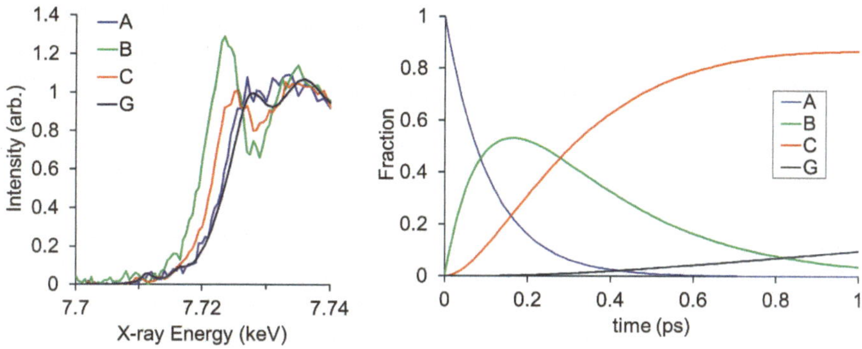

Fig. 11 (*Left*) Calculated X-ray absorption spectra for cyanocobalamin using the kinetic model in Fig. 9. (*Right*) Calculated populations of the 4 cyanocobalamin species using this model. (Left) *Reproduced with permission from Miller, N. A., Deb, A., Alonso-Mori, R., Glownia, J. M., Kiefer, L. M., Konar, A., et al. (2018). Ultrafast X-ray absorption near edge structure reveals ballistic excited state structural dynamics.* The Journal of Physical Chemistry A, *122, 4963–4971. https://doi.org/10.1021/acs.jpca.8b04223. Copyright American Chemical Society.*

B must have bond-length elongations, relative to **G**, that are at least twice as large as those for **C**. In fact, a ligand at such a long distance from the cobalt (2.5–2.6 Å) would make almost no contribution to the bonding. Simulations using such long distances are unable to reproduce the spectrum for **B** shown in Fig. 11 and instead give a predicted spectrum similar to that seen for cob(*II*)alamin.

The reason that the calculated spectrum for **B** is so unusual can be understood by considering the relative populations of each of the species (Fig. 11, right). Because **A** and **B** have similar lifetimes, a significant population of **B** never accumulates. In this model, the spectral differences seen at ~200 fs must reflect a large change in X-ray absorption coefficient at ~7.72 keV. While this model fits the data, it can only do so by making unrealistic assumptions about the X-ray absorption coefficient of **B**. This illustrates an important benefit of tr-XAS: absorption coefficients in X-ray absorption can be constrained in ways that are not possible for UV–visible spectra. The structured absorption in the XANES region arises from scattering of the X-ray excited photoelectron (see Section 6), and thus provides *direct* structural information that is not available from UV–visible spectroscopy. This allows one to constrain the possible changes in X-ray absorption coefficient in ways that are not possible for UV–visible spectroscopy.

The error behind the physically unrealistic predicted spectra for **B** (Fig. 11) is the assumption that the early time behavior of cyanocobalamin can be described by an exponential model. Inherent in this model (Eq. 2) is the assumption that the time required to go from **A** → **B** and from **B** → **C** is short in comparison with lifetimes of **A** and **B**. For ~100 fs lifetimes, this is probably not a good assumption, particularly if substantial structural rearrangement is required, such as the ~0.36 Å elongation of the Co—C distance in **C**. In the harmonic oscillator model, a typical metal-ligand stretch has a period of ~100 fs and a displacement of <0.2 Å. At this velocity it is plausible to stretch the Co-CN distance by 0.36 Å in hundreds of fs, but it is *not* plausible for this elongation to happen so fast that it is negligible in comparison with the calculated lifetimes. A similar logic is likely to apply to all excited states that involve substantial (>~0.1 Å) changes in bond lengths on ultrafast (~100 fs) time scales.

In order to account for the observed spectra, we adopted a different model (Miller et al., 2018). Rather than assume that the **A** → **B** → **C** conversion can be modeled as two sequential exponentials, we asked whether this could be modeled as the evolution of a single coherent wavepacket, going from **A** → → → → **C** over the course of ~250 fs. According to this model, the maximum in the XANES difference at ~190 fs reflects a structure that evolves as a slightly underdamped harmonic oscillator, rather than the formation of a discrete, short-lived intermediate. Continuous elongation in the Co—C distance with a slight overshoot followed by recovery to the equilibrium structure for **C** (see Fig. 12) can explain the observed data without requiring unrealistic XANES spectra for the intermediates.

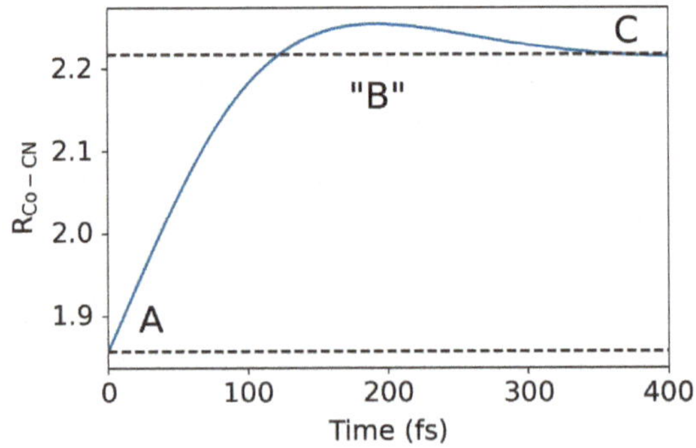

Fig. 12 Schematic illustration of one possible model for the structural evolution of photoexcited cyanocobalamin from **A** to **C**, showing the continuous evolution in Co-C distance that initially slightly overshoots the distance for **C**, giving rise to the maximum in XANES difference at ~190 fs.

This raises the question—when can pump-probe XANES be used to distinguish between the coherent evolution of a wave packet from **A** → **B** and the first-order exponential decay of **A** into **B**? In fortunate cases, it is possible to detect the presence of coherent oscillations, as seen for example in photoexcited $Fe(bpy)_3^{2+}$ (Lemke et al., 2017). In the more general case, however, the absence of oscillations cannot be taken as evidence for exponential kinetics. A critically damped harmonic oscillator (i.e., one with no overshoot) will have exponential conversion of the bond length:

$$R(t) = R(\infty) - \Delta R e^{-kt} \tag{17}$$

where ΔR is the difference between R(∞) and R(0). To the extent that the XANES is a linear function of distance, it would be impossible to distinguish between these models, since the average distance would be identical. XANES is not a linear function of distance, particularly for multishell ligands with significant outer-shell scattering (see below), thus it should be possible to distinguish between these. However, the deviations are small, and it does not appear that anyone has yet collected time-resolved XANES with a sufficient signal/noise ratio to allow for this distinction.

6. Comparison of theory and experiment

While a great deal of useful information can be obtained from the experimental lifetimes and spectra that are deduced from transient absorption measurements, it is often desirable to support measurements with theoretical calculations of the expected spectra. It is (relatively) straightforward to obtain accurate calculations of the ground-state electronic structure, using methods such as density functional theory (DFT). Nevertheless, for molecules as large as cobalamin, these calculations can become quite expensive, especially for compounds with unpaired electrons (e.g., cob(II)alamin). In order to predict the spectrum of a compound, one needs information about both the ground and the excited state electronic structure, and the situation is generally worse for excited state electronic structure calculations. These often require larger basis sets and, in many cases, the need for multireference methods to account for configuration interaction.

There has been significant progress in the use of time-dependent DFT (TD-DFT) methods for calculating ground-state absorption spectra, where the electronic structure is solved in the presence of a time-varying applied electric field. Using TD-DFT, it is often possible to find at least qualitative agreement between measured and calculated ground-state absorption spectra, even for molecules as large as cobalamin (Garabato, Kumar, Lodowski, Jaworska, & Kozlowski, 2016; Kornobis et al., 2011; Lodowski, Jaworska, Kornobis, Andruniow, & Kozlowski, 2011; Pallares & Brunold, 2014; Solheim, Kornobis, Ruud, & Kozlowski, 2011; Stich, Brooks, Buan, & Brunold, 2003; Stich, Buan, & Brunold, 2004; Stich, Buan, Escalante-Semerena, & Brunold, 2005; Wiley et al., 2018). However, in order to describe transient absorption data, one needs not only the ground-state absorbance, but also the excited-state absorbance. The latter remains extremely challenging to calculate since one needs to describe not only the electronic structure of the excited state (i.e., the state formed by the pump pulse) but also the interaction of this excited state with the oscillating electric field of probe pulse. There are not, at present, quantitatively accurate calculations of excited state UV–visible absorption spectra for cobalamins.

6.1 X-ray absorption spectra

At a high level, calculation of the X-ray absorption spectrum for a molecule is the same problem as calculation of UV–visible absorption spectra—one needs to describe the interaction of an oscillating electric field with the electronic structure of the molecule. The X-ray photo-absorption cross-section, $\sigma(\omega)$ can be written, using Fermi's golden rule, as

$$\sigma(\omega) = 4\pi^2 \frac{\omega}{c} \sum_f \left| i \middle| \hat{O} \middle| f \right|^2 \delta\left(\omega - E_i - E_f\right) \tag{18}$$

where $\langle i|$ and $|f\rangle$ are shorthand notation for the initial and final-state wave functions, $\hat{O}$ is an operator reflecting the interaction of the X-ray electric field with the molecule, ω is the energy of the X-ray photon, E_i and E_f are the energies of the initial and final states, and the delta function guarantees conservation of energy. (For details, see e.g., Joly & Grenier, 2016). While this equation is formally equivalent to the equations used to calculate UV–visible absorption spectra, applications in the X-ray region are significantly more difficult for several reasons. One is that, in contrast with most UV–visible spectroscopy, where only electric dipole transitions are important, higher-order (e.g., electric quadrupole transitions) are important for the electric field operator in the X-ray range. Worse, from a computational perspective, the high energy of X-rays means that many more final states are accessible, and thus the sum in Eq. (18) must be extended over a wider range. It is no longer sufficient to consider only bound states for $|f\rangle$, where (in a one-electron picture) the electron is excited from a filled to an empty molecular orbital; rather, one must also consider final states where there is a hole in the 1 s shell of the X-ray excited atom (i.e., the Co for cobalamin Co K-edge spectra) and an electron wave with an arbitrary energy Ω, such that $E_f = E_{Co(1s^1)} + \Omega$. Finally, description of the final-state wave function is complicated by the presence of the core-hole. Ejection of a 1 s electron leaves behind a Co atom with a higher charge, which can change the electronic structure of the cobalamin.

As noted above, X-ray absorption spectra can be divided into three regions: pre-edge, XANES, and EXAFS. Theoretical description in the EXAFS region is the best developed. Unfortunately, it is also the most difficult region experimentally, and there are only a few published ultrafast time-resolved EXAFS studies, none of which involve cobalamins. Theoretical description of pre-edge and XANES regions generally take different approaches.

6.1.1 Pre-edge spectra

In the pre-edge region, the incident X-ray does not have sufficient energy to eject the core-electron, and thus probes bound states—empty molecular orbitals that are below the ionization threshold. The probed orbitals need to overlap with the initial state, thus limiting pre-edge spectra to molecular orbitals with some Co character. For Co K-edges, this means, in practice, that transitions are seen either to the empty Co 3d orbitals or to the molecular orbitals with substantial Co 4p character. Both of these tend to be quite weak (see all Fig. 3). In the case of 1s-3d transitions, they are weak because this transition is forbidden by dipole selection rules and mostly gains intensity due to the much weaker quadrupole coupling (Hahn et al., 1982). The latter, which are formally Co(1 s) → ligand charge-transfer transitions tend to be weak due to small amount of Co character (typically Co 4p) in the relevant molecular orbitals. As a consequence, it is challenging to measure pre-edge difference spectra with a high signal/noise ratio (see Fig. 13). Data obtained following the excitation of CNCbl exhibit rapid bleaching of two pre-edge transitions and the appearance of new transitions at lower energy.

Theoretical approaches to pre-edge transitions generally use either a semi-empirical approach of ligand-field multiplet theory (de Groot, 2005) or TD-DFT calculations as used for UV–visible spectroscopy (e.g., Neese, 2012). One of the challenges in accurately describing time-resolved spectra in this region is accounting for the changes in electronic structure caused by both the pump and the probe beams. This is an important area for future development in time-resolved studies of cobalamins.

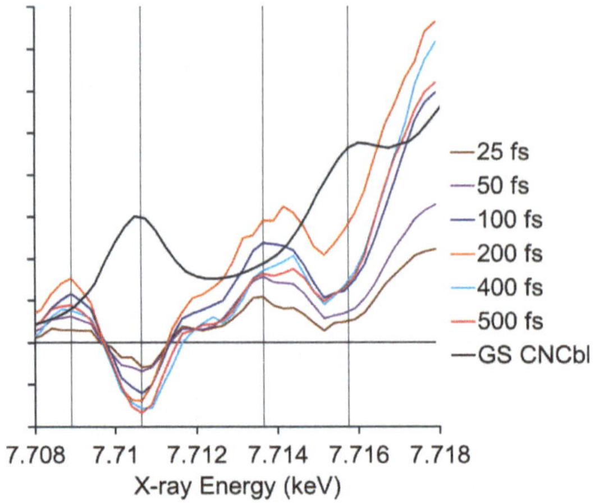

Fig. 13 Pre-edge region for ground state CNCbl (*black*) and time-resolved difference spectra at the indicated time delays.

6.1.2 XANES spectra

In the XANES region, the energy of the incident X-ray photon exceeds the binding energy of the core electron. Although it is empirically well established that the energy of the absorption threshold is correlated with metal-ion oxidation state, and this has often been used to assign oxidation states, this correlation is complicated by the fact that bond-length also changes with oxidation state. Indeed, some of the earliest attempts to explain trends in XANES spectra noted that the energy of the edge varies as $1/R^2$, where R is the metal-ligand distance (Bianconi, Dell'Ariccia, Gargano, & Natoli, 1983; Natoli, 1983). This effect, which is sometimes referred to as a "shape resonance," can be rationalized as X-ray excited photoelectron being temporarily trapped in a potential well defined by the potential of the neighboring atoms, in analogy with the classic particle-in-a-box model (Stöhr, 1992). Thus, time-resolved changes in the edge energy provide at least qualitative insight into the bond lengths around the X-ray absorbing atom. Importantly, XANES spectra are not explicitly sensitive to the electronic structure of the molecule and thus can be modeled, at least to a first approximation, by considering only the molecular structure.

In order to be quantitative, one would like to perform theoretical calculations of XANES spectra. While this could, in principle, be done using TD-DFT, the size of the basis set needed to describe the final state is unrealistically large. For this reason, most theoretical descriptions of the XANES region rely on some sort of scattering treatment. One of the most common programs is FEFF (Rehr, Kas, Vila, Prange, & Jorissen, 2010), which utilizes a real-space Green's function approach to model the X-ray absorption cross-section. This is an efficient code; by avoiding the need to describe explicitly the final-state wave function, FEFF is quite fast. It does an excellent job of modeling the EXAFS region and is one of the most widely used programs for simulating molecular X-ray absorption spectra. In order to obtain reliable XANES spectra, it is necessary to use the full multiple-scattering (FMS) option of FEFF since the multiple-scattering path expansion that FEFF normally uses to speed up calculations often fails to converge near the edge threshold. With FMS, FEFF can give a good description of the XANES region and was used to explain the time-resolved XANES spectra of methylcobalamin (Subramanian et al., 2018).

One of the weaknesses of FEFF for descriptions of the XANES region is that FEFF models the electron density using the "muffin-tin" approximation, where each atomic potential is spherically averaged and a constant interstitial potential is used to describe the areas between atoms. While the muffin-tin approximation works well at higher energies, it can lead to errors in the XANES region. One alternative is the program FDMNES, which uses a finite-difference method to calculate the potential over a grid of points centered on the absorbing atom (Joly, 2001). Although this is computationally much slower than FEFF, we have found this to give better simulations of the edge (Michocki et al., 2019; Miller et al., 2017, 2018).

Regardless of the program that is used, the computational cost increases rapidly as the size of the cluster increases. For this reason, it is common to truncate the structure, typically keeping atoms within 6–8 Å of the absorbing atom. It is common to utilize published crystal structures for atomic coordinates. It is important to note, however, that crystallographic C—H distances are often set to be significantly shorter than the true C—H distance, which can affect the calculated XANES.

6.2 Interpretation of time-dependent changes in cobalamin spectra

To the extent that XANES spectra depend solely on nearest-neighbor distance, one might expect them to be insensitive to *which* distance changes. This is largely the case, as illustrated in Fig. 14. This shows the error that is introduced by changing the axial bond lengths from a starting position of 2.1 Å (R_{lower}:Co-$N_{benzimidazole}$ distance) and 2.0 Å (R_{upper}:Co-$C_{cyanide}$ distance). Over a fairly wide range of distances, one finds that for the isotropic XANES, elongation of either axial bond length can be approximately

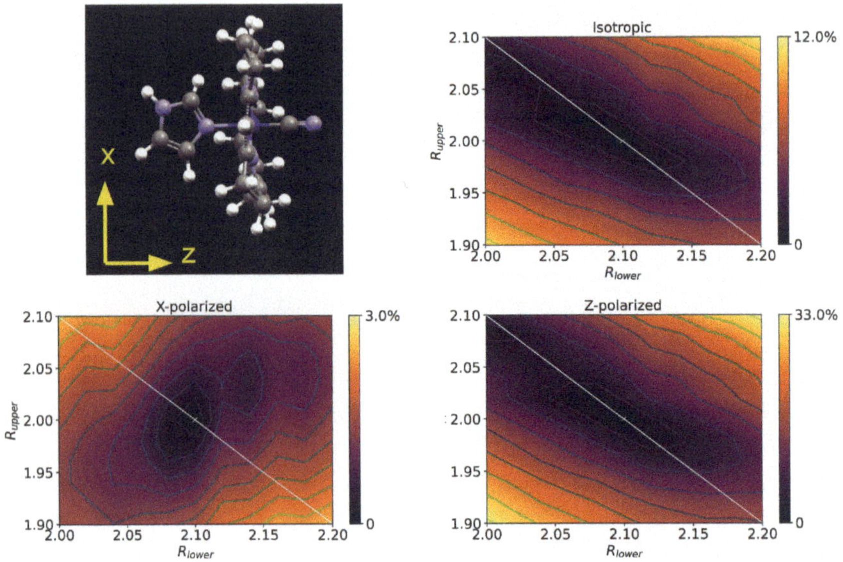

Fig. 14 Effect of geometric distortion on calculated XANES. Molecular axes are defined in upper left. Upper right shows the root-mean-square change in the isotropic XANES, normalized to the edge jump, as the upper and lower Co-ligand distances are changed. Bottom row shows the calculated errors for x and z polarized spectra.

compensated by contracting the other distance, giving rise to a diagonal valley with only small errors. These calculations are for noise-free simulations; the sensitivity is even lower for noisy experimental data.

However, this does not mean that elongation of the two different bond is indistinguishable, at least not if polarized spectra can be measured. For XANES spectra polarized along the x axis (see Fig. 1 for definition of the orientations), there is little ambiguity in R_{lower}. This reflects the fact that the benzimidazole ligand lies in the x-z plane so that changes in Co-$N_{benzimidazole}$ alter not only the z -polarized XANES, but also the x polarized XANES. In isotropic spectra, this sensitivity is overwhelmed by the much larger variation in the z polarized XANES (note the difference in scale in Fig. 14), and can only be reliably resolved using polarized measurements.

Finally, we return to the question of whether XANES alone can be used to distinguish between different structural models for the evolution of the excited state. As noted above, if the XANES spectrum were a linear function of distance, it would be impossible to distinguish between exponential decay of $A \rightarrow B$ and critically-damped harmonic motion from $A \rightarrow B$ since these

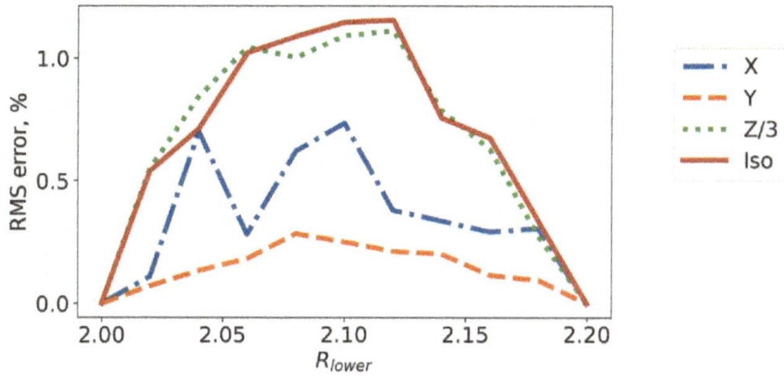

Fig. 15 RMS error, normalized to the edge jump, for interpolating spectra with different R_{lower}. Equivalent results are seen for interpolation of R_{upper}, although with less sensitivity to polarization.

would have the same average Co-ligand distance. Fortunately, XANES spectra are not linear in distance, as illustrated in Fig. 15. A spectrum S_{test} with R_{test} cannot be modeled as a linear combination $S_0(R_0)$ and $S_1(R_1)$:

$$S_{test} = \frac{R_1 - R_{test}}{R_1 - R_0} S_0 + \frac{R_{test} - R_0}{R_1 - R_0} S_1 \tag{19}$$

This doesn't guarantee that it will be possible to distinguish between different structural models, particularly in cases where the excited state wavepacket evolves along multiple different trajectories, but it does point to the potential of XANES, if available with a sufficiently good signal to noise ratio, to answer critical questions about excited state structural evolution.

Acknowledgments

This work was supported by the National Science Foundation NSF-CHE 1464584 and NSF-CHE 1836435 to R.J.S. P.D. was supported by an REU award (NSF-PHY 1757574). Time-resolved XANES spectra were measured at the Linac Coherent Light Source (LCLS), SLAC National Accelerator Laboratory, which is supported by the U.S. Department of Energy, Office of Science, Office of Basic Energy Sciences under Contract No. DE-AC02-76SF00515. In particular, we thank Tim B. van Driel, Roberto Alonso-Mori and Marco Reinhard at LCLS for their assistance with the power dependence measurements in Fig. 6–8.

References

Bacellar, C., Kinschel, D., Mancini, G. F., Ingle, R. A., Rouxel, J., Cannelli, O., et al. (2020). Spin cascade and doming in ferric haems: Femtosecond X-ray absorption and X-ray emission studies. *Proceedings of the National Academy of Science*, *117*, 21914–21920. https://doi.org/10.1073/pnas.2004490117.

Bianconi, A., Dell'Ariccia, M., Gargano, A., & Natoli, C. R. (1983). Bond length determination using XANES. In A. Bianconi, L. Incoccia, & S. Stipcich (Eds.), *EXAFS and near edge structure* (pp. 57–61). Berlin Heidelberg, Berlin, Heidelberg: Springer. https://doi.org/10.1007/978-3-642-50098-5_8.

Britz, A., Abraham, B., Biasin, E., van Driel, T. B., Gallo, A., Garcia-Esparza, A. T., et al. (2020). Resolving structures of transition metal complex reaction intermediates with femtosecond EXAFS. *Physical Chemistry Chemical Physics*, *22*, 2660–2666. https://doi.org/10.1039/C9CP03483H.

Cole, A. G., Yoder, L. M., Shiang, J. J., Anderson, N. A., Walker, L. A., II, Banaszak Holl, M. M., et al. (2002). Time-resolved spectroscopic studies of B_{12} coenzymes: A comparison of the primary photolysis mechanism in methyl-, ethyl-, n-propyl-, and 5′-deoxyadenosylcobalamin. *Journal of the American Chemical Society*, *124*, 434–441. https://doi.org/10.1021/ja011628s.

de Groot, F. (2005). Multiplet effects in X-ray spectroscopy. *Coordination Chemistry Reviews*, *249*, 31–63. https://doi.org/10.1016/j.ccr.2004.03.018.

Garabato, B. D., Kumar, N., Lodowski, P., Jaworska, M., & Kozlowski, P. M. (2016). Electronically excited states of cob(II) alamin: Insights from CASSCF/XMCQDPT2 and TD-DFT calculations. *Physical Chemistry Chemical Physics*, *18*, 4513–4526. https://doi.org/10.1039/c5cp06439b.

Hahn, J. E., Scott, R. A., Hodgson, K. O., Doniach, S., Desjardins, S. R., & Solomon, E. I. (1982). Observation of an electric quadrupole transition in the x-ray absorption-spectrum of a cu(II) complex. *Chemical Physics Letters*, *88*, 595–598. https://doi.org/10.1016/0009-2614(82)85016-1.

Hannibal, L., Smith, C. A., Smith, J. A., Axhemi, A., Miller, A., Wang, S., et al. (2009). High resolution crystal structure of the methylcobalamin analogues ethylcobalamin and butylcobalamin by X-ray synchrotron diffraction. *Inorganic Chemistry*, *48*, 6615–6622. https://doi.org/10.1021/ic900590p.

Joly, Y. (2001). X-ray absorption near-edge structure calculations beyond the muffin-tin approximation. *Physical Review B*, *63*, 125120. https://doi.org/10.1103/PhysRevB.63.125120.

Joly, Y., & Grenier, S. (2016). Theory of X-ray absorption near edge structure. In J. A. van Bokhoven, & C. Lamberti (Eds.), *X-ray absorption and X-ray emission spectroscopy: Theory and applications* (pp. 73–97). Incorporated, New York: John Wiley & Sons.

Jones, A. R., Russell, H. J., Greetham, G. M., Towrie, M., Hay, S., & Scrutton, N. S. (2012). Ultrafast infrared spectral fingerprints of vitamin B_{12} and related cobalamins. *Journal of Physical Chemistry A*, *116*, 5586–5594. https://doi.org/10.1021/jp304594d.

Katayama, T., Northey, T., Gawelda, W., Milne, C. J., Vanko, G., Lima, F. A., et al. (2019). Tracking multiple components of a nuclear wavepacket in photoexcited cu(I)-phenanthroline complex using ultrafast X-ray spectroscopy. *Nature Communications*, *10*, 3606. https://doi.org/10.1038/s41467-019-11499-w.

Kinschel, D., Bacellar, C., Cannelli, O., Sorokin, B., Katayama, T., Mancini, G. F., et al. (2020). Femtosecond X-ray emission study of the spin cross-over dynamics in haem proteins. *Nature Communications*, *11*, 4145. https://doi.org/10.1038/s41467-020-17923-w.

Kornobis, K., Kumar, N., Wong, B. M., Lodowski, P., Jaworska, M., Andruniow, T., et al. (2011). Electronically excited states of vitamin B12: Benchmark calculations including time-dependent density functional theory and correlated ab initio methods. *Journal of Physical Chemistry A*, *115*, 1280–1292. https://doi.org/10.1021/jp110914y.

Kratky, C., Farber, G., Gruber, K., Wilson, K., Dauter, Z., Nolting, H.-F., et al. (1995). Accurate structural data demystify B12: High-resolution solid-state structure of aquocobalamin perchlorate and structure analysis of the aquocobalamin ion in solution. *Journal of the American Chemical Society*, *117*, 4654–4670. https://doi.org/10.1021/ja00121a022.

Lemke, H. T., Kjaer, K. S., Hartsock, R., van Driel, T. B., Chollet, M., Glowina, J. M., & Cammarata, M. (2017). Coherent structural trapping through wave packet dispersion during photoinduced spin state switching. *Nature Communications*, *8*(2), 2–8. https://doi.org/10.1038/ncomms15342.

Lodowski, P., Jaworska, M., Kornobis, K., Andruniow, T., & Kozlowski, P. M. (2011). Electronic and structural properties of low-lying excited states of vitamin B_{12}. *Journal of Physical Chemistry B*, *115*, 13304–13319. https://doi.org/10.1021/jp200911y.

Lott, W. B., Chagovetz, A. M., & Grissom, C. B. (1995). Alkyl radical geometry controls geminate cage recombination in alkylcobalamins. *Journal of the American Chemical Society*, *117*, 12194–12201. https://doi.org/10.1021/ja00154a020.

Mebs, S., Henn, J., Dittrich, B., Paulmann, C., & Luger, P. (2009). Electron densities of three B_{12} vitamins. *Journal of Physical Chemistry A*, *113*, 8366–8378. https://doi.org/10.1021/jp902433x.

Michocki, L. B., Miller, N. A., Alonso-Mori, R., Britz, A., Deb, A., Glownia, J. M., et al. (2019). Probing the excited state of Methylcobalamin using polarized time-resolved X-ray absorption spectroscopy. *Journal of Physical Chemistry B*, *123*, 6042–6048. https://doi.org/10.1021/acs.jpcb.9b05854.

Miller, N. A., Deb, A., Alonso-Mori, R., Garabato, B. D., Glownia, J. M., Kiefer, L. M., et al. (2017). Polarized XANES monitors femtosecond structural evolution of Photoexcited vitamin B-12. *Journal of the American Chemical Society*, *139*, 1894–1899. https://doi.org/10.1021/jacs.6b11295.

Miller, N. A., Deb, A., Alonso-Mori, R., Glownia, J. M., Kiefer, L. M., Konar, A., et al. (2018). Ultrafast X-ray absorption near edge structure reveals ballistic excited state structural dynamics. *Journal of Physical Chemistry A*, *122*, 4963–4971. https://doi.org/10.1021/acs.jpca.8b04223.

Miller, N. A., Kaneshiro, A. K., Konar, A., Alonso-Mori, R., Britz, A., Deb, A., et al. (2020). The photoactive excited state of the B12-based photoreceptor CarH. *The Journal of Physical Chemistry B*, *124*, 10732–10738. https://doi.org/10.1021/acs.jpcb.0c09428.

Miller, N. A., Michocki, L. B., Alonso-Mori, R., Britz, A., Deb, A., DePonte, D. P., et al. (2019). Antivitamins B-12 in a microdrop: The excited-state structure of a precious sample using transient polarized X-ray absorption near-edge structure. *Journal of Physical Chemistry Letters*, *10*, 5484–5489. https://doi.org/10.1021/acs.jpclett.9b02202.

Miller, N. A., Michocki, L. B., Konar, A., Alonso-Mori, R., Deb, A., Glownia, J. M., et al. (2020). Ultrafast XANES monitors femtosecond sequential structural evolution in Photoexcited coenzyme B-12. *Journal of Physical Chemistry B*, *124*, 199–209. https://doi.org/10.1021/acs.jpcb.9b09286.

C.R. Natoli, Near Edge Absorption Structure in the Framework of the Multiple Scattering Model. Potential resonance or barrier effects?, Springer, Berlin, 1983, pp. 43–56. doi: https://doi.org/10.1007/978-3-642-50098-5_7.

Neese, F. (2012). The ORCA program system. *Wiley Interdisciplinary Reviews: Computational Molecular Science*, *2*, 73–78. https://doi.org/10.1002/wcms.81.

Ouyang, L. Z., Rulis, P., Ching, W. Y., Slouf, M., Nardin, G., & Randaccio, L. (2005). Electronic structure and bonding in hydroxocobalamin. *Spectrochimica Acta Part A–Molecular and Biomolecular Spectroscopy*, *61*, 1647–1652. https://doi.org/10.1016/j.saa.2004.11.066.

Padmanabhan, S., Jost, M., Drennan, C. L., & Elias-Arnanz, M. (2017). A new facet of vitamin B_{12}: Gene regulation by cobalamin-based photoreceptors. *Annual Review of Biochemistry*, *86*, 485–514. https://doi.org/10.1146/annurev-biochem-061516-044500.

Padmanabhan, S., Perez-Castano, R., & Elias-Arnanz, M. (2019). B-12-based photoreceptors: From structure and function to applications in optogenetics and synthetic biology. *Current Opinion in Structural Biology*, *57*, 47–55. https://doi.org/10.1016/j.sbi.2019.01.020.

Pallares, I. G., & Brunold, T. C. (2014). Spectral and electronic properties of Nitrosylcobalamin. *Inorganic Chemistry*, *53*, 7676–7691. https://doi.org/10.1021/ic500986x.

Peng, J. A., Tang, K. C., McLoughlin, K., Yang, Y., Forgach, D., & Sension, R. J. (2010). Ultrafast excited-state dynamics and photolysis in base-off B-12 coenzymes and analogues: Absence of the trans-nitrogenous ligand opens a channel for rapid nonradiative decay. *Journal of Physical Chemistry B*, *114*, 12398–12405. https://doi.org/10.1021/jp104641u.

Prieto, L., Neuburger, M., Spingler, B., & Zelder, F. (2016). Inorganic cyanide as protecting Group in the stereospecific Reconstitution of vitamin B12 from an artificial green secocorrinoid. *Organic Letters*, *18*, 5292–5295. https://doi.org/10.1021/acs.orglett.6b02611.

Rehr, J. J., Kas, J. J., Vila, F. D., Prange, M. P., & Jorissen, K. (2010). Parameter-free calculations of X-ray spectra with FEFF9. *Physical Chemistry Chemical Physics*, *12*, 5503–5513. https://doi.org/10.1039/B926434E.

Salerno, E. V., Miller, N. A., Konar, A., Li, Y., Kieninger, C., Kräutler, B., et al. (2020). Ultrafast excited state dynamics and fluorescence from vitamin B12 and organometallic [Co]–C≡C–R cobalamins. *The Journal of Physical Chemistry B*, *124*, 6651–6656. https://doi.org/10.1021/acs.jpcb.0c04886.

Salerno, E. V., Miller, N. A., Konar, A., Salchner, R., Kieninger, C., Wurst, K., et al. (2020). Exceptional photochemical stability of the co–C bond of alkynyl cobalamins, potential antivitamins B12 and Core elements of B12-based biological vectors. *Inorganic Chemistry*, *59*, 6422–6431. https://doi.org/10.1021/acs.inorgchem.0c00453.

Sension, R. J., Cole, A. G., Harris, A. D., Fox, C. C., Woodbury, N. W., Lin, S., et al. (2004). Photolysis and recombination of adenosylcobalamin bound to glutamate mutase. *Journal of the American Chemical Society*, *126*, 1598–1599. https://doi.org/10.1021/ja0396910.

Sension, R. J., Harris, D. A., & Cole, A. G. (2005). Time-resolved spectroscopic studies of B_{12} coenzymes: A comparison of the influence of solvent on the primary photolysis mechanism and geminate recombination of methyl-, ethyl-, n-propyl-, and 5′-deoxyadenosylcobalamin. *Journal of Physical Chemistry B*, *109*, 21954–21962. https://doi.org/10.1021/jp053202w.

Sension, R. J., Harris, D. A., Stickrath, A., Cole, A. G., Fox, C. C., & Marsh, E. N. G. (2005). Time-resolved measurements of the photolysis and recombination of adenosylcobalamin bound to glutamate mutase. *Journal of Physical Chemistry B*, *109*, 18146–18152. https://doi.org/10.1021/jp052492d.

Shiang, J. J., Cole, A. G., Sension, R. J., Hang, K., Weng, Y. X., Trommel, J. S., et al. (2006). Ultrafast excited-state dynamics in vitamin B-12 and related cob(III)alamins. *Journal of the American Chemical Society*, *128*, 801–808. https://doi.org/10.1021/ja054374.

Solheim, H., Kornobis, K., Ruud, K., & Kozlowski, P. M. (2011). Electronically excited states of vitamin B12 and methylcobalamin: Theoretical analysis of absorption, CD, and MCD data. *Journal of Physical Chemistry B*, *115*, 737–748. https://doi.org/10.1021/jp109793r.

Stich, T. A., Brooks, A. J., Buan, N. R., & Brunold, T. C. (2003). Spectroscopic and computational studies of Co^{3+}-corrinoids: Spectral and electronic properties of the B_{12} cofactors and biologically relevant precursors. *Journal of the American Chemical Society*, *125*, 5897–5914. https://doi.org/10.1021/ja029328d.

Stich, T. A., Buan, N. R., & Brunold, T. C. (2004). Spectroscopic and computational studies of Co2+corrinoids: Spectral and electronic properties of the biologically Relevant Base-on and base-off forms of Co2+cobalamin. *Journal of the American Chemical Society*, *126*, 9735–9749. https://doi.org/10.1021/ja0481631.

Stich, T. A., Buan, N. R., Escalante-Semerena, J. C., & Brunold, T. C. (2005). Spectroscopic and computational studies of the ATP: Corrinoid adenosyltransferase (CobA) from salmonella enterica: Insights into the mechanism of adenosylcobalamin biosynthesis. *Journal of the American Chemical Society, 127*, 8710–8719. https://doi.org/10.1021/ja042142p.

Stickrath, A., Carroll, E. C., Dai, X., Harris, D. A., Rury, A., Smith, B., et al. (2009). Solvent-dependent cage dynamics of small nonpolar radicals: Lessons from the photodissociation and geminate recombination of alkylcobalamins. *Journal of Physical Chemistry A, 113*, 8513–8522. https://doi.org/10.1021/jp9017986.

Stöhr, J. (1992). *Experimental and calculated K-shell spectra of simple free molecules, NEXAFS spectroscopy* (pp. 79–113). Berlin, Heidelberg: Springer Berlin Heidelberg. https://doi.org/10.1007/978-3-662-02853-7_4.

Subramanian, G., Zhang, X., Kodis, G., Kong, Q., Liu, C., Chizmeshya, A., et al. (2018). Direct structural and chemical characterization of the photolytic intermediates of methylcobalamin using time-resolved X-ray absorption spectroscopy. *Journal of Physical Chemistry Letters, 9*, 1542–1546. https://doi.org/10.1021/acs.jpclett.8b00083.

Taylor, R. T., Smucker, L., Hanna, M. L., & Gill, J. (1973). Aerobic photoiysis of alkylcobalamins: Quantum yields and light-action spectra. *Archives of Biochemistry and Biophysics, 156*, 521–533. https://doi.org/10.1016/0003-9861(73)90301-9.

Wiley, T. E., Miller, N. A., Miller, W. R., Sofferman, D. L., Lodowski, P., Toda, M. J., et al. (2018). Off to the races: Comparison of excited state dynamics in vitamin B_{12} derivatives hydroxocobalamin and aquocobalamin. *The Journal of Physical Chemistry A, 122*, 6693–6703. https://doi.org/10.1021/acs.jpca.8b06103.

CHAPTER FOURTEEN

Electronic structure studies of free and enzyme-bound B_{12} species by magnetic circular dichroism and complementary spectroscopic techniques

Laura D. Elmendorf and Thomas C. Brunold*
Department of Chemistry, University of Wisconsin—Madison, Madison, WI, United States
*Corresponding author: e-mail address: brunold@chem.wisc.edu

Contents

Abstract

Electronic absorption (Abs) and circular dichroism (CD) spectroscopic techniques have been used successfully for over half a century in studies of free and enzyme-bound B_{12} species. More recently, magnetic circular dichroism (MCD) spectroscopy and other complementary techniques have provided an increasingly detailed understanding of the electronic structure of cobalamins. While CD spectroscopy measures the difference in the absorption of left- and right-circularly polarized light, MCD spectroscopy adds the application of a magnetic field parallel to the direction of light propagation. Transitions that are formally forbidden according to the Abs and CD selection rules, such as ligand field (or $d \rightarrow d$) transitions, can gain MCD intensity through spin-orbit coupling. As such, MCD spectroscopy provides a uniquely sensitive probe of the different binding modes, Co oxidation states, and axial ligand environments of B_{12} species in enzyme active sites, and thus the distinct reactivities displayed by these species. This chapter summarizes representative MCD studies of free and enzyme-bound B_{12} species,

Methods in Enzymology, Volume 669
ISSN 0076-6879
https://doi.org/10.1016/bs.mie.2022.02.002

including those present in adenosyltransferases, isomerases, and reductive dehalogenases. Complementary spectroscopic and computational data are also presented and discussed where appropriate.

1. Overview

The rich history of electronic absorption (Abs) and circular dichroism (CD) spectroscopic studies of B_{12} species goes back over half a century, and by the 1960s much of the nomenclature used to describe characteristic B_{12} spectral features had been established (Firth, Hill, Pratt, Williams, & Jackson, 1967; Toohey, 1965). Since then, magnetic circular dichroism (MCD) spectroscopy and other complementary techniques have provided an increasingly detailed understanding of the electronic structure of cobalamins. CD spectroscopy measures the difference in the absorption of left- and right-circularly polarized light (Johnson, 2000). MCD spectroscopy adds the application of a magnetic field parallel to the direction of light propagation. MCD spectroscopy is an extremely powerful tool for studying enzymes and cofactors containing transition metal ions; it does not require isotope enrichment, works with paramagnetic and diamagnetic species, probes both the ground and excited states, and, because all electronic transitions in the visible/near-IR region originate from the metalloenzyme active site or metallocofactor, is highly site selective (Johnson, 2000). Additionally, the intensities of the electronic transitions dominating CD and MCD spectra are governed by different selection rules than those dominating Abs spectra, often allowing for electronic transitions that are too weak or too close in energy to be resolved in Abs spectra to be probed using these complementary methods (Johnson, 2000; Solomon, Kirk, Gamelin, & Pulver, 1995; Solomon, Neidig, & Schenk, 2004).

For an electronic transition from the ground state to an excited state, the Abs intensity is proportional to the square of the transition moment integral (Solomon et al., 1995):

$$I^{Abs} \propto \langle \Psi_{GS} | \hat{\mu} | \Psi_{ES} \rangle^2$$

Ψ_{GS} and Ψ_{ES} are the ground and excited state wavefunctions, respectively, and $\hat{\mu}$ is the transition moment operator. This can be broken down into two terms of significance, electric dipole intensity (I^{ED}) and magnetic dipole intensity (I^{MD}) (Johnson, 2000):

$$I^{ED} \propto \langle \Psi_{GS} | \hat{\mu}_e | \Psi_{ES} \rangle^2$$
$$I^{MD} \propto \langle \Psi_{GS} | \hat{\mu}_m | \Psi_{ES} \rangle^2$$

$\hat{\mu}_e$ and $\hat{\mu}_m$ are the electric dipole and magnetic dipole moment operators, respectively. Electric dipole transitions are usually several orders of magnitude more intense than magnetic dipole transitions, so these dominate Abs spectra. Evaluating the transition moment integral for electric dipole transitions leads to the following selection rules: (i) transitions between states of different spin are forbidden (spin selection rule) and (ii) transitions between wavefunctions of the same parity are forbidden, i.e., only $g \rightarrow u$ and $u \rightarrow g$ transitions are allowed (orbital or Laporte selection rule). Importantly, ligand field (LF or $d \rightarrow d$) transitions are parity forbidden $g \rightarrow g$ excitations and thus have low Abs intensity (Johnson, 2000).

The CD signal intensity is governed by the rotational strength (R), which is proportional to the imaginary (Im) part of the dot product of the electric and magnetic dipole transition moments (Johnson, 2000):

$$R \propto \mathrm{Im}\{\langle \Psi_{GS} | \hat{\mu}_e | \Psi_{ES} \rangle \cdot \langle \Psi_{GS} | \hat{\mu}_m | \Psi_{ES} \rangle\}$$

Obtaining rotational strength therefore requires transitions that are both electric and magnetic dipole allowed, and the corresponding transition dipole moments must be colinear. These requirements are only fulfilled by certain low-symmetry point groups in which the electric dipole and magnetic dipole operators transform as the same irreducible representation. A CD signal therefore requires that the molecule being investigated lacks a mirror plane (more precisely, an improper rotation axis).

While the presence of a CD signal requires molecular chirality, an MCD signal is observed for all matter as a consequence of the Faraday effect (Johnson, 2000). Transitions that are formally forbidden according to the Abs and CD selection rules, such as LF transitions, can gain MCD intensity through spin-orbit coupling. Three different terms contribute to the MCD signal intensity, which is measured as the difference in extinction coefficients for left-circularly polarized (LCP) and right-circularly polarized (RCP) light, $\Delta\varepsilon = \varepsilon_{LCP} - \varepsilon_{RCP}$ (Johnson, 2000; Neese & Solomon, 1999):

$$\frac{\Delta\varepsilon}{E} = \frac{\varepsilon_{LCP} - \varepsilon_{RCP}}{E} \propto B\left\{-A_1\left(\frac{\partial f(E)}{\partial E}\right) + \left(B_0 + \frac{C_0}{kT}\right)f(E)\right\}$$

where E is the energy of the incident radiation, B is the magnetic flux density, f(E) is the absorption bandshape, k is the Boltzmann constant, T is the temperature, and the A_1, B_0, and C_0 terms depend on the transition under study. The magnetic field results in a Zeeman splitting of orbitally degenerate excited states (A_1, or the "A-term"), field-induced mixing of all electronic states (B_0, or "B-term"), and a Zeeman splitting of the ground state of paramagnetic species, causing a population difference among the spin sublevels (C_0, or "C-term"). The temperature-dependent C-term contributions dominate low-temperature MCD spectra of paramagnetic species.

MCD spectroscopy is particularly effective when used in combination with other spectroscopic techniques and computational tools, such as electron paramagnetic resonance (EPR) spectroscopy and density functional theory calculations, both discussed in other chapters. Another technique that nicely complements MCD spectroscopy is resonance Raman (rR) spectroscopy, which involves laser excitation in the visible or near-UV spectral region to enhance vibrational modes along which the chromophore distorts when undergoing an electronic transition in resonance with the laser wavelength. Because the intense bands in the visible region of the Abs spectra of B_{12} species arise from corrin $\pi \rightarrow \pi^*$ transitions, rR spectroscopy provides an excellent probe of conformational differences of the corrin ring of cobalamins in different enzyme environments (Dong, Padmakumar, Banerjee, & Spiro, 1996, 1998, 1999; Dong, Padmakumar, Maiti, Banerjee, & Spiro, 1998; Park, Mera, Escalante-Semerena, & Brunold, 2008, 2012, 2016; Park, Mera, Moore, Escalante-Semerena, & Brunold, 2015). The information it yields on vibronic coupling and the polarization of electronic transitions has been extremely helpful for assigning electronic transitions associated with key Abs, CD, and MCD spectral features displayed by B_{12} species in all three accessible Co oxidation states (Brunold, Conrad, Liptak, & Park, 2009; Liptak & Brunold, 2006; Park & Brunold, 2013; Stich, Brooks, Buan, & Brunold, 2003; Stich, Buan, & Brunold, 2004).

This chapter summarizes representative MCD studies of free and enzyme-bound B_{12} species. Complementary spectroscopic and computational data are also presented and discussed where appropriate.

2. Spectroscopic studies of free B_{12} species

2.1 Co^{III}Corrinoids

The Abs, CD, and MCD spectra of Co^{III}Cobalamins (Co^{III}Cbls; Fig. 1), which are six-coordinate, low spin ($S=0$) $3d^6$ species, are dominated by corrin $\pi \rightarrow \pi^*$ transitions. Because they are diamagnetic, Co^{III}Cbls lack the intense, temperature-dependent C-term MCD features observed

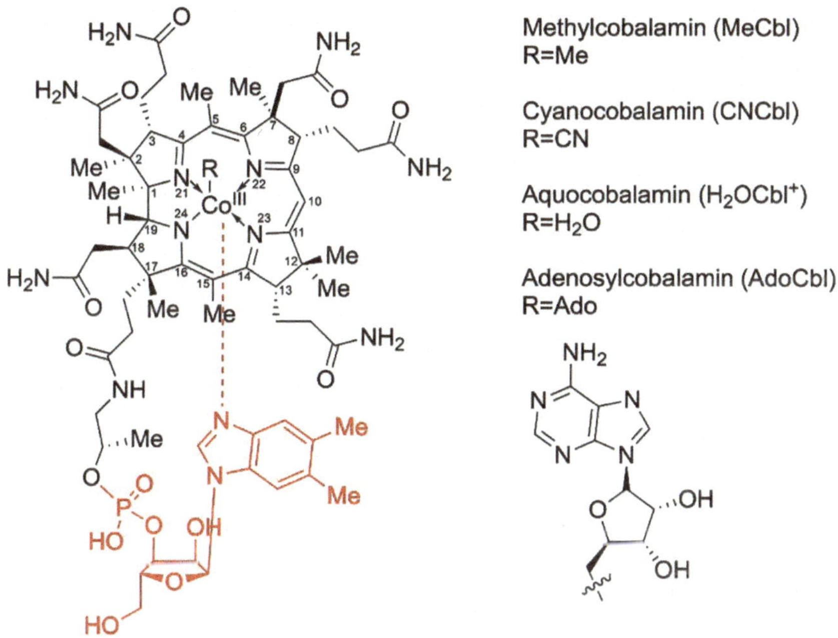

Fig. 1 Chemical structure of Co^{III}Cbls, where **R** indicates the upper axial ligand. Relevant atom numbers are included. Note that in Co^{III}cobinamides, the nucleotide loop (shown in *red*) including the coordinating dimethylbenzimidazole (DMB) is absent and a water molecule occupies the lower axial position.

for paramagnetic species. Nonetheless, their MCD spectra are still quite informative. Electronic communication between the axial ligands and corrin macrocycle causes the Abs, CD, and MCD spectra of Co^{III}Cbls to be sensitive to the axial coordination environment (Stich et al., 2003). The identity of the upper axial ligand is particularly impactful; early spectroscopic studies led to a distinction between "typical" Abs spectra displayed by the majority of Co^{III}Cbls, such as CNCbl and H_2OCbl^+, and "unique" spectra exhibited primarily by alkylcobalamins like AdoCbl and MeCbl (Firth et al., 1967).

Co^{III}Cbl Abs spectra (Fig. 2) can be divided into three regions: (i) the α/β bands at lowest energy in the visible region, (ii) the D/E bands toward higher energy, and (iii) the γ band in the UV region (Pratt, 1999; Stich et al., 2003). The α band was definitively assigned as a corrin $\pi \rightarrow \pi^*$ transition polarized along the long corrin axis (the $C^5 \cdots C^{15}$ vector in Fig. 1), while the identity of the nearby β band—whether a vibronic sideband of the α transition or a separate electronic transition—has been a matter of debate (Pratt, 1999; Salama & Spiro, 1977). Time-dependent density functional theory (TD-DFT) computations failed to definitively settle this question, as the choice of functional greatly impacted the number of electronic transitions predicted in this region (Andruniów et al., 2008; Kornobis et al., 2011; Solheim, Kornobis, Ruud, & Kozlowski, 2011). However, recent experimental studies using triply resonant sum frequency (TRSF) spectroscopy (Handali et al., 2018) and ultrafast Abs and fluorescence spectroscopy (Salerno et al., 2020) confirmed that the β feature in "typical" Abs spectra arises from a vibrational progression rather than a distinct electronic transition. This assignment is also supported by the presence of a single weak feature coinciding with the α band in the MCD spectra of CNCbl and H_2OCbl^+ (Stich et al., 2003). While the α/β peaks in the "unique" Abs spectra observed for MeCbl and AdoCbl were also attributed to the origin and vibrational sideband of a single corrin $\pi \rightarrow \pi^*$ transition (Stich et al., 2003), at least one additional electronic transition must be present in close energetic proximity, as the corresponding MCD spectra exhibit two oppositely signed features in this region (Fig. 2). Notably, "typical" and "unique" Abs spectra differ with respect to the energy of the α/β bands, which red-shift as the σ-donor strength of the upper axial ligand increases from H_2OCbl^+ to CNCbl to MeCbl (Fig. 2).

The D/E and γ regions of typical $Co^{III}Cbl$ Abs spectra further demonstrate the power of CD and MCD spectroscopy to resolve overlapping Abs features.

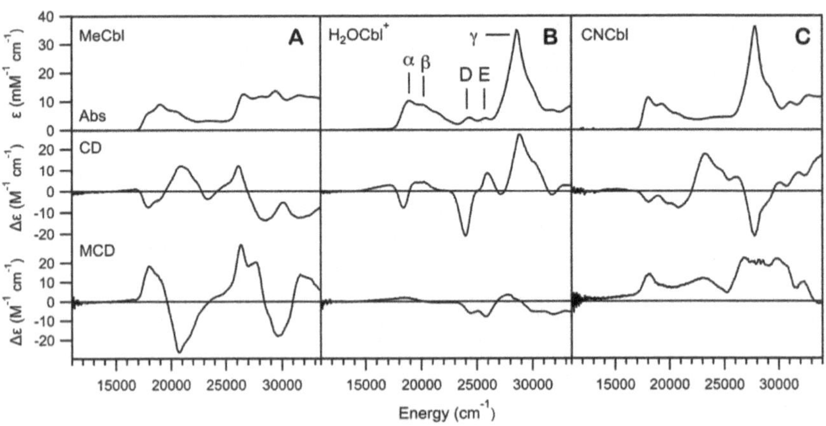

Fig. 2 Absorption (*top*), CD (*center*), and 7T MCD (*bottom*) spectra recorded at 4.5 K of (A) MeCbl, (B) H_2OCbl^+, and (C) CNCbl. Band designations are given for H_2OCbl^+. *Adapted from Stich, T. A., Brooks, A. J., Buan, N. R., & Brunold, T. C. (2003). Spectroscopic and computational studies of Co^{3+} corrinoids: Spectral and electronic properties of the B_{12} cofactors and biologically relevant precursors.* Journal of the American Chemical Society, 125, *5897–5914.*

The two prominent bands in the D/E region of the Abs spectrum of H_2OCbl^+ arise from two separate electronic transitions, as their spacing ($\sim$1500 cm^{-1}) exceeds the energy of the corrin-based vibrational modes coupled to these electronic transitions (Stich et al., 2003). The presence of two intense CD features offset from those bands indicates that at least four transitions actually contribute to this region. TD-DFT calculations revealed that the most intense Abs peak in the D/E region can be attributed to a corrin HOMO-1 → LUMO ($\pi \rightarrow \pi^*$) transition (Stich et al., 2003). Meanwhile, the features in the γ region are particularly distinctive for different Co^{III}Cbls. In "typical" Abs spectra, the γ region is dominated by a single intense band arising from a corrin $\pi \rightarrow \pi^*$ transition polarized along the short corrin axis (the Co···C^{10} vector in Fig. 1) (Salama & Spiro, 1977). In "unique" Abs spectra, the intensity of the γ band is redistributed over several weaker features of similar intensity in the UV spectral region.

The remarkable differences observed between the Abs, CD, and MCD spectra of different Co^{III}Cbl species are the result of subtle but important contributions of the axial ligands to the corrin-based frontier orbitals. TD-DFT computations, which accurately reproduced the key experimental features and trends (Fig. 3), revealed that the distinction between "typical" and "unique" Abs spectra can be understood in terms of differences in σ-donation from the upper axial ligand to the Co^{III} ion (Stich et al., 2003). With increasing σ-donor strength of the upper axial ligand, the contribution from the formally unoccupied Co $3d_{z^2}$ orbital to the HOMO increases, which induces a stronger σ-antibonding interaction between the Co^{III} ion and lower axial base. This raises the energy of the HOMO and red-shifts the HOMO → LUMO transition. Increased σ-donation also decreases the effective nuclear charge of the Co^{III} ion, raising the relative energies of the filled Co 3d orbitals and thus leading to more extensive mixing of these orbitals with the corrin π-based frontier orbitals. As a result, the number of filled MOs directly beneath the HOMO increases, causing the appearance of multiple, similarly intense features in the γ region of "unique" Abs spectra (Stich et al., 2003).

Substitution of the lower axial DMB ligand also has a significant effect on the Abs, CD, and MCD spectra. This is nicely illustrated by Co^{III}cobinamide ($Co^{III}Cbi^+$) species, natural precursors of cobalamins that lack the nucleotide loop including the dimethylbenzimidazole (DMB) moiety and instead bind H_2O as the lower axial ligand in aqueous solution, which makes them ideal models of "base-off" Co^{III}Cbls at neutral pH (Stich et al., 2003). In the Abs spectrum of $AdoCbi^+$, the α band is

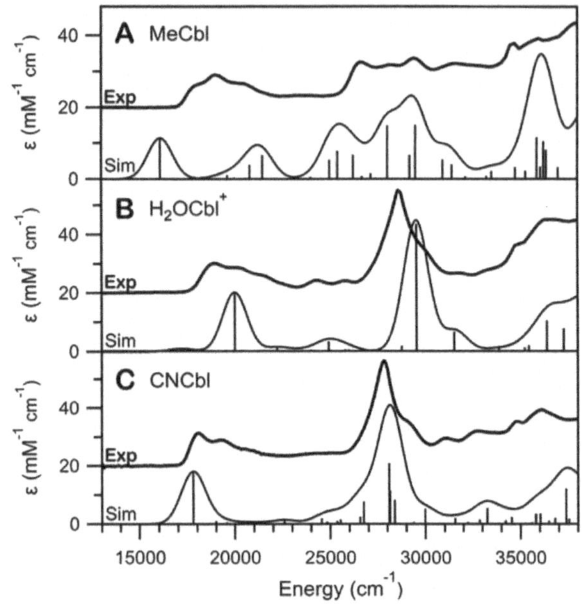

Fig. 3 Experimental and TD-DFT computed Abs spectra for (A) MeCbl, (B) H_2OCbl^+, and (C) CNCbl. The computed spectra were uniformly red shifted by 5500 cm^{-1} to facilitate a comparison with the experimental data. *Adapted from Stich, T. A., Brooks, A. J., Buan, N. R., & Brunold, T. C. (2003). Spectroscopic and computational studies of Co^{3+} corrinoids: Spectral and electronic properties of the B_{12} cofactors and biologically relevant precursors.* Journal of the American Chemical Society, 125, *5897–5914.*

blue-shifted by ~3000 cm^{-1} from its counterpart in the AdoCbl Abs spectrum (Fig. 4) (Reig, Conrad, & Brunold, 2012). This shift was attributed to the elimination of the moderate Co–N(DMB) σ* interaction in the HOMO upon replacement of the DMB ligand with a more weakly σ-donating water molecule, which results in a sizeable stabilization of the donor MO for this transition (Stich et al., 2003).

2.2 Co^{II}Corrinoids

Co^{II}corrinoids can be formed in several different ways, such as via (i) homolytic cleavage of the Co—C bond of alkylcobalamins, as is the case in the catalytic cycles of AdoCbl-dependent enzymes, (ii) one-electron reduction of a Co^{III}corrinoid species, and (iii) oxidation of Co^{I}Cbl, chemically or accidentally in the catalytic cycles of MeCbl-dependent enzymes (Stich et al., 2004). Co^{II}corrinoids are low-spin ($S=1/2$) 3d^7 species that are usually

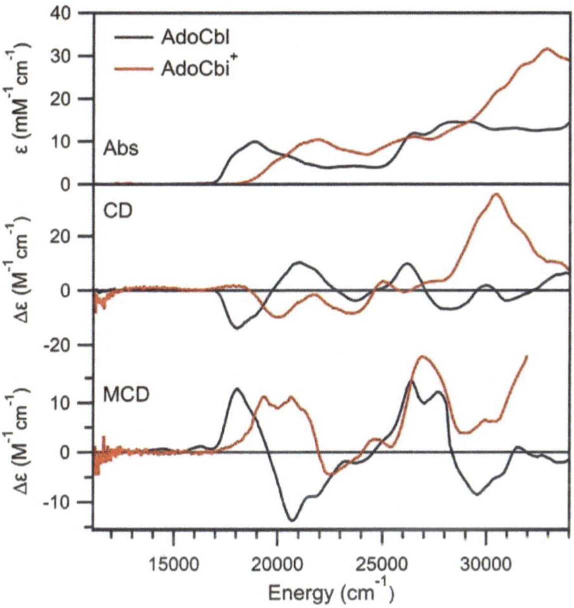

Fig. 4 Comparison of the 4.5 K Abs (*top*), CD (*center*), and 7 T MCD (*bottom*) spectra of AdoCbl and AdoCbi+. *Adapted from Stich, T. A., Brooks, A. J., Buan, N. R., & Brunold, T. C. (2003). Spectroscopic and computational studies of Co^{3+} corrinoids: Spectral and electronic properties of the B_{12} cofactors and biologically relevant precursors.* Journal of the American Chemical Society, 125, *5897–5914.*

five-coordinate, lacking an upper axial ligand. Because of their paramagnetic ground states, Co^{II}corrinoids are excellent targets for MCD studies. Distinct transitions dominate the different regions of the Abs, CD, and MCD spectra of Co^{II}corrinoids (Fig. 5): $d \rightarrow d$ transitions below ~17,000 cm^{-1}, metal-to-ligand charge transfer (MLCT) transitions from ~17,000 to 20,000 cm^{-1}, and corrin $\pi \rightarrow \pi^*$ transition above ~20,000 cm^{-1} (Brooks, Vlasie, Banerjee, & Brunold, 2005). The Co $d \rightarrow d$ transitions are parity forbidden and therefore contribute minimally to the Abs spectra but gain significant MCD intensity through spin-orbit coupling. The opposite effect is observed for corrin $\pi \rightarrow \pi^*$ transitions; e.g., the transition responsible for the dominant Abs band in the visible region with a maximum at ~21,000 cm^{-1} carries very little intensity in the MCD spectrum, a testament to the different selection rules for these techniques (Park et al., 2016; Park & Brunold, 2013; Stich et al., 2004).

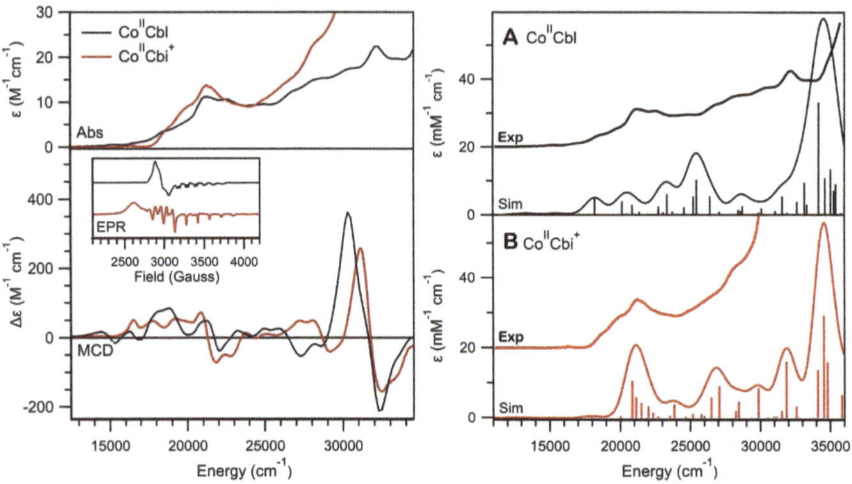

Fig. 5 (*Left*): Abs and 7 T MCD spectra at 4.5 K of $Co^{II}Cbl$ (*black*) and $Co^{II}Cbi^+$ (*red*). The corresponding X-band EPR spectra at 40 K are shown in the inset. (*Right*): Experimental 4.5 K and TD-DFT-computed Abs spectra of (A) $Co^{II}Cbl$ and (B) $Co^{II}Cbi^+$. *Adapted from Stich, T. A., Buan, N. R., & Brunold, T. C. (2004). Spectroscopic and computational studies of Co^{2+} corrinoids: Spectral and electronic properties of the biologically relevant base-on and base-off forms of Co^{2+} cobalamin.* Journal of the American Chemical Society, 126, *9735–9749.*

The prominent feature at ~21,000 cm^{-1} in the $Co^{II}Cbl$ Abs spectrum originates from a corrin $\pi \rightarrow \pi^*$ transition involving the same donor and acceptor orbitals as the transition associated with the α/β bands of $Co^{III}Cbls$. However, because the donor MO for this transition contains less Co 3d character in $Co^{II}Cbl$ than in $Co^{III}Cbl$, the transition energy is less affected by changes in the lower axial ligand (Stich et al., 2004). As a result, the Abs spectra of $Co^{II}Cbl$ and $Co^{II}Cbi^+$ are quite similar. A much greater difference is observed, however, for the LF transitions, only discernable in the MCD spectra. The weaker σ-donation of H_2O compared to the DMB ligand leads to a sizable stabilization of the occupied Co $3d_{z^2}$-based MO, causing a significant blue-shift of the LF transitions originating from this orbital. MCD spectroscopy is therefore much better suited than Abs spectroscopy for studying the axial ligand environment of $Co^{II}Cbls$. EPR spectroscopy complements MCD spectroscopy particularly well in this context, as it is also sensitive to the relative energies of the Co 3d-based MOs and the composition of the singly occupied, Co $3d_{z^2}$-based MO (Fig. 5, *inset*).

2.3 Co^{I}Corrinoids

Co^{I}Cbl, the "superreduced" form of B_{12}, is formed (i) via reduction of Co^{II}Cbl (e.g. using Zn dust), (ii) as a part of the catalytic cycles of MeCbl-dependent enzymes and reductive dehalogenases, and (iii) during AdoCbl biosynthesis (Liptak & Brunold, 2006). Co^{I}Cbl is a four-coordinate, low-spin ($S=0$) $3d^8$ species in which the Co^{I} ion is ligated solely by the corrin macrocycle; consequently, Co^{I}Cbl and Co^{I}cobinamide are indistinguishable by Abs, CD, and MCD spectroscopy. As in the case of Co^{III}Cbls, the diamagnetic ground state of Co^{I}Cbl results in a temperature-independent and relatively weak, but nonetheless quite well-resolved MCD spectrum (Fig. 6) (Liptak & Brunold, 2006). The Abs spectrum of Co^{I}Cbl contains four moderately intense bands in the near-IR/visible region (at ~12,500, 15,000, 18,000, and 22,000 cm^{-1}) and a much more intense feature in the UV region (at ~26,000 cm^{-1}). TD-DFT computations revealed that the HOMO of Co^{I}Cbl has primarily Co $3d_{z}^{2}$ orbital character, while the three lowest unoccupied MOs are corrin π^*-based (Liptak & Brunold, 2006). As a result, the four low-energy transitions observed in the near-IR/visible regions of the Co^{I}Cbl Abs spectrum (bands i–iv in Fig. 6) formally arise from MLCT transitions, though they also possess significant $\pi \rightarrow \pi^*$ character due to the mixing between the Co 3d and corrin π frontier orbitals. Toward higher energy, the dominant Abs feature at ~26,000 cm^{-1} arises from a similar corrin $\pi \rightarrow \pi^*$ transition as the one responsible for the α/β bands in the Co^{III}Cbl Abs spectra. Interestingly, the energy of this transition increases from MeCbl to Co^{II}Cbl and Co^{I}Cbl. Computational studies revealed that the blue-shift of this transition with decreasing formal oxidation state of the Co ion can be attributed to an increase in π-backbonding from the filled Co 3d orbitals to the empty corrin π^*-based acceptor MO and the consequent destabilization of this orbital (Liptak & Brunold, 2006).

The different binding modes, Co oxidation states, and axial ligand environments are key to the distinct reactivities displayed by different B_{12} species, and MCD spectroscopy provides a uniquely sensitive probe of these characteristics. The detailed understanding of the correlation that exists between structural perturbations and spectral changes provided the necessary foundation for carrying out Abs, CD, and MCD spectroscopic studies of corrinoids in various enzyme active sites. The following section outlines examples of how this approach was used to obtain molecular-level insight into the reaction mechanisms of enzymes involved in the biosynthesis of B_{12} and B_{12}-dependent enzymes.

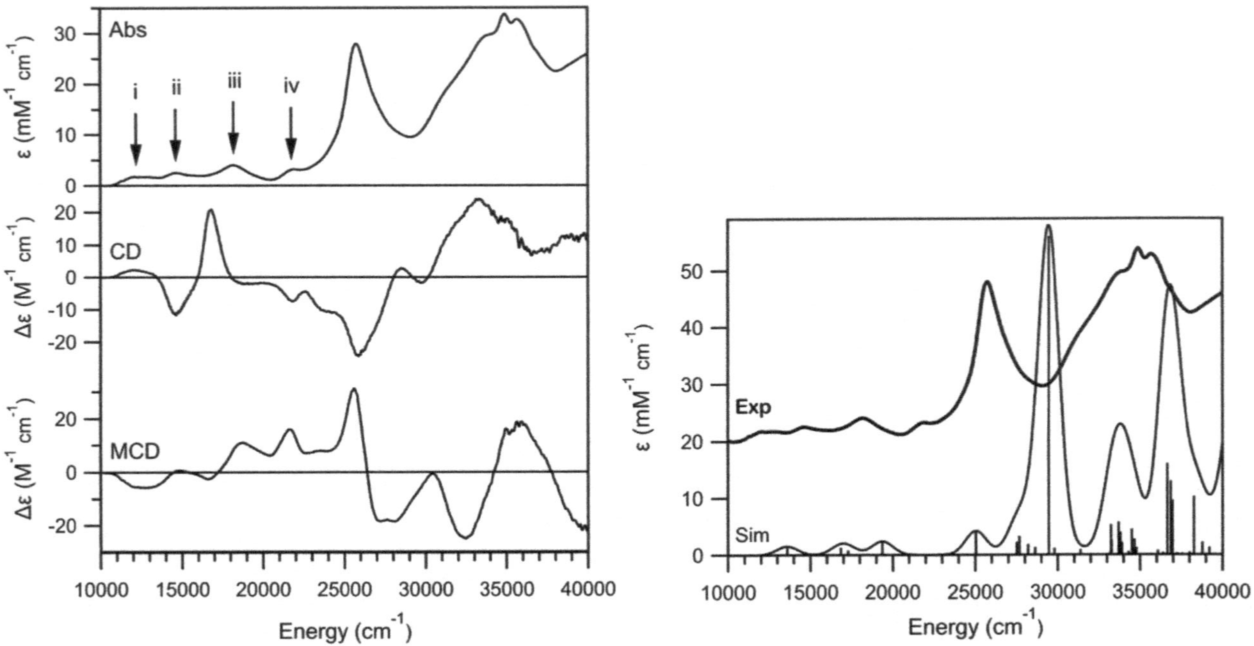

Fig. 6 (*Left*): Absorption (*top*), CD (*center*), and 7 T MCD (*bottom*) spectra recorded at 4.5 K of $Co^{I}Cbl$. The four low-energy Abs features in the near-IR/visible region are labeled i–iv. (*Right*): Experimental 4.5 K and TD-DFT-computed Abs spectra of $Co^{I}Cbl$. *Adapted from Liptak, M. D., & Brunold, T. C. (2006). Spectroscopic and computational studies of* Co^{1+} *Cobalamin: Spectral and electronic properties of the "superreduced"* B_{12} *cofactor.* Journal of the American Chemical Society, 128, *9144–9156.*

3. Application of MCD spectroscopy to the study of enzyme-bound B_{12} species

3.1 ATP:Co^{I}Corrinoid adenosyltransferases (ACATs)

Because AdoCbl-dependent enzymes are susceptible to inactivation via oxidation of the Co^{II}Cbl intermediate generated during turnover (Banerjee & Ragsdale, 2003), all organisms that employ these enzymes in their metabolism are believed to produce ATP:Co^{I}corrinoid adenosyltransferases (ACATs) (Warren, Raux, Schubert, & Escalante-Semerena, 2002). ACATs catalyze the replacement of the upper axial ligand of exogenous cobalamins or incomplete corrinoids with the 5′-deoxyadenosyl (Ado) group of ATP (Mera & Escalante-Semerena, 2010a; Warren et al., 2002). Three non-homologous families of ACATs have been identified and classified according to their biological roles in *Salmonella enterica* (Mera & Escalante-Semerena, 2010a). The CobA ACAT is involved in the de novo synthesis of AdoCbl and in the scavenging of exogenous corrinoids, whereas the PduO and EutT ACATs are required for the growth of *S. enterica* on 1,2-propanediol and ethanolamine (via supply of AdoCbl to diol dehydratase and ethanolamine ammonia-lyase, respectively) (Buan, Suh, & Escalante-Semerena, 2004; Johnson et al., 2001; Johnson, Buszko, & Bobik, 2004; Moore, Mera, & Escalante-Semerena, 2014).

The general mechanism proposed for ACATs involves the formation of a Co^{I}corrinoid intermediate that performs a nucleophilic attack on the 5′-carbon of co-substrate ATP (Fig. 7) (Fonseca & Escalante-Semerena, 2000; Mera & Escalante-Semerena, 2010a, 2010b; Park et al., 2008; Warren et al., 2002). A particularly intriguing step in this mechanism is the thermodynamically challenging $Co^{II} \rightarrow Co^{I}$corrinoid reduction (step ii → iii) that must precede the adenosylation (step iii → iv), given that the reduction potentials of free Co^{II}corrinoids (e.g., E^0 vs NHE of −610 mV for Co^{II}Cbl and −490 mV for $Co^{II}Cbi^{+}$) (Faure, Lexa, & Savéant, 1982) are well below those of readily available reducing agents in the cell. However, when bound to an ACAT active site in the presence of ATP, Co^{II}corrinoids can be reduced by free dihydroflavins ($E^0 = -228$ mV for FMN at pH 7.5) (Mera & Escalante-Semerena, 2010b; Moore et al., 2014).

Our initial ACAT target was the PduO-type human adenosyltransferase *h*ATR (Stich, Yamanishi, Banerjee, & Brunold, 2005). The MCD spectrum of Co^{II}Cbl bound to *h*ATR in the absence of co-substrate ATP was found to exhibit only small differences from that of free Co^{II}Cbl (cf. Fig. 8C and A).

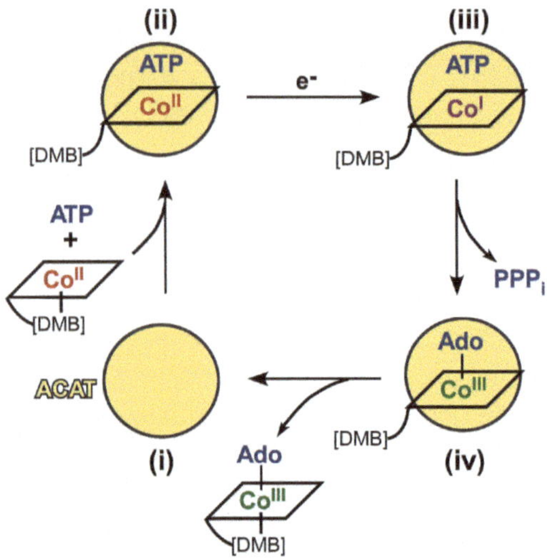

Fig. 7 Generalized reaction cycle of ACATs. Binding of ATP and Co^{II}Cbl to substrate-free enzyme (i) yields an unusual four-coordinate Co^{II}Cbl species (ii) that is activated for reduction to the Co^{I} state (iii). This strongly nucleophilic Co^{I}Cbl intermediate then attacks the 5′-carbon of ATP to form the Co—C bond of AdoCbl (iv) and triphosphate (PPP_i). Release of AdoCbl product or transfer to the respective acceptor enzyme completes the catalytic cycle.

Likewise, our X-band EPR spectra of free and *h*ATR-bound Co^{II}Cbl were similar (cf. Fig. 9C and A), though a quantitative analysis indicated a sizeable (~40%) contribution from base-off Co^{II}Cbl (Fig. 9B) to the protein spectrum. Much more dramatic changes were observed upon Co^{II}Cbl binding to *h*ATR complexed with ATP (Fig. 8D). Most noticeably, the lowest-energy transition exhibited a large red-shift and a substantial increase in intensity, signaling the formation of an unprecedented Co^{II}Cbl species. Similarly, a very unusual EPR spectrum was obtained for Co^{II}Cbl bound to the *h*ATR/ATP complex (Fig. 9D). A fit of the EPR spectrum associated with the newly formed Co^{II}Cbl species revealed unusually large $g_\perp$ and ^{59}Co hyperfine values. On the basis of these results, we concluded that *h*ATR accomplishes the thermodynamically challenging $Co^{II} \rightarrow Co^{I}$Cbl reduction by generating an effectively 4-coordinate (4c), square-planar Co^{II} species that lacks any significant axial bonding interactions. Formation of this unusual species only in the presence of co-substrate ATP suggested that the enzyme exerts exquisite control over the timing of Co^{II}Cbl reduction, thereby protecting itself from deleterious side-reactions by the transiently formed Co^{I}Cbl "supernucleophile" (Schrauzer & Deutsch, 1969).

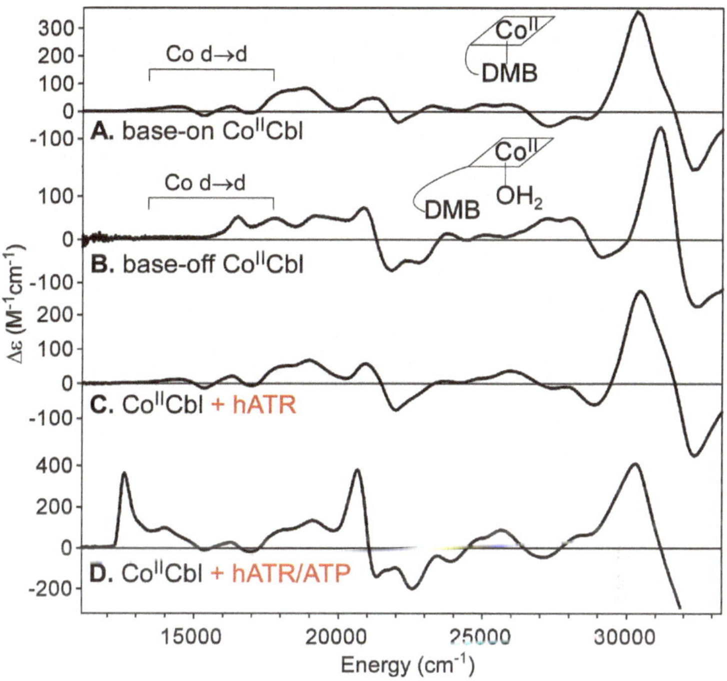

Fig. 8 MCD spectra at 4.5 K and 7 T of (A) base-on Co^{II}Cbl, (B) base-off Co^{II}Cbl, (C) Co^{II}Cbl in the presence of *h*ATR, and (D) Co^{II}Cbl in the presence of *h*ATR and ATP. *Adapted from Stich, T. A., Yamanishi, M., Banerjee, R., & Brunold, T. C. (2005). Spectroscopic evidence for the formation of a four-coordinate Co^{2+} cobalamin species upon binding to the human ATP:Cobalamin adenosyltransferase.* Journal of the American Chemical Society, 127, *7660–7661.*

Even though no evolutionary relationship exists among the CobA, PduO, and EutT families of ACATs (Mera & Escalante-Semerena, 2010a), spectroscopic studies of CobA from *Salmonella enterica* (*Se*CobA) (Stich, Buan, Escalante-Semerena, & Brunold, 2005), the PduO-type ACAT from *Lactobacillus reuteri* (*Lr*PduO) (Park et al., 2008, 2012, 2016), and the EutT-type ACAT from *S. enterica* (*Se*EutT) (Pallares, Brunold, Moore, & Escalante-Semerena, 2016; Park et al., 2015) provided compelling evidence that all three families use the same basic mechanism for the reduction of their Co^{II}corrinoid substrates, as outlined above for *h*ATR. However, our results also revealed some intriguing mechanistic differences among the ACAT families, especially with regards to the order of substrate binding and the effectiveness of the conversion of Co^{II}Cbl and $Co^{II}Cbi^{+}$ to 4c species. While *Se*CobA and *Lr*PduO achieve higher 5c→4c conversion yields with $Co^{II}Cbi^{+}$

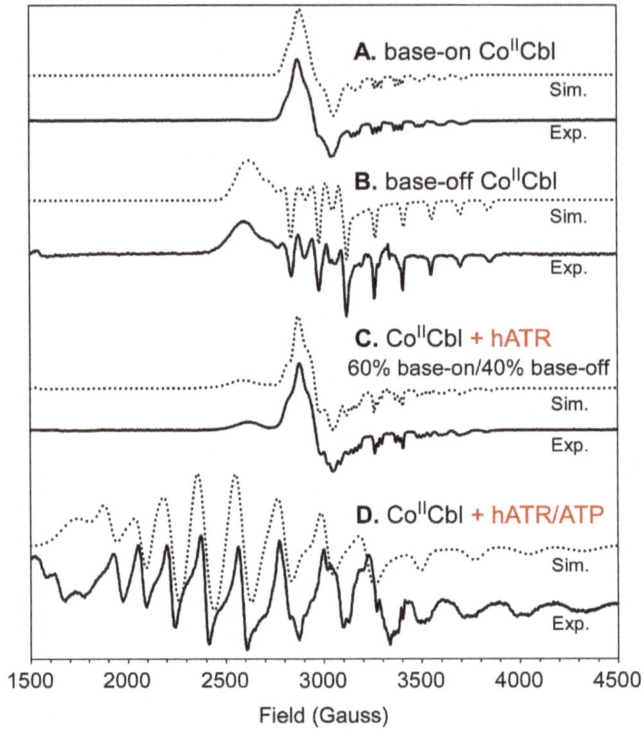

Fig. 9 X-band EPR spectra at 40 K of (A) base-on Co^{II}Cbl, (B) base-off Co^{II}Cbl, (C) Co^{II}Cbl in the presence of *h*ATR, and (D) Co^{II}Cbl in the presence of *h*ATR and ATP. Simulated spectra (dotted lines) are shown above the experimental spectra (solid lines). *Adapted from Stich, T. A., Yamanishi, M., Banerjee, R., & Brunold, T. C. (2005). Spectroscopic evidence for the formation of a four-coordinate Co^{2+} cobalamin species upon binding to the human ATP:Cobalamin adenosyltransferase.* Journal of the American Chemical Society, *127, 7660–7661.*

(Pallares, Moore, Escalante-Semerena, & Brunold, 2014; Park et al., 2012), which is expected given the fact that the Co–OH_2 bond is considerably weaker than the Co–N(DMB) bond, *Se*EutT fails to generate 4c Co^{II}Cbi$^+$ and is thus unable to adenosylate this cobalamin precursor (Pallares et al., 2016; Park et al., 2015).

X-ray crystallographic studies of Co^{II}Cbl bound to *Se*CobA (Moore, Newmister, Rayment, & Escalante-Semerena, 2012) (Fig. 10) and *Lr*PduO (St. Maurice et al., 2008) confirmed our prediction that 4c Co^{II}Cbl species are formed in these enzymes when ATP is present. In both structures, co-substrate ATP occupies a position near the vacant "upper" axial coordination site of the Co^{II}Cbl, while hydrophobic amino acid residues block access

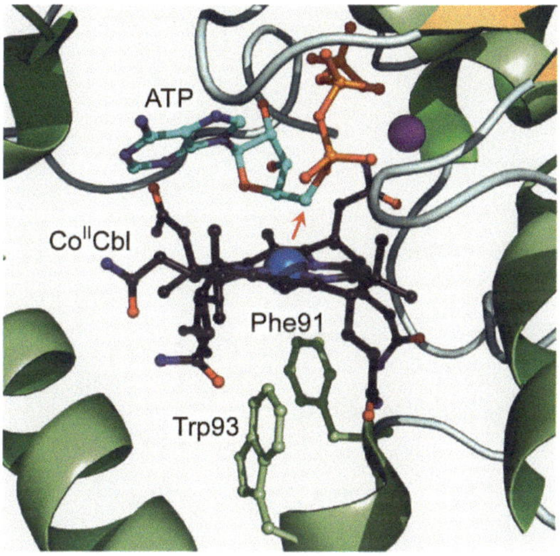

Fig. 10 X-ray crystal structure of 4c Co^{II}Cbl (*black*) in the active site of *Se*CobA, with Mg-ATP located above the corrin ring. The Co and Mg atoms are shown as *blue* and *purple spheres*, respectively. Key hydrophobic residues that block off the lower axial position are depicted in *pale green*. The 5′-carbon of ATP is highlighted by a *red arrow*.

to the "lower" site to prevent the coordination of a water molecule. Although EutT has thus far eluded characterization by X-ray crystallography, kinetic and spectroscopic studies of the *Salmonella enterica* EutT revealed that this enzyme exhibits an additional metal cofactor requirement not observed for the other ACAT families (Moore et al., 2014; Pallares et al., 2016). Intriguingly, though, some EutT enzymes from other organisms, including *Listeria monocytogenes*, lack this divalent metal cofactor (Stracey, Costa, Escalante-Semerena, & Brunold, 2018).

3.2 AdoCbl-dependent enzymes: Isomerases and eliminases

AdoCbl-dependent enzymes use the unique reactivity of the Co–C(Ado) bond to perform a range of isomerization reactions (Banerjee, 2003; Banerjee & Ragsdale, 2003; Greenhalgh & Brunold, 2021; Toraya, 2003, 2014). Differences in their catalytic mechanisms allow for the classification of these enzymes into two subclasses: (i) Class I isomerases, which catalyze the 1,2-exchange of a functional group, such as an alcohol, amine, or carbon-containing fragment, and (ii) Class II eliminases, for which substrate rearrangement is followed by the spontaneous elimination of a leaving group, such as water or ammonia (Fig. 11) (Banerjee & Ragsdale, 2003; Toraya, 2003). Interestingly, the two subclasses differ in their modes of

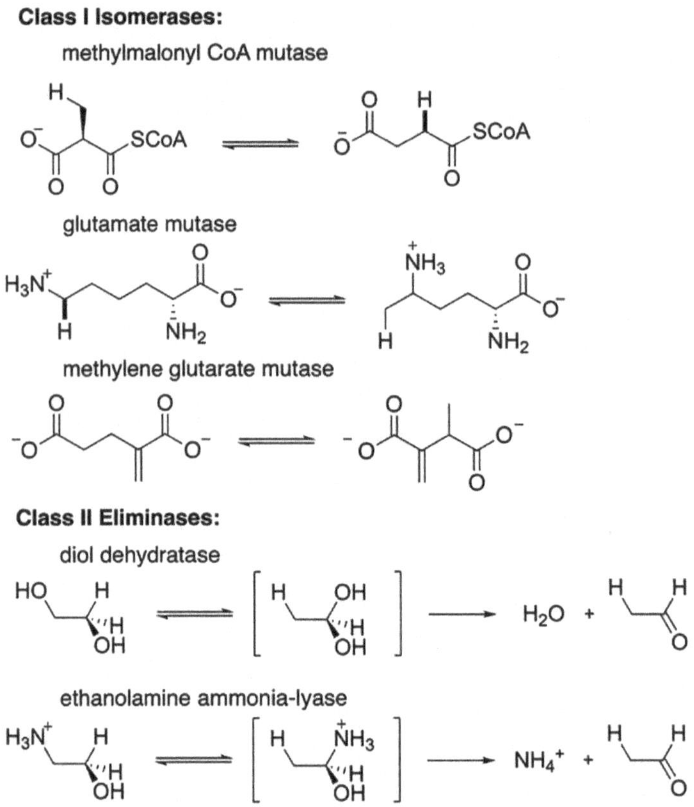

Fig. 11 Classification of representative AdoCbl-dependent enzymes with their substrates and products.

AdoCbl binding. Crystal structures and EPR spectroscopic studies revealed that Class I isomerases bind AdoCbl in the base-off/His-on mode, where the DMB ligand is displaced by a protein-derived histidine (Gruber, Reitzer, & Kratky, 2001; Mancia, Smith, & Evans, 1999; Padmakumar, Taoka, Padmakumar, & Banerjee, 1995; Zelder, Beatrix, Kroll, & Buckel, 1995). In contrast, Class II eliminases bind AdoCbl base-on, with the DMB ligand still bound to the Co^{III} ion (Abend et al., 1999; Shibata et al., 2010; Shibata, Sueyoshi, Higuchi, & Toraya, 2018). Class I isomerases include the only AdoCbl-dependent enzyme found in humans, methylmalonyl coenzyme-A mutase (MMCM), which catalyzes the conversion of methylmalonyl-CoA to succinyl-CoA. The role this reaction plays in the catabolism of branched-chain amino acids, odd-chain fatty acids, and cholesterol makes MMCM vital to human health (Banerjee & Ragsdale, 2003; Flavin, Ortiz, & Ochoa, 1955; Katz & Chaikoff, 1955).

Following substrate binding, the catalytic cycles of all AdoCbl-dependent enzymes are initiated through homolytic cleavage of the cofactor's Co—C bond, which generates Co^{II}Cbl and an organic radical centered on the 5′-carbon of the Ado moiety (Ado$^{\bullet}$) (Fig. 12). Remarkably, the rate of Co—C bond homolysis is increased by 12 orders of magnitude (corresponding to a ~17 kcal/mol decrease in activation energy) when the cofactor is bound to the enzyme and substrate is added (Chowdhury & Banerjee, 2000; Hay & Finke, 1987). Elucidating the mechanism by which AdoCbl-dependent enzymes accomplish this trillion-fold rate enhancement for Co—C bond homolysis while suppressing potentially harmful side reactions involving the post-homolysis radical species has been an area of research for decades (Banerjee, 2003; Kräutler, Keller, & Kratky, 1989; Toraya, 2003). In general, this Co—C bond activation could be achieved via destabilization of the AdoCbl ground state, stabilization of the post-homolysis products Ado$^{\bullet}$ and Co^{II}Cbl, or a combination of the two. MCD spectroscopy is well-suited for evaluating these possibilities, as it permits a detailed assessment of the cofactor/enzyme active site interactions in both the AdoCbl and Co^{II}Cbl states. With our understanding of the origin of the key spectral features of Cbls, changes in the MCD spectra can be correlated with geometric and electronic perturbations of the cofactor by the active site.

This MCD-based approach was originally used for studying the Class I isomerase MMCM. Abs and MCD spectra were collected for MMCM-bound AdoCbl and Co^{II}Cbl both in the absence and presence of substrate (analogue) and compared to the spectra obtained for the free cofactors (Brooks et al., 2005; Brooks, Vlasie, et al., 2005; Brooks, Vlasie, Banerjee, & Brunold, 2004). Importantly, only minor spectral changes were observed upon incorporation of AdoCbl into the enzyme active site and the subsequent addition of substrate (Fig. 13) (Brooks et al., 2004). The energies of all features in the α/β and γ regions remained largely unchanged, and the slight

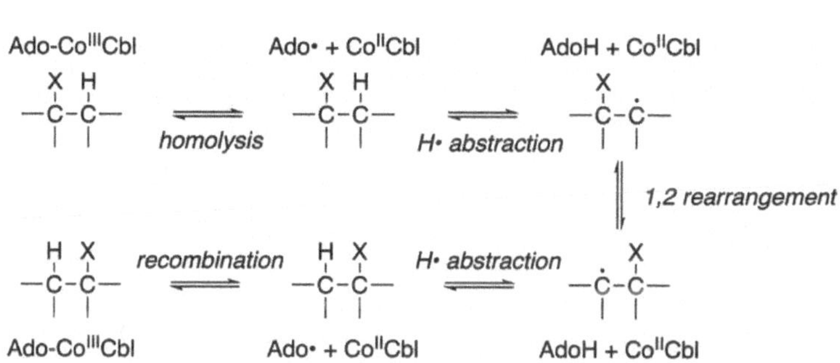

Fig. 12 Basic mechanism of substrate rearrangement catalyzed by AdoCbl-dependent enzymes.

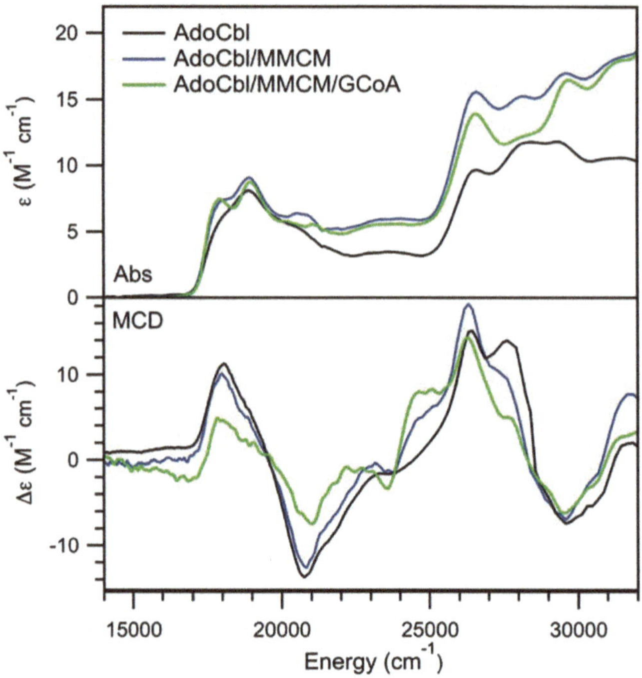

Fig. 13 Abs spectra at 4.5 K (*top*) and MCD spectra (*bottom*) at 25 K and 7 T of free AdoCbl, AdoCbl bound to MMCM (AdoCbl/MMCM), and AdoCbl bound to MMCM complexed with the "slow" substrate glutaryl-CoA (AdoCbl/MMCM/GCoA). *Adapted from Brooks, A. J., Vlasie, M., Banerjee, R., & Brunold, T. C. (2004). Spectroscopic and computational studies on the adenosylcobalamin-dependent methylmalonyl-CoA mutase: Evaluation of enzymatic contributions to Co-C bond activation in the Co^{3+} ground state.* Journal of the American Chemical Society, 126, *8167–8180.*

increase in Abs intensity of the γ features could be attributed to a change in the dielectric environment of the cofactor. These data argued against a significant enzyme-induced weakening of the Co—C bond in the AdoCbl ground state. In contrast, the electronic structure of Co^{II}Cbl was found to be notably perturbed upon protein binding and, to an even greater extent, following the addition of substrate. MCD spectra revealed a substantial blue-shift of the Co 3d → corrin π* MLCT transitions in the ~17,000–20,000 cm^{-1} region, while the LF and corrin-centered π → π* transitions were only weakly affected (Fig. 14). An analysis of these results in the framework of DFT computations led to the proposal that MMCM induces a fairly uniform stabilization of the Co 3d-based MOs of Co^{II}Cbl and, because these orbitals are filled, a stabilization of the enzyme-bound cofactor as a whole (Brooks, Fox, et al., 2005; Brooks, Vlasie, et al., 2005). Nearly identical results were obtained in a

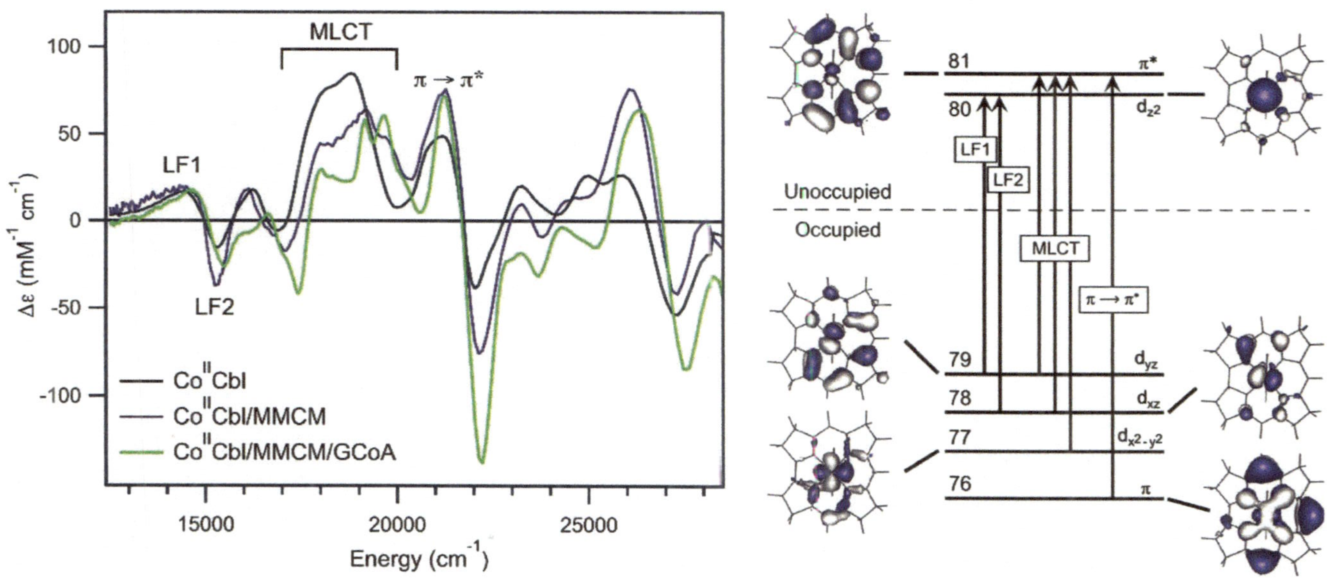

Fig. 14 (*Left*): MCD spectra at 4.5 K and 7 T of free $Co^{II}Cbl$, $Co^{II}Cbl$ bound to MMCM ($Co^{II}Cbl$/MMCM), and $Co^{II}Cbl$ bound to MMCM complexed with the "slow" substrate glutaryl-CoA ($Co^{II}Cbl$/MMCM/GCoA). (*Right*): Relative energies and isosurface plots of the relevant spin-down MOs of $Co^{II}Cbl$ based on a spin-unrestricted DFT calculation. Electronic transitions that have been correlated to bands in the experimental $Co^{II}Cbl$ MCD spectra on the *left* are indicated by *vertical arrows*. *Adapted from Brooks, A. J., Vlasie, M., Banerjee, R., & Brunold, T. C. (2004). Spectroscopic and computational studies on the adenosylcobalamin-dependent methylmalonyl-CoA mutase: Evaluation of enzymatic contributions to Co-C bond activation in the* Co^{3+} *ground state.* Journal of the American Chemical Society, 126, *8167–8180.*

subsequent study of another member of the Class I isomerase family, glutamate mutase (GM), which catalyzes the interconversion of L-glutamate and L-threo-3-methylaspartate (Brooks, Fox, et al., 2005).

X-ray crystal structures reported for MMCM and GM provide clues as to how these enzymes stabilize the $Co^{II}Cbl$ post-homolysis product to promote Co—C bond homolysis. Displacement of the DMB ligand by a histidine side-chain offers the Class I isomerases a means of influencing the electronic properties of $Co^{II}Cbl$. Additional tuning of the Co–N(His) bond is facilitated through a conserved DXHXXG motif, which encodes a network of H-bonding residues involving the coordinating His and conserved aspartate and lysine/asparagine residues (Gruber et al., 2001; Mancia & Evans, 1998; Marsh & Holloway, 1992). Substituting any of these residues dramatically reduces catalytic activity, emphasizing their importance for enzyme function (Chen & Marsh, 1997; Vlasie, Chowdhury, & Banerjee, 2002). Coupling proton uptake by the H-bonding triad to Co—C bond homolysis would be expected to reduce charge donation from the His ligand in the $Co^{II}Cbl$ state, causing a stabilization of the Co 3d orbitals and thus the cofactor as a whole, as suggested by our MCD results (Brooks, Vlasie, et al., 2005).

Further evidence that the mechanism of Co—C bond activation employed by Class I isomerases relies primarily on a stabilization of the post-homolysis products, rather than a destabilization of the AdoCbl ground state, was obtained by a number of experimental and computational studies. A rR spectroscopic study of MMCM revealed that vibrational modes involving the Co–C(Ado) bond of AdoCbl were essentially unaffected upon cofactor binding to MMCM, even following the addition of substrate, ruling out any significant Co—C bond lengthening prior to homolysis (Dong et al., 1999; Dong, Padmakumar, Maiti, et al., 1998). Similarly, a computational study of MMCM led to the conclusion that AdoCbl binding to the enzyme active site does not induce any significant changes in Co—C bond length or corrin ring conformation (Freindorf & Kozlowski, 2004). Other computational studies of MMCM and GM showed that electrostatic interactions between $Co^{II}Cbl$ and the ribose oxygens of the Ado moiety contribute to the stabilization of the post-homolysis products, along with favorable interactions between the Ado moiety and nearby protein residues (Dölker, Maseras, & Siegbahn, 2004; Jensen & Ryde, 2005; Kwiecien et al., 2006; Sharma, Chu, Olsson, & Warshel, 2007). Conformational changes of the Ado moiety prior to Co—C bond homolysis were noted in one of these studies, but the corrin ring was found to be minimally perturbed (Kwiecien et al., 2006).

Interestingly, the Co—C bond activation mechanism proposed for Class I isomerases cannot be the same for Class II eliminases such as ethanolamine ammonia-lyase (EAL), which catalyzes the conversion of ethanolamine to acetaldehyde and ammonia (Bradbeer, 1965; Chang & Chang, 1975; Mori et al., 2014; Toraya, 2003). Because Class II eliminases bind AdoCbl base-on, they are unable to activate the Co—C bond through preferential stabilization of the Co^{II}Cbl state via H-bonding interactions involving the lower axial ligand or proton uptake. Advanced EPR techniques were used to compare the reaction intermediates and key structural features of the active sites of isomerases and eliminases in some detail, which revealed interesting differences between these two subclasses (Bandarian & Reed, 2002; Bender, Poyner, & Reed, 2008; Ke, Torrent, Museav, Morokuma, & Warncke, 1999; Ke & Warncke, 1999; LoBrutto et al., 2001; Padmakumar et al., 1995; Robertson, Wang, & Warncke, 2011; Warncke, Schmidt, & Ke, 1999; Warncke & Utada, 2001; Zelder et al., 1995). For example, in Class I isomerases, the substrate binds ~6–7 Å from the Co ion, and following Co—C bond cleavage the Ado moiety undergoes a ribose "pseudo-rotation" to orient itself toward the substrate (Bothe et al., 1998; Gruber et al., 2001). In Class II eliminases, the substrate binds farther from the Co ion, at a distance of ~10–12 Å, and the ribosyl moiety rotates about the glycosidic bond in order to span the greater distance (Bandarian & Reed, 2002; Bender et al., 2008; Ke & Warncke, 1999; LoBrutto et al., 2001; Warncke & Utada, 2001). Possible protein-cofactor interactions that contribute to this mechanism have been explored for EAL through the use of cofactor analogues with modifications to the upper and lower axial ligands (Chen et al., 2013; Fukuoka et al., 2002). Despite these efforts, the mechanism of Co—C bond activation employed by Class II eliminases remains unclear.

4. Reductive dehalogenases

Organisms that produce reductive dehalogenases (RDases) utilize halogenated aromatic and aliphatic substances as terminal electron acceptors in a process termed organohalide respiration (John et al., 2009). These organisms can couple the reduction of halogenated substances with the production of ATP. Early EPR studies of as-isolated and reduced RDases revealed the presence of Fe—S clusters and a Co^{II}corrinoid cofactor (Schumacher, Holliger, Zehnder, & Hagen, 1997; van de Pas et al., 1999). Indeed, all dehalogenases examined so far contain two Fe—S clusters, either a pair of [4Fe—4S] clusters or one [4Fe—4S] and one [4Fe—3S] cluster, which are expected to serve as electron-transfer centers (Krasotkina, Walters, Maruya, & Ragsdale, 2001; Maillard et al., 2003; Neumann et al., 2002; van de Pas et al., 1999; Wohlfarth & Diekert, 1997). Moreover, all but one of the characterized RDases contain a corrinoid cofactor, the exception being the *Desulfomonile tiedjei* DCB-1 enzyme, which appears to possess a heme cofactor (Ni, Fredrickson, & Xun, 1995).

Tetrachloroethylene reductive dehalogenase (PceA) uses norpseudo-Co^{II}Cbl (Co^{II}Cbl*)—an unusual cobalamin that lacks the methyl group found on the intramolecular loop adjacent to the phosphate moiety and features an adenine base in lieu of the DMB—to catalyze the reductive dehalogenation of per- and trichloroethylenes (PCE and TCE, respectively) to primarily *cis*-dichloroethylene (DCE) (Bommer et al., 2014). The enzymatic conversion of PCE to TCE (and subsequently DCE) was proposed to proceed via a mechanism in which the first step involves one of several options: single electron transfer; nucleophilic addition followed by chloride elimination or protonation; or direct attack at the halogen (Fig. 15) (Greenhalgh, Kunze, Schubert, Diekert, & Brunold, 2021). Difficulties with producing adequate quantities of PceA greatly hampered direct experimental studies of the reaction mechanism. To overcome these challenges, we collected MCD spectra of PceA in the absence and presence of substrate, generated computational models of these states using the quantum mechanics/molecular mechanics (QM/MM) method, and validated our models on the basis of our MCD data and X-ray crystal structures (Greenhalgh et al., 2021). Computational models of viable reaction intermediates were then constructed using the same computational methodology.

Consistent with published X-ray crystal structures (Bommer et al., 2014), the MCD spectrum of as-isolated PceA indicated that Co^{II}Cbl* is five-coordinate, with the intramolecular base replaced by a water molecule in one of the axial positions (Fig. 16) (Greenhalgh et al., 2021). The lowest energy feature, which arises from a LF transition that is especially sensitive to perturbations of the axial bonding interaction (Stich et al., 2004), was found to be red-shifted by $\sim$320 cm^{-1} from its counterpart in the MCD spectrum of $Co^{II}Cbi^{+}$, indicating a weaker Co–O(H_2) bond in the protein-bound Co^{II}Cbl*. Addition of TCE to PceA caused only minor changes to the MCD spectrum, ruling out any major conformational changes of the Co^{II}Cbl* species (Fig. 16). Most importantly, this result provided compelling evidence that the PceA-bound Co^{II}Cbl* species remained five-coordinate even in the presence of substrate. Thus, the mechanism for the thermodynamically challenging $Co^{II} \rightarrow Co^{I}$Cbl* reduction used by

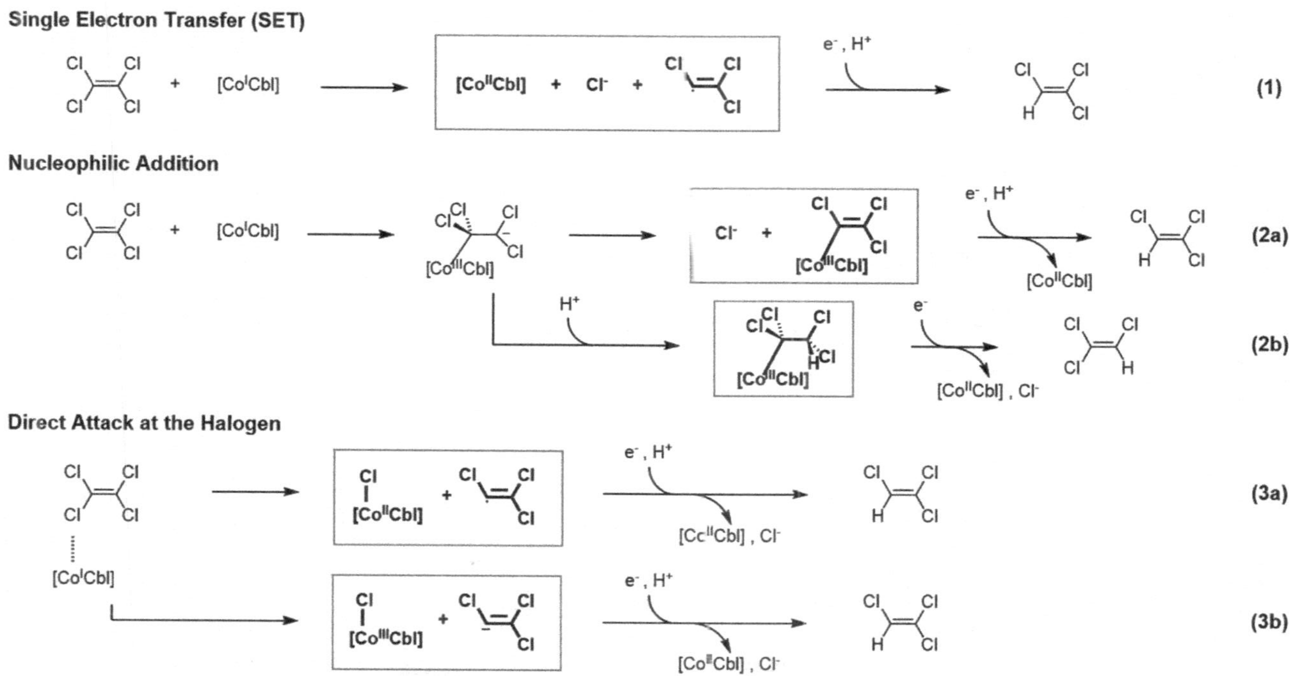

Fig. 15 Possible mechanisms for the initial step of the conversion of PCE to TCE catalyzed by PceA: (1) single electron transfer; (2a) nucleophilic addition followed by Cl^- elimination; (2b) nucleophilic addition followed by protonation; and direct attack at the halogen with the Cl—C bond cleaved (3a) homolytically or (3b) heterolytically. The intermediates highlighted in bold and shown in a box were evaluated computationally. *Adapted from Greenhalgh, E. D., Kunze, C., Schubert, T., Diekert, G., & Brunold, T. C. A. (2021). Spectroscopically validated computational investigation of viable reaction intermediates in the catalytic cycle of the reductive dehalogenase PceA.* Biochemistry, 60, *2022–2032.*

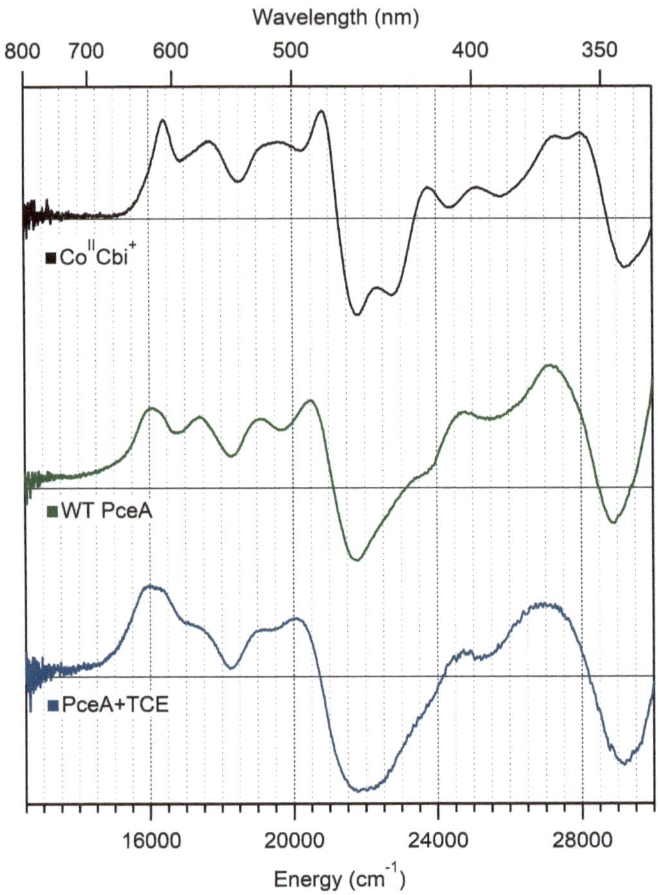

Fig. 16 MCD spectra at 4.5 K and 7 T of (from top to bottom) $Co^{II}Cbi^{+}$; as-isolated PceA, and PceA incubated with TCE. *Adapted from Greenhalgh, E. D., Kunze, C., Schubert, T., Diekert, G., & Brunold, T. C. A. (2021). Spectroscopically validated computational investigation of viable reaction intermediates in the catalytic cycle of the reductive dehalogenase PceA.* Biochemistry, 60, *2022–2032.*

PceA must differ fundamentally from that utilized by the ACATs (discussed above). In contrast to ACATs, PceA contains two low potential [4Fe—4S] clusters that serve as an internal reducing system (Schubert, Adrian, Sawers, & Diekert, 2018). While the physiological reductant remains unknown, these $[4Fe—4S]^{1+}$ clusters are sufficiently potent to reduce PceA bound $Co^{II}Cbl^{*}$ (Kräutler et al., 2003).

Importantly, our QM/MM optimized models of substrate-free and TCE-bound PceA both featured five-coordinate $Co^{II}Cbl^{*}$ with a water molecule occupying the lower axial position, as stipulated by the close resemblance of the MCD spectra obtained for these species and free $Co^{II}Cbi^+$ (Fig. 16) (Greenhalgh et al., 2021). The same QM/MM computational methodology was then applied to viable reaction intermediates in the catalytic cycle of PceA (Fig. 15). The intermediate predicted to possess the lowest energy was the one resulting from single electron transfer (SET) from $Co^{I}Cbl^*$ to the substrate to yield $Co^{II}Cbl^*$, a chloride ion, and a vinylic radical (Fig. 15). While earlier studies led to conflicting conclusions as to the mechanism of enzymatic reductive dehalogenation, we were able to rule out mechanisms invoking a Co—Cl adduct. Instead, our computations strongly favored the SET initiated reaction.

Acknowledgments

T.C.B. thanks his current and former graduate students who have contributed to the cobalamin project, his superb collaborators on this project, and the National Science Foundation (grant CHE-1710339) for financial support.

References

Abend, A., Bandarian, V., Nitsche, R., Stupperich, E., Rétey, J., & Reed, G. H. (1999). Ethanolamine ammonia-lyase has a 'base-on' binding mode for coenzyme B_{12}. *Archives of Biochemistry and Biophysics, 370*, 138–141.

Andruniów, T., Jaworska, M., Lodowski, P., Zgierski, M. Z., Dreos, R., Randaccio, L., et al. (2008). Time-dependent density functional theory study of cobalt corrinoids: Electronically excited states of methylcobalamin. *The Journal of Chemical Physics, 129*, 085101.

Bandarian, V., & Reed, G. H. (2002). Analysis of the electron paramagnetic resonance spectrum of a radical intermediate in the coenzyme B_{12}-dependent ethanolamine ammonia-lyase catalyzed reaction of S-2-aminopropanol. *Biochemistry, 41*, 8580–8588.

Banerjee, R. (2003). Radical carbon skeleton rearrangements: Catalysis by coenzyme B_{12}-dependent mutases. *Chemical Reviews, 103*, 2083–2094.

Banerjee, R., & Ragsdale, S. W. (2003). The many faces of vitamin B_{12}: Catalysis by cobalamin-dependent enzymes. *Annual Review of Biochemistry, 72*, 209–247.

Bender, G., Poyner, R. R., & Reed, G. H. (2008). Identification of the substrate radical intermediate derived from ethanolamine during catalysis by ethanolamine ammonia-lyase. *Biochemistry, 47*, 11360–11366.

Bommer, M., Kunze, C., Fesseler, J., Schubert, T., Diekert, G., & Dobbek, H. (2014). Structural basis for organohalide respiration. *Science, 346*, 455–458.

Bothe, H., Darley, D. J., Albracht, S. P. J., Gerfen, G. J., Golding, B. T., & Buckel, W. (1998). Identification of the 4-glutamyl radical as an intermediate in the carbon skeleton rearrangement catalyzed by coenzyme B_{12}-dependent glutamate mutase from *Clostridium cochlearium*. *Biochemistry, 37*, 4105–4113.

Bradbeer, C. (1965). The clostridial fermentations of choline and ethanolamine I. Preparation and properties of cell-free extracts. *The Journal of Biological Chemistry, 240*, 4669–4674.

Brooks, A. J., Fox, C. C., Marsh, E. N. G., Vlasie, M., Banerjee, R., & Brunold, T. C. (2005). Electronic structure studies of the adenosylcobalamin cofactor in glutamate mutase. *Biochemistry*, *44*, 15167–15181.

Brooks, A. J., Vlasie, M., Banerjee, R., & Brunold, T. C. (2004). Spectroscopic and computational studies on the adenosylcobalamin-dependent methylmalonyl-CoA mutase: Evaluation of enzymatic contributions to Co-C bond activation in the Co^{3+} ground state. *Journal of the American Chemical Society*, *126*, 8167–8180.

Brooks, A. J., Vlasie, M., Banerjee, R., & Brunold, T. C. (2005). Co-C bond activation in methylmalonyl-CoA mutase by stabilization of the post-homolysis product Co^{2+} cobalamin. *Journal of the American Chemical Society*, *127*, 16522–16528.

Brunold, T. C., Conrad, K. S., Liptak, M. D., & Park, K. (2009). Spectroscopically validated density functional theory studies of the B_{12} cofactors and their interactions with enzyme active sites. *Coordination Chemistry Reviews*, *253*, 779–794.

Buan, N. R., Suh, S. J., & Escalante-Semerena, J. C. (2004). The eutT gene of *Salmonella enterica* encodes an oxygen-labile, metal-containing ATP:Corrinoid adenosyltransferase enzyme (vol 186, pg 5711, 2004). *Journal of Bacteriology*, *186*, 7826.

Chang, G. W., & Chang, J. T. (1975). Evidence for the B_{12}-dependent enzyme ethanolamine deaminase in *Salmonella*. *Nature*, *254*, 150–151.

Chen, H. P., & Marsh, E. N. G. (1997). How enzymes control the reactivity of adenosylcobalamin: Effect on coenzyme binding and catalysis of mutations in the conserved histidine-aspartate pair of glutamate mutase. *Biochemistry*, *36*, 7884–7889.

Chen, Z.-G. G., Zietek, M. A., Russell, H. J., Tait, S., Hay, S., Jones, A. R., et al. (2013). Dynamic, electrostatic model for the generation and control of high-energy radical intermediates by a coenzyme B_{12}-dependent enzyme. *Chembiochem*, *14*, 1529–1533.

Chowdhury, S., & Banerjee, R. (2000). Thermodynamic and kinetic characterization of Co−C bond homolysis catalyzed by coenzyme B_{12}-dependent methylmalonyl-CoA mutase. *Biochemistry*, *39*, 7998–8006.

Dölker, N., Maseras, F., & Siegbahn, P. E. M. (2004). Stabilization of the adenosyl radical in coenzyme B_{12}—A theoretical study. *Chemical Physics Letters*, *386*, 174–178.

Dong, S., Padmakumar, R., Banerjee, R., & Spiro, T. G. (1996). Resonance Raman Co−C stretching frequencies reflect bond strength changes in alkyl cobalamins, but are unaffected by trans ligand substitution. *Journal of the American Chemical Society*, *118*, 9182–9183.

Dong, S., Padmakumar, R., Banerjee, R., & Spiro, T. G. (1998). Co-C force constants from resonance Raman spectra of alkylcobalamins: Insensitivity to dimethylbenzylimidazole coordination. *Inorganica Chimica Acta*, *270*, 392–398.

Dong, S., Padmakumar, R., Banerjee, R., & Spiro, T. G. (1999). Co-C bond activation in B_{12}-dependent enzymes: Cryogenic resonance Raman studies of methylmalonyl-coenzyme a mutase. *Journal of the American Chemical Society*, *121*, 7063–7070.

Dong, S., Padmakumar, R., Maiti, N., Banerjee, R., & Spiro, T. G. (1998). Resonance raman spectra show that coenzyme B_{12} binding to methylmalonyl-coenzyme a mutase changes the corrin ring conformation but leaves the Co-C bond essentially unaffected. *Journal of the American Chemical Society*, *120*, 9947–9948.

Faure, D., Lexa, D., & Savéant, J.-M. (1982). Electrochemistry of vitamin B_{12}. *Journal of Electroanalytical Chemistry and Interfacial Electrochemistry*, *140*, 297–309.

Firth, R. A., Hill, H. A. O., Pratt, J. M., Williams, R. J. P., & Jackson, W. R. (1967). The circular dichroism and absorption spectra of some vitamin B_{12} derivatives. *Biochemistry*, *6*, 2178–2189.

Flavin, M., Ortiz, P. J., & Ochoa, S. (1955). Metabolism of propionic acid in animal tissues. *Nature*, *176*, 823–826.

Fonseca, M. V., & Escalante-Semerena, J. C. (2000). Reduction of cob(III)alamin to cob(II) alamin in *Salmonella enterica* serovar typhimurium LT2. *Journal of Bacteriology*, *182*, 4304–4309.

Freindorf, M., & Kozlowski, P. M. (2004). A combined density functional theory and molecular mechanics study of the relationship between the structure of coenzyme B_{12} and its binding to methylmalonyl-CoA mutase. *Journal of the American Chemical Society, 126*, 1928–1929.

Fukuoka, M., Yamada, S., Miyoshi, S., Yamashita, K., Yamanishi, M., Zou, X., et al. (2002). Functions of the D-ribosyl moiety and the lower axial ligand of the nucleotide loop of coenzyme B_{12} in diol dehydratase and ethanolamine ammonia-lyase reactions. *Journal of Biochemistry, 132*, 935–943.

Greenhalgh, E. D., & Brunold, T. C. (2021). Biorelevant chemistry of cobalamin. In L. Que Jr.,, & Y. Lu (Eds.), *Comprehensive coordination chemistry III* (pp. 812–824). Elsevier.

Greenhalgh, E. D., Kunze, C., Schubert, T., Diekert, G., & Brunold, T. C. (2021). A spectroscopically validated computational investigation of viable reaction intermediates in the catalytic cycle of the reductive dehalogenase PceA. *Biochemistry, 60*, 2022–2032.

Gruber, K., Reitzer, R., & Kratky, C. (2001). Radical shuttling in a protein: Ribose pseudorotation controls alkyl-radical transfer in the coenzyme B_{12} dependent enzyme glutamate mutase. *Angewandte Chemie, International Edition, 40*, 3377–3380.

Handali, J. D., Sunden, K. F., Thompson, B. J., Neff-Mallon, N. A., Kaufman, E. M., Brunold, T. C., et al. (2018). Three dimensional triply resonant sum frequency spectroscopy revealing vibronic coupling in cobalamins: Toward a probe of reaction coordinates. *The Journal of Physical Chemistry. A, 122*, 9031–9042.

Hay, B. P., & Finke, I. R. G. (1987). Thermolysis of the Co-C bond in adenosylcorrins: 3: Quantification of the axial base effect in adenosylcobalamin by the synthesis and thermolysis of axial base-free adenosylcobinamide: insights into the energetics of enzyme-assisted cobalt-carbon bond homolysis. *Journal of the American Chemical Society, 109*, 8012–8018.

Jensen, K. P., & Ryde, U. (2005). How the Co-C bond is cleaved in coenzyme B_{12} enzymes: A theoretical study. *Journal of the American Chemical Society, 127*, 9117–9128.

John, M., Rubick, R., Schmitz, R. P., Rakoczy, J., Schubert, T., & Diekert, G. (2009). Retentive memory of bacteria: Long-term regulation of dehalorespiration in *Sulfurospirillum multivorans*. *Journal of Bacteriology, 191*, 1650–1655.

Johnson, M. K. CD and MCD spectroscopy. in Physical methods in bioinorganic chemistry: Spectroscopy and magnetism (ed. Que, L. J.) 233–285 (University Science Books, 2000).

Johnson, C. L. V., Buszko, M. L., & Bobik, T. A. (2004). Purification and initial characterization of the *Salmonella enterica* PduO ATP:cob(I)alamin adenosyltransferase. *Journal of Bacteriology, 186*, 7881–7887.

Johnson, C. L. V., Pechonick, E., Park, S. D., Havemann, G. D., Leal, N. A., & Bobik, T. A. (2001). Functional genomic, biochemical, and genetic characterization of the *Salmonella* pduO gene, an ATP:çob(I)alamin adenosyltransferase gene. *Journal of Bacteriology, 183*, 1577–1584.

Katz, J., & Chaikoff, I. L. (1955). The metabolism of propionate by rat liver slices and the formation of isosuccinic acid. *Journal of the American Chemical Society*, 77, 2659–2660.

Ke, S. C., Torrent, M., Museav, D. G., Morokuma, K., & Warncke, K. (1999). Identification of dimethylbenzimidazole axial coordination and characterization of ^{14}N superhyperfine and nuclear quadrupole coupling in cob(II)alamin bound to ethanolamine deaminase in a catalytically-engaged substrate radical-cobalt(II) biradical state. *Biochemistry, 38*, 12681–12689.

Ke, S. C., & Warncke, K. (1999). Interactions of substrate and product radicals with Co(II) in cobalamin and with the active site in ethanolamine deaminase, characterized by ESE-EPR and ^{14}N ESEEM spectroscopies. *Journal of the American Chemical Society, 121*, 9922–9927.

Kornobis, K., Kumar, N., Wong, B. M., Lodowski, P., Jaworska, M., Andruniów, T., et al. (2011). Electronically excited states of vitamin B_{12}: Benchmark calculations including time-dependent density functional theory and correlated ab initio methods. *The Journal of Physical Chemistry. A, 115*, 1280–1292.

Krasotkina, J., Walters, T., Maruya, K. A., & Ragsdale, S. W. (2001). Characterization of the B_{12}- and iron-sulfur-containing reductive dehalogenase from *Desulfitobacterium chlororespirans*. *The Journal of Biological Chemistry*, *276*, 40991–40997.

Kräutler, B., Fieber, W., Ostermann, S., Fasching, M., Ongania, K.-H. H., Gruber, K., et al. (2003). The cofactor of tetrachloroethene reductive dehalogenase of *Dehalospirillum multivorans* is norpseudo-B_{12}, a new type of a natural corrinoid. *Helvetica Chimica Acta*, *86*, 3698–3716.

Kräutler, B., Keller, W., & Kratky, C. (1989). Coenzyme B12 chemistry: The crystal and molecular structure of Cob(II)alamin. *Journal of the American Chemical Society*, *111*, 8936–8938.

Kwiecien, R. A., Khavrutskii, I. V., Musaev, D. G., Morokuma, K., Banerjee, R., & Paneth, P. (2006). Computational insights into the mechanism of radical generation in B_{12}-dependent methylmalonyl-CoA mutase. *Journal of the American Chemical Society*, *128*, 1287–1292.

Liptak, M. D., & Brunold, T. C. (2006). Spectroscopic and computational studies of Co^{1+}cobalamin: Spectral and electronic properties of the "superreduced" B_{12} cofactor. *Journal of the American Chemical Society*, *128*, 9144–9156.

LoBrutto, R., Bandarian, V., Magnusson, O. T., Chen, X., Schramm, V. L., & Reed, G. H. (2001). 5′-Deoxyadenosine contacts the substrate radical intermediate in the active site of ethanolamine ammonia-lyase: ^2H and ^{13}C electron nuclear double resonance studies. *Biochemistry*, *40*, 9–14.

Maillard, J., Schumacher, W., Vazquez, F., Regeard, C., Hagen, W. R., & Holliger, C. (2003). Characterization of the corrinoid iron-sulfur protein tetrachloroethene reductive dehalogenase of *Dehalobacter restrictus*. *Applied and Environmental Microbiology*, *69*, 4628–4638.

Mancia, F., & Evans, P. R. (1998). Conformational changes on substrate binding to methylmalonyl CoA mutase and new insights into the free radical mechanism. *Structure*, *6*, 711–720.

Mancia, F., Smith, G. A., & Evans, P. R. (1999). Crystal structure of substrate complexes of methylmalonyl-CoA mutase. *Biochemistry*, *38*, 7999–8005.

Marsh, E. N. G., & Holloway, D. E. (1992). Cloning and sequencing of glutamate mutase component S from *Clostridium tetanomorphum* homologies with other cobalamin-dependent enzymes. *FEBS Letters*, *310*, 167–170.

Mera, P. E., & Escalante-Semerena, J. C. (2010a). Multiple roles of ATP:cob(I)alamin adenosyltransferases in the conversion of B12 to coenzyme B12. *Applied Microbiology and Biotechnology*, *88*, 41–48.

Mera, P. E., & Escalante-Semerena, J. C. (2010b). Dihydroflavin-driven adenosylation of 4-coordinate Co(II) corrinoids: Are cobalamin reductases enzymes or electron transfer proteins? *The Journal of Biological Chemistry*, *285*, 2911–2917.

Moore, T. C., Mera, P. E., & Escalante-Semerena, J. C. (2014). The EutT enzyme of *Salmonella enterica* is a unique ATP:Cob(I)alamin adenosyltransferase metalloprotein that requires ferrous ions for maximal activity. *Journal of Bacteriology*, *196*, 903–910.

Moore, T. C., Newmister, S. A., Rayment, I., & Escalante-Semerena, J. C. (2012). Structural insights into the mechanism of four-coordinate Cob(II)alamin formation in the active site of the *Salmonella enterica* ATP:Co(I)rrinoid adenosyltransferase enzyme: Critical role of residues Phe91 and Trp93. *Biochemistry*, *51*, 9647–9657.

Mori, K., Oiwa, T., Kawaguchi, S., Kondo, K., Takahashi, Y., & Toraya, T. (2014). Catalytic roles of substrate-binding residues in coenzyme B_{12}-dependent ethanolamine ammonia-lyase. *Biochemistry*, *53*, 2661–2671.

Neese, F., & Solomon, E. I. (1999). MCD C-term signs, saturation behavior, and determination of band polarizations in randomly oriented systems with spin $S \geq 1/2$. Applications to $S = 1/2$ and $S = 5/2$. *Inorganic Chemistry*, *38*, 1847–1865.

Neumann, A., Siebert, A., Trescher, T., Reinhardt, S., Wohlfarth, G., & Diekert, G. (2002). Tetrachloroethene reductive dehalogenase of *Dehalospirillum multivorans*: Substrate specificity of the native enzyme and its corrinoid cofactor. *Archives of Microbiology, 177*, 420–426.

Ni, S., Fredrickson, J. K., & Xun, L. (1995). Purification and characterization of a novel 3-chlorobenzoate-reductive dehalogenase from the cytoplasmic membrane of *Desulfomonile tiedjei* DCB-1. *Journal of Bacteriology, 177*, 5135–5139.

Padmakumar, R., Taoka, S., Padmakumar, R., & Banerjee, R. (1995). Coenzyme B_{12} is coordinated by histidine and not dimethylbenzimidazole on methylmalonyl-CoA mutase. *Journal of the American Chemical Society, 117*, 7033–7034.

Pallares, I. G., Brunold, T. C., Moore, T. C., & Escalante-Semerena, J. C. (2016). Spectroscopic studies of the EutT adenosyltransferase from *Salmonella enterica*: Mechanism of four-coordinate Co(II)Cbl formation. *Journal of the American Chemical Society, 138*, 3694–3704.

Pallares, I. G., Moore, T. C., Escalante-Semerena, J. C., & Brunold, T. C. (2014). Spectroscopic studies of the *Salmonella enterica* adenosyltransferase enzyme SeCobA: Molecular-level insight into the mechanism of substrate Cob(II)alamin activation. *Biochemistry, 53*, 7969–7982.

Park, K., & Brunold, T. C. (2013). Combined spectroscopic and computational analysis of the vibrational properties of vitamin B_{12} in its Co^{3+}, Co^{2+}, and Co^{1+} oxidation states. *The Journal of Physical Chemistry B, 117*, 5397–5410.

Park, K., Mera, P. E., Escalante-Semerena, J. C., & Brunold, T. C. (2008). Kinetic and spectroscopic studies of the ATP:corrinoid adenosyltransferase PduO from *Lactobacillus reuteri*: Substrate specificity and insights into the mechanism of Co(II)corrinoid reduction. *Biochemistry, 47*, 9007–9015.

Park, K., Mera, P. E., Escalante-Semerena, J. C., & Brunold, T. C. (2012). Spectroscopic characterization of active-site variants of the PduO-type ATP:Corrinoid adenosyltransferase from *Lactobacillus reuteri*: Insights into the mechanism of four-coordinate Co(II)corrinoid formation. *Inorganic Chemistry, 51*, 4482–4494.

Park, K., Mera, P. E., Escalante-Semerena, J. C., & Brunold, T. C. (2016). Resonance Raman spectroscopic study of the interaction between Co(II)rrinoids and the ATP:corrinoid adenosyltransferase PduO from *Lactobacillus reuteri*. *Journal of Biological Inorganic Chemistry, 21*, 669–681.

Park, K., Mera, P. E., Moore, T. C., Escalante-Semerena, J. C., & Brunold, T. C. (2015). Unprecedented mechanism employed by the *Salmonella enterica* EutT ATP:Co(I)rrinoid adenosyltransferase precludes adenosylation of incomplete Co(II)rrinoids. *Angewandte Chemie, International Edition, 54*, 7158–7161.

Pratt, J. M. (1999). The roles of Co, corrin, and protein. II. Electronic spectra and structure of the corrin ligand: Molecular machinery of the protein. In R. Banerjee (Ed.), *Chemistry and biochemistry of B12* (pp. 113–164). Wiley.

Reig, A. J., Conrad, K. S., & Brunold, T. C. (2012). Combined spectroscopic/computational studies of vitamin B_{12} precursors: Geometric and electronic structures of cobinamides. *Inorganic Chemistry, 51*, 2867–2879.

Robertson, W. D., Wang, M., & Warncke, K. (2011). Characterization of protein contributions to cobalt−carbon bond cleavage catalysis in adenosylcobalamin-dependent ethanolamine ammonia-lyase by using photolysis in the ternary complex. *Journal of the American Chemical Society, 133*, 6968–6977.

Salama, S., & Spiro, T. G. (1977). Visible and near-ultraviolet resonance Raman spectra of photolabile vitamin B_{12} derivatives with a rapid-flow technique. *Journal of Raman Specroscopy, 6*, 57–60.

Salerno, E. V., Miller, N. A., Konar, A., Li, Y., Kieninger, C., Kraütler, B., et al. (2020). Ultrafast excited state dynamics and fluorescence from vitamin B_{12} and organometallic [Co]-C≡C-R cobalamins. *The Journal of Physical Chemistry B, 124*, 6651–6656.

Schrauzer, G. N., & Deutsch, E. (1969). Reactions of cobalt(I) supernucleophiles. The alkylation of vitamin B_{12}s, cobaloximes(I), and related compounds. *Journal of the American Chemical Society, 91*, 3341–3350.

Schubert, T., Adrian, L., Sawers, R. G., & Diekert, G. (2018). Organohalide respiratory chains: Composition, topology and key enzymes. *FEMS Microbiology Ecology, 94*.

Schumacher, W., Holliger, C., Zehnder, A. J., & Hagen, W. R. (1997). Redox chemistry of cobalamin and iron-sulfur cofactors in the tetrachloroethene reductase of *Dehalobacter restrictus*. *FEBS Letters, 409*, 421–425.

Sharma, P. K., Chu, Z. T., Olsson, M. H. M., & Warshel, A. (2007). A new paradigm for electrostatic catalysis of radical reactions in vitamin B_{12} enzymes. *Proceedings of the National Academy of Sciences of the United States of America, 104*, 9661–9666.

Shibata, N., Sueyoshi, Y., Higuchi, Y., & Toraya, T. (2018). Direct participation of a peripheral side chain of a corrin ring in coenzyme B_{12} catalysis. *Angewandte Chemie, International Edition, 57*, 7830–7835.

Shibata, N., Tamagaki, H., Hieda, N., Akita, K., Komori, H., Shomura, Y., et al. (2010). Crystal structures of ethanolamine ammonia-lyase complexed with coenzyme B_{12} analogs and substrates. *The Journal of Biological Chemistry, 285*, 26484–26493.

Solheim, H., Kornobis, K., Ruud, K., & Kozlowski, P. M. (2011). Electronically excited states of vitamin B_{12} and methylcobalamin: Theoretical analysis of absorption, CD, and MCD data. *The Journal of Physical Chemistry B, 115*, 737–748.

Solomon, E. I., Kirk, M. L., Gamelin, D. R., & Pulver, S. (1995). Bioinorganic spectroscopy. *Methods in Enzymology, 246*, 71–110.

Solomon, E. I., Neidig, M. L., & Schenk, G. (2004). Magnetic circular dichroism of paramagnetic species. In J. A. McCleverty, & T. J. Meyer (Eds.), *Vol. 2. Comprehensive coordination chemistry II* (pp. 339–349). Elsevier Ltd.

St. Maurice, M., Mera, P., Park, K., Brunold, T. C., Escalante-Semerena, J. C., & Rayment, I. (2008). Structural characterization of a human-type corrinoid adenosyltransferase confirms that coenzyme B_{12} is synthesized through a four-coordinate intermediate. *Biochemistry, 47*, 5755–5766.

Stich, T. A., Brooks, A. J., Buan, N. R., & Brunold, T. C. (2003). Spectroscopic and computational studies of Co^{3+}corrinoids: Spectral and electronic properties of the B_{12} cofactors and biologically relevant precursors. *Journal of the American Chemical Society, 125*, 5897–5914.

Stich, T. A., Buan, N. R., & Brunold, T. C. (2004). Spectroscopic and computational studies of Co^{2+}corrinoids: Spectral and electronic properties of the biologically relevant base-on and base-off forms of Co^{2+}cobalamin. *Journal of the American Chemical Society, 126*, 9735–9749.

Stich, T. A., Buan, N. R., Escalante-Semerena, J. C., & Brunold, T. C. (2005). Spectroscopic and computational studies of the ATP:corrinoid adenosyltransferase (CobA) from *Salmonella enterica*: Insights into the mechanism of adenosylcobalamin biosynthesis. *Journal of the American Chemical Society, 127*, 8710–8719.

Stich, T. A., Yamanishi, M., Banerjee, R., & Brunold, T. C. (2005). Spectroscopic evidence for the formation of a four-coordinate Co^{2+}cobalamin species upon binding to the human ATP:cobalamin adenosyltransferase. *Journal of the American Chemical Society, 127*, 7660–7661.

Stracey, N. G., Costa, F. G., Escalante-Semerena, J. C., & Brunold, T. C. (2018). Spectroscopic study of the EutT adenosyltransferase from *Listeria monocytogenes*: Evidence for the formation of a four-coordinate Cob(II)alamin intermediate. *Biochemistry, 57*, 5088–5095.

Toohey, J. I. (1965). A vitamin B_{12} compound containing no cobalt. *Proceedings of the National Academy of Sciences of the United States of America, 54*, 934–942.

Toraya, T. (2003). Radical catalysis in coenzyme B_{12}-dependent isomerization (eliminating) reactions. *Chemical Reviews*, *103*, 2095–2127.
Toraya, T. (2014). Cobalamin-dependent dehydratases and a deaminase: Radical catalysis and reactivating chaperones. *Archives of Biochemistry and Biophysics*, *544*, 40–57.
van de Pas, B. A., Smidt, H., Hagen, W. R., van der Oost, J., Schraa, G., Stams, A. J., et al. (1999). Purification and molecular characterization of ortho-chlorophenol reductive dehalogenase, a key enzyme of halorespiration in *Desulfitobacterium dehalogenans*. *The Journal of Biological Chemistry*, *274*, 20287–20292.
Vlasie, M., Chowdhury, S., & Banerjee, R. (2002). Importance of the histidine ligand to coenzyme B_{12} in the reaction catalyzed by methylmalonyl-CoA mutase. *The Journal of Biological Chemistry*, *277*, 18523–18527.
Warncke, K., Schmidt, J. C., & Ke, S. C. (1999). Identification of a rearranged-substrate, product radical intermediate and the contribution of a product radical trap in vitamin B_{12} coenzyme-dependent ethanolamine deaminase catalysis. *Journal of the American Chemical Society*, *121*, 10522–10528.
Warncke, K., & Utada, A. S. (2001). Interaction of the substrate radical and the 5′-deoxyadenosine-5′-methyl group in vitamin B_{12} coenzyme-dependent ethanolamine deaminase. *Journal of the American Chemical Society*, *123*, 8564–8572.
Warren, M. J., Raux, E., Schubert, H. L., & Escalante-Semerena, J. C. (2002). The biosynthesis of adenosylcobalamin (vitamin B_{12}). *Natural Product Reports*, *19*, 390–412.
Wohlfarth, G., & Diekert, G. (1997). Anaerobic dehalogenases. *Current Opinion in Biotechnology*, *8*, 290–295.
Zelder, O., Beatrix, B., Kroll, F., & Buckel, W. (1995). Coordination of a histidine residue of the protein-component S to the cobalt atom in coenzyme B_{12}-dependent glutamate mutase from *Clostridium cochlearium*. *FEBS Letters*, *369*, 252–254.

Kolibri 360 GmbH
Ferdinand-Jühlke-Straße 7
99095 Erfurt, Deutschland
produktsicherheit@kolibri360.de